State of the Art in Ethiopian Church Forests and Restoration Options

Mengistie Kindu · Thomas Schneider ·
Alemayehu Wassie · Mulugeta Lemenih ·
Demel Teketay · Thomas Knoke
Editors

State of the Art in Ethiopian Church Forests and Restoration Options

Springer

Editors
Mengistie Kindu
Technical University of Munich
Freising, Bayern, Germany

Thomas Schneider
Technical University of Munich
Freising, Bayern, Germany

Alemayehu Wassie
Organization for Rehabilitation
and Development
Bahir Dar, Ethiopia

Mulugeta Lemenih
Unique Forestry and Land Use GmbH
Addis Ababa, Ethiopia

Demel Teketay
Department of Range and Forest Resources
Botswana University of Agriculture
and Natural Resources
Gaborone, Botswana

Thomas Knoke
Fachgebiet für Waldinventur und
nachhaltige Nutzung
Technische Universität München
Freising, Bayern, Germany

ISBN 978-3-030-86625-9 ISBN 978-3-030-86626-6 (eBook)
https://doi.org/10.1007/978-3-030-86626-6

This Springer imprint is published by the registered company Springer Nature Switzerland AG
The registered company address is: Gewerbestrasse 11, 6330 Cham, Switzerland

Contents

Editors and Contributors

About the Editors

Mengistie Kindu is a scientific staff member at the Technical University of Munich (TUM) in Germany. He is working in the field of geo-informatics applications to resource management, with particular attention to a novel multidisciplinary perspective. He received Ph.D. in Natural Science, specialized geo-informatics application to resources management from TUM, M.Sc. in Geo-information Science from Wageningen University (The Netherlands) and B.Sc. in Forestry from Debub University (Ethiopia). He develops methodological frameworks to characterize the changes to Earth's LULC and integrate remote sensing observations with model-based approaches to understand the impacts retrospectively and prospectively. His research focuses on object-based image analysis, big data analysis, land use/land cover change mapping, time series analysis, machine learning, drivers of LULC changes, value chain mapping, land suitability and valuing ecosystem services. He has been actively involved in several research projects and taught different courses. He has published over 50 papers in peer-reviewed journals, as chapters and in proceedings. He is a member of several professional associations and serves as a reviewer of over 25 international journals.

Thomas Schneider is a scientific staff member of the Technical University of Munich (TUM) in Germany. He is serving at various positions within the university as the head of Geo-informatics (remote sensing and GIS) working groups of the Institute of Forest Management (IFM) and the chair for Aquatic Systems Biology (Limnological Research Station), a project leader and a scientist for over 14 national and international projects as well as a lecturer in remote sensing and image analysis and applied geo-informatics. He obtained Ph.D. in Forest Science and Diploma in Geology/Paleontology from Ludwig-Maximilians Universität Munich (Germany). His research focuses on the field and laboratory spectroscopy in forestry, agriculture and limnology, BRDF approximation by field goniometry, physical models and parameter estimation by inversion as well as application-oriented remote sensing and GIS-based expert system development in agriculture, forestry and limnology. He also serves as a reviewer of over 17 international journals.

Alemayehu Wassie has attended Alemaya University of Agriculture (Ethiopia) where he received a B.Sc. degree in Forestry in 1992. He then received a M.Sc. in Farm Forestry from the Swedish University of Agricultural Science (Sweden) for his research on the roles of Ethiopian Orthodox Tewahido churches in the conservation of forest biodiversity. He continued his research in church forest and further earned a Ph.D. in 2007 from Wageningen University (the Netherlands). He has worked at different capacity in government offices and NGOs. Currently, he is an executive director for a local NGO called Organization for Rehabilitation and Development in Amhara (ORDA) and an adjunct professor at Bahir Dar University (Ethiopia). He is a dedicated researcher, expert and conservationist for Ethiopian Orthodox Church Forests and named as "Unsung Hero" by African Geographic Magazine. Far beyond his academic research, he is also a writer in the world of literature. He has published mystic nobles named "Emegua", "Ziguara", "Merbebit" and "Sebez", which are among the best-selling books in Ethiopia.

Mulugeta Lemenih received his B.Sc. from Alemaya University in Ethiopia, M.Sc. from ITC in the Netherlands and Ph.D. from Swedish University of Agricultural Sciences, Sweden, all related to natural resources management. He has served in various organizations in various capacities. Currently, he is serving as Ethiopia country representative of UNIQUE land use and forestry, a German-based international consulting organization. He has published over 100 scientific articles and technical reports. He is a fellow of the Ethiopian Academy of Science and a member of the African Forest Forum and the Ethiopian Foresters Association.

Demel Teketay received his Ph.D. in Forest Vegetation Ecology from the Swedish University of Agricultural Sciences (Sweden) and M.Sc. in Plant Taxonomy from Reading University (UK). He has served at various capacities and countries as the professor in Forest Science (Botswana), an executive manager (Kenya), an associate professor and a research fellow (Botswana), a regional director (Ghana), a postdoctoral fellow (Sweden) and a postgraduate fellow (Netherlands) as well as the director general, the director of Forest Research, an assistant professor, a graduate assistant, an assistant lecturer, a lecturer and an extension agent (Ethiopia). He is currently serving as a professor of Forest Sciences and the dean of the Faculty of Natural Resources at the Botswana University of Agriculture and Natural Resources. He has published over 265 scientific articles and technical reports, including more than 165 peer-reviewed articles in national and international scientific journals. He has also taught different courses in different universities and supervised/co-supervised 20 Ph.D. and 20 M.Sc. students, examined 22 Ph.D. students in universities of different countries. He is a fellow of The World Academy of Sciences, an international fellow of the Royal Swedish Academy of Agriculture and Forestry, a fellow of the African Academy of Sciences, an associate fellow of the Ethiopian Academy of Sciences, and a member of the Botswana Academy of Science.

Thomas Knoke leads the Institute of Forest Management (IFM) at the Technical University of Munich (TUM) and explores sustainable use concepts for forest ecosystems. His particular interest lies in risk modeling and diversification strategy planning. Methodologies drawn from decision theory, operations research and modern finance theory are applied to forest science issues and land use problems in general. After studying forest science at Munich's Ludwig Maximilian University, he completed his doctorate in 1998. He did his lecturer qualification at TUM in 2003. In 2013, he was assigned to Professorship "Forstökonomie und Forstplanung" (W3), at Albert-Ludwigs-Universität Freiburg, which he declined. His research interests led him to the Institute for Commercial Forestry Research (South Africa), the Instituto Forestal (Chile) and the Chinese Academy of Forestry. He also traveled to Ecuador as a member of a multidisciplinary research group. His other activities include acting as a court expert, giving expert advice to forest owners and giving talks on international forestry investment.

Contributors

Abdu Abdelkadir Ethiopian Environment and Forest Research Institute (EEFRI), Addis Ababa, Ethiopia

Abrham Abiyu World Agroforestry (ICRAF), C/O ILRI Campus, Addis Ababa, Ethiopia

Enyew Adgo Department of Natural Resource Management, Bahir Dar University, Bahir Dar, Ethiopia

Mehari Alebachew Central Ethiopia Environment and Forest Research Centre, Addis Ababa, Ethiopia

Shiferaw Alem Central Ethiopia Environment and Forest Research Center, Addis Ababa, Ethiopia

Mekuria Argaw Center for Environmental Science, College of Natural and Computational Sciences, Addis Ababa University, Addis Ababa, Ethiopia

Dessie Assefa Department of Natural Resources Management, Bahir Dar University, Bahir Dar, Ethiopia

Shimeles Aynalem Department of Wildlife and Ecotourism Management, Bahir Dar University, Bahir Dar, Ethiopia

Merga Belina Addis Ababa University, Addis Ababa, Ethiopia

Achim Bräuning Institute of Geography, Friedrich-Alexander-University Erlangen-Nuremberg, Erlangen, Germany

Fikirte Demissie Ethiopian Institute of Architecture, Building Construction, and City Development (EiABC), Addis Ababa University, Addis Ababa, Ethiopia; Department of Architecture, Adama Science andTechnology University, Adama, Ethiopia

Richard P. Dick Ohio State University, Columbus, OH, USA

Martin Döllerer Institute of Forest Management, TUM School of Life Sciences Weihenstephan, Technical University of Munich, Freising, Germany

Eyasu Elias Addis Ababa University, Addis Ababa, Ethiopia

Abeje Eshete Ethiopian Environment and Forest Research Institute, Addis Ababa, Ethiopia

Zewdu Eshetu Addis Ababa University, Addis Ababa, Ethiopia

Amaury Frankl Department of Geography, Ghent University, Ghent, Belgium

Aster Gebrekirstos World Agroforestry (ICRAF), United Nations Avenue, Nairobi, Kenya

Melaku Getachew Ethiopian Environment and Forest Research Institute, Addis Ababa, Ethiopia

Douglas Godbold Institute of Forest Ecology, University of Natural Resources and Life Science, Vienna, Austria

David K. Goodin McGill School of Religious Studies, Montreal, Canada

Kiflu Gudeta Ministry of Agriculture, Addis Ababa, Ethiopia

Kiros Hadgu World Agroforestry (ICRAF), C/O ILRI Campus, Addis Ababa, Ethiopia

Niguse Hagazi World Agroforestry (ICRAF), C/O ILRI Campus, Addis Ababa, Ethiopia

Nesru Hassen Central Ethiopia Environment and Forest Research Center, Addis Ababa, Ethiopia

Aleksandra Ivanova Institute of Forest Management, TUM School of Life Sciences Weihenstephan, Technical University of Munich, Freising, Germany

Getachew Kebede Ethiopian Environment and Forest Research Institute, Addis Ababa, Ethiopia

Asia Khamzina Department of Environmental Science and Ecological Engineering, Korea University, Seoul, Korea

Eguale Tadesse Kifle Department of Environmental Science and Ecological Engineering, Korea University, Seoul, Korea;
Ethiopian Environment and Forest Research Institute (EEFRI), Addis Ababa, Ethiopia

Mengistie Kindu Institute of Forest Management, TUM School of Life Sciences Weihenstephan, Technical University of Munich, Freising, Germany

Thomas Knoke Institute of Forest Management, TUM School of Life Sciences Weihenstephan, Technical University of Munich, Freising, Germany

Gudina Legessa Addis Ababa University, Addis Ababa, Ethiopia

Mulugeta Lemenih Unique Forestry and Land Use GmbH, Addis Ababa, Ethiopia

Luc Lens Terrestrial Ecology Unit, Ghent University, Ghent, Belgium

Seyoum Leta Addis Ababa University, Addis Ababa, Ethiopia

Ferehiwot Mequanint Department of Wildlife and Ecotourism Management, Bahir Dar University, Bahir Dar, Ethiopia;
Terrestrial Ecology Unit, Ghent University, Ghent, Belgium

Mulugeta Mokira World Agroforestry (ICRAF), C/O ILRI Campus, Addis Ababa, Ethiopia

Bart Muys Division Forest, Nature and Landscape, KU Leuven, Leuven, Belgium

Demeke Nigussie Ethiopian Institute of Agricultural Research, Addis Ababa, Ethiopia

Florent Noulèkoun Department of Environmental Science and Ecological Engineering, Korea University, Seoul, Korea

Jan Nyssen Department of Geography, Ghent University, Ghent, Belgium

Boris Rewald Institute of Forest Ecology, University of Natural Resources and Life Science, Vienna, Austria

Mesfin Sahle Wolkite Univeristy, Wolkite, Ethiopia

Hans Sandén Institute of Forest Ecology, University of Natural Resources and Life Science, Vienna, Austria

Thomas Schneider Institute of Forest Management, TUM School of Life Sciences Weihenstephan, Technical University of Munich, Freising, Germany

Kibruyesfa Sisay Central Ethiopia Environment and Forest Research Center, Addis Ababa, Ethiopia

Yowhan Son Department of Environmental Science and Ecological Engineering, Korea University, Seoul, Korea

Teshome Soromessa Addis Ababa University, Addis Ababa, Ethiopia

Bernd Stimm Ecosystem Dynamics & Forest Management in Mountain Landscapes, TUM School of Life Sciences Weihenstephan, Technical University of Munich, Freising, Germany

Diederik Strubbe Terrestrial Ecology Unit, Ghent University, Ghent, Belgium

Wubalem Tadesse Ethiopian Environment and Forest Research Institute (EEFRI), Addis Ababa, Ethiopia

Demel Teketay Department of Range and Forest Resources, Botswana University of Agriculture and Natural Resources, Gaborone, Botswana

Zelalem Teshager Bahir Dar Environment and Forest Research Center, Bahir Dar, Ethiopia

Mindaye Teshome Central Ethiopia Environment and Forest Research Centre, Addis Ababa, Ethiopia

Degefie Tibebe Ethiopian Institute of Agricultural Research, Addis Ababa, Ethiopia

Tulu Tolla Kotebe Metropolitan University, Addis Ababa, Ethiopia

Alemayehu Wassie Organization for Rehabilitation and Development in Amhara, Bahir Dar, Ethiopia

Nesibu Yahya Ethiopian Environment and Forest Research Institute, Addis Ababa, Ethiopia;
Central Ethiopia Environment and Forest Research Center, Addis Ababa, Ethiopia

Kumelachew Yeshitela Ethiopian Institute of Architecture, Building Construction, and City Development (EiABC), Addis Ababa University, Addis Ababa, Ethiopia

Haileab Zegeye Department of Biology, Faculty of Natural and Computational Sciences, Debre Tabor University, Debre Tabor, Ethiopia

Chapter 1
Ethiopian Church Forests and Restoration Options—An Introduction

Mengistie Kindu, Thomas Schneider, Alemayehu Wassie, Mulugeta Lemenih, Demel Teketay, and Thomas Knoke

Abstract The Ethiopian church forests are now considered as an effective socio-ecological system, which contributes to the conservation of biodiversity. They provide habitat as critical sanctuaries for many endangered and endemic plant as well as invertebrate taxa. They are also considered as blueprints and seed repositories for native species to assist restoration of degraded landscapes in the country. This introductory chapter, written by the editors, provides an overview of background information and an insight into the different chapters of the book entitled "**State of the Art in Ethiopian Church Forests and Restoration Options**".

Keywords Church forests · Ecosystem services · Degradation · Restoration · Changes · Drivers · Ethiopia

M. Kindu (✉) · T. Schneider · T. Knoke
Institute of Forest Management, TUM School of Life Sciences Weihenstephan, Technical University of Munich, Hans-Carl-von-Carlowitz-Platz 2, 85354 Freising, Germany
e-mail: mengistie@tum.de

T. Schneider
e-mail: tomi.schneider@tum.de

T. Knoke
e-mail: knoke@tum.de

A. Wassie
Organization for Rehabilitation and Development in Amhara, PO Box 204, Bahir Dar, Ethiopia

M. Lemenih
Unique Forestry and Land Use GmbH, Addis Ababa, Ethiopia

D. Teketay
Department of Range and Forest Resources, Botswana University of Agriculture and Natural Resources, Private Bag 0027, Gaborone, Botswana
e-mail: dteketay@buan.ac.bw

M. Kindu et al. (eds.), *State of the Art in Ethiopian Church Forests and Restoration Options*, https://doi.org/10.1007/978-3-030-86626-6_1

1.1 Background

Tropical forests, including Ethiopian forests, contain over half of Earth's species (Lewis et al. 2015). They are among the terrestrial biomes with the largest flow of ecosystem services (Foley et al. 2007; Guariguata and Balvanera 2009; Kindu et al. 2016; Carrasco et al. 2017; Gashaw et al. 2018; Knoke et al. 2020). However, these ecosystems are rapidly declining in spatial cover and quality, and this has been driven by multiple social, economic and ecological factors.

Ethiopia, which once hosted a significant tropical forest cover, has witnessed loss of over 97% of its historic forest cover (Darbyshire et al. 2003; Nyssen et al. 2004; Kindu et al. 2013), a process that continued to date (Kindu et al. 2018). Central and northern parts of the country have experienced ages of continued deforestation that today few scattered patches of tropical forests remained to witness the heritage of the past. The remnant patches in these parts of the country are mostly confined around the Ethiopian Orthodox Tewahido Churches (EOTC), monasteries and other holy sites (Wassie 2002; Tilahun et al. 2015; Aerts et al. 2016). The Ethiopian Orthodox Tewahido Church is one of the oldest Christian churches in the world, with some of the earliest church buildings dating to 300 A.D (Wassie, 2002). For centuries, EOTC and followers of EOTC have conserved patches of many indigenous trees and shrubs around churches (Wassie 2002; Wassie et al. 2005; Woldemedhin and Teketay 2016).

The church forests have much spiritual and religious value and represent important cultural heritage sites (Wassie 2007; Woldemedhin and Teketay 2016). In addition, many studies are being conducted focusing on the contribution of Ethiopian Church Forests (ECFs) to the conservation of biodiversity and ecosystem services (Votrin 2005; Wassie and Teketay 2006; Wassie et al. 2009a, b; Aynekulu et al. 2016; Woldemedhin and Teketay 2016; Cardelús et al. 2017; Yilma and Derero 2020). Specifically, the ECFs provide habitat as critical sanctuaries for many endangered and endemic plant as well as invertebrate taxa in Ethiopia (Bongers et al. 2006). They can be used as important reserves for species and biodiversity conservation and as seed sources for restoration of degraded areas with composition closely similar to the past vegetation. They are also important storehouse of biomass carbon, hence playing climate mitigation role. Overall, the ECFs are considered as in situ conservation sites for biodiversity, in particular for indigenous trees and shrubs that found refugee from human destruction (Wassie 2002; Woldemedhin and Teketay 2016).

The ECFs also serve as sources of past climate information, since climate change is also a global phenomenon, but its impacts are heterogeneous across different parts of the world (IPCC 2014). Tree rings from old grown trees in ECFs and woods used in church constructions are among the most useful and readily available proxy data sources of past hydroclimate information of various sites in Ethiopia (Wils et al. 2011; Gebrekirstos et al. 2014; Mokria et al. 2015; Gebregeorgis et al. 2020).

The above facts show that ECFs are within complex social–ecological and climate systems that have provided a variety of cultural, ecological and economic benefits for churches and the surrounding communities for centuries (Wassie et al. 2010; Woldemedhin and Teketay 2016). The positive attitude to the resources sheltered by the church and the acceptance of the church tradition are also an opportunity for forest ecosystem conservation and restoration (Wassie 2002). However, the remnants of native forest fragments continue to decline driven by both natural and anthropogenic factors. Thus, the value of these church forests in continuing to offer ecosystem services will depend on their long-term sustainable management. This calls for immediate invention to persevere these biodiversity hot spots of indigenous/endemic flora and fauna. In addition, using ECFs as blueprint for restoration of degraded landscapes in the country is an innovative concept to improve the landscape capacity to provide ecosystem services.

A study from the Ethiopian highland showed that the estimated rate of soil loss from cultivated fields is up to 79 Mg ha^{-1} yr^{-1}, which by all measures exceeds the rate of soil formation (Bewket and Sterk 2003). Such ongoing land degradation is also, globally, a very serious problem (Willemen et al. 2020). In this regard, the ECFs can serve as "stepping stones" for restoration. However, this needs identification of hot spot areas of erosion and prioritizing areas of intervention for reducing further degradation and restoring the degraded areas. Once prioritized, restoration approaches can be in place. The choice of methods for restoration may depend on the severity of degradation and will significantly affect the speed within which the restoration process proceeds (Lemenih and Teketay 2004).

Despite valuable findings from several studies on ECFs and restoration options of degraded landscapes, comprehensive compilation and synthesis of such works in a form of book are uncommon. Such demanding task requires numerous lenses to bear on the subject matter and understand fully the state of church forests for their management and contribution in restoring degraded landscapes from multidisciplinary perspective.

This book, thus, aims to present and discuss the state of the art in ECFs and the associated potential restoration options, which are compiled in different chapters. The book brings together a collection of original research and review chapters that deal with the different aspects of the ECFs and potential restoration options. In the following sections, highlights of the different chapters included in this book are presented.

1.2 Overview of Chapters

The book is organized into seventeen chapters. The chapters contribute to a more evidence-based view of the topics, which are instrumental for informed decisions. The book starts with this introductory chapter. The main body is subdivided accordingly into four main parts: "General Overview of Ethiopian Church Forests", "Present Role and Future Challenges of Ethiopian Church Forests", "Structure and Diversity of Ethiopian Church Forests" "Restoration Options to the Surrounding Landscapes of Ethiopian Church Forests". The book closes with a synthesis and conclusion chapter.

The chapters in **Part One** focus on the general overview of ECFs, emphasizing their importance inside the society to understand sociocultural context and the changes in church forests.

The chapter on *Understanding Land Use/Land Cover Dynamics in and the surroundings of Ethiopian Church Forests* by Kindu et al. provides a review of the state of the art of scientific knowledge on land use/land cover (LULC) dynamics within and in the surrounding areas of ECFs. In this emerging research field on LULC change in Ethiopia, the authors reviewed and explained how the changes are captured (approaches used) and presented the states of overall LULC changes and underlying reasons/drivers of the changes. The authors also outline further research challenges and opportunities within and surrounding the ECFs. Similarly, Yahya et al. focus on LULC changes and fragmentations of four selected church forests as a case study site in the chapter on *Land Use Land Cover Changes and Forest Fragmentation on the Surrounding of Selected Church Forests in Ethiopia.* This chapter provides the rate of changes and level of fragmentations and major recommendations based on the findings.

Besides overview of changes and their drivers, Goodin provides an overview in his chapter on *Sacred Texts and Environmental Ethics: Lessons in Sustainability from Ethiopia* about the resilience of ECFs from the diverse religious history that comes together in the Tewahedo Eco-theology. The author examines the cultural importance of ECFs with respect to environmental ethics and eco-theology, both within Tewahedo tradition for restoration projects in Ethiopia and for the potential to inform and enrich Christian traditions throughout the developed and developing world. This chapter presents specific recommendations for improving project success by identifying how the sacred forests of Ethiopia are protected by Tewahedo theology, and this cultural protection can be increased or lost in the eyes of the people.

The chapters in **Part Two**: "Present Role and Future Challenges of Ethiopian Church Forests" provide and discuss contributions of ECFs and associated challenges in order to find ways and means to conserve the remaining forests.

The first chapter by Assefa et al. focuses on *Soil Carbon Stocks and Dynamics of Church Forests in Northern Ethiopia.* The authors examined soil carbon storage of three selected church forests in comparison with adjacent land use systems (eucalypt plantations, grazing land and cropland). The authors estimated and presented the amount of loss of soil organic carbon (SOC) stock due to conversion of church forests to other land use types. While considering these challenges, they call for urgent attention to restore essential ecosystem functions and soil carbon sequestrations. Similarly, Tolla et al. focus on the trends of carbon stock in the chapter on *Estimation and Mapping of Asabot Monastery Dry Afromontane Forest Carbon Stock Under Diverse Land Use Scenarios.* This chapter provides the carbon stock estimates under different scenarios and describes their importance in Dry Afromontane forests. In another related chapter, Alebachew et al. concentrate on about *Aboveground and Belowground Carbon Pools for Some Selected Native and Introduced Tree Species of Abune Teklehayimanot Church Forest, Welayita Sodo, Southern Ethiopia.* Using established methods, the authors presented above and belowground biomass and the mean carbon stock density of aboveground biomass recorded in the natural forests of the church. In light of their results, the authors recommended forest carbon-related awareness creation for the local people and promotion of the local conservation knowledge as a possible option for sustainable forest management.

In **Part Three**: "Structure and Diversity of Ethiopian Church Forests", the chapters assess and describe the composition, diversity structure and regeneration status of woody species in church forests.

The first chapter by Alem et al. deals with *Floristic Composition, Diversity, Population Structure and Regeneration Status of Woody Species in Four Church Forests in Ethiopia*. The authors documented the number of woody species, their diversity and densities as well as number of seedlings. They assessed the regeneration status of the woody species as well as causes of observed hampered regeneration and recommended appropriate silvicultural treatments and management practices to enhance the regeneration of native woody species in the studied church forests. Similarly, Zegeye provides interesting results about *Diversity, Regeneration Status and Socio-Economic Importance of Tara Gedam, Abebaye and Fach Forests, Libokemkem District of South Gondar Administrative Zone, Northwestern Ethiopia.* The author used a combination of methods for data collection and analysis. The results showed status of the church forests, their socio-economic importance and anthropogenic factors affecting them. The author further forwarded recommendations to ensure the long-term maintenance of the church forests.

Mequanint et al. deliver another study on *Woody Vegetation Composition and Structure of Church Forests in Southeast of Lake Tana, Northwest Ethiopia.* The authors documented the woody species composition, presented population structural features and diversity and regeneration status covering 24 church forests in southeast of Lake Tana. They also outlined existing problems and possible options of management. In a similar chapter, Kifle et al. also present *Woody Species Composition, Diversity, Structure and Uses of Selected Church Forests in the Central Ethiopia.* Apart from assessing species composition and diversity, the authors also documented threatened species according to the IUCN Red List. Subsequently, they recommended possible conservation strategies especially by giving priorities to the IUCN red-listed species.

The chapters in **Part Four**: "Restoration Options to the Surrounding Landscapes of Ethiopian Church Forests" focus on restoration options for the surrounding landscapes. This part explores various options to restore degraded lands in Ethiopia, which have been disturbed by humans over the last decades.

The first chapter by Stimm et al. focuses on *Church Forests as Repositories for Forest Reproductive Material of Native Species and their Possible Role as Starting Points for the Restoration of Degraded Areas.* The authors pointed out the possible purpose of church forests as sources for reproductive material and highlighted the potential role of new church nurseries in high-quality planting stock production. They concluded with recommendations for a conceptual framework on tree seed procurement and its implementation for the purpose of successful restoration of Ethiopian forests.

Demissie et al. illustrated another applicable concept that focuses on *Ecological Status and Plan for Connectivity of Fragmented Forests as a Means of Degraded*

Land Restoration in South Gonder, Ethiopia. The authors evaluated the state of land use fragmentation and proposed ecological connectivity in a case study site that can be used as a model. They also outlined recommendations to put the proposed development strategies into practice. Another dimension of the restoration approach is highlighted in the chapter of Gudeta et al. on *Identifying Priority Areas for Conservation in Mojo River Watershed of Ethiopia using GIS-Based Erosion Risk Evaluation.* In this case, the authors used revised universal soil loss equations and multicriteria evaluation approaches in GIS environment for prioritization of degraded areas for undertaking the required conservation measures. Through such approach, they outlined critical areas under very high and severe category and recommended for immediate conservation interventions. Similarly, Schneider et al. focus on *Rehabilitation Sites Prioritization on Base of Remote Sensing Time Series, Erosion Risk and Woody Biomass Modelling* to present a concept for a land use management system adapted to the rural landscapes of the Ethiopian Highlands. The authors used multiple data sets and combination of techniques for developing the concept.

Based on a framework for using church forests as evidence-based climate-smart restoration efforts, Mokira et al. discuss specific paths of reconstructing multi-century climate information, in their chapter on *Ethiopian Church Forests as Monitoring Towers in Reconstructing Climate Change and Its Impacts and to Make Evidence-Based Climate-Smart Restoration Efforts.* They explain the role of dendrochronological data to better understand forest growth dynamics, tree to forest level responses to changing climate and to reconstruct multi-century climate information and characterize extreme climate events beyond instrumental periods. The authors address opportunities and challenges of dendroecological application and summarized results from different studies on *Juniperus procera* growing in church forests, as they contain old standing trees that are less exposed to anthropogenic disturbances.

1.3 Conclusions

Ethiopia's landscapes have seen decades of deforestation, biodiversity depletion and degradation of their capacity to provide ecosystem services. Such processes are clear indicators for an unsustainable development pathway of a country (Carrasco et al. 2017; Kindu et al. 2018). Now it is high time to revert the detrimental landscape depletion and degradation processes. In this light, ECFs provide an invaluable asset. Hosting of an extremely rich biodiversity, their direct value is high (Paul et al. 2020). In addition, ECF provides high indirect value as a library to inform landscape restoration concepts. We hope that our book can contribute to conserve and utilize the high value inherent in ECFs.

References

Aynekulu E, Aerts R, Denich M, Negussie A, Friis I, Demissew S, Boehmer HJ (2016) Plant diversity and regeneration in a disturbed isolated dry Afromontane forest in northern Ethiopia. Folia Geobot, 51:115–127

Aerts R, van Overtveld K, November E, Wassie A, Abiyu A, Demissew S, Daye DD, Giday K, Haile M, TewoldeBerhan S, Teketay D, Teklehaimanot Z, Binggeli P, Deckers J, Friis I, Gratzer G, Hermy M, Heyn M, Honnay O, Paris M, Sterck FJ, Muys B, Bongers F, Healey JR (2016) Conservation of the Ethiopian church forests: threats, opportunities and implications for their management. Sci Total Environ 551–552:404–414

Bongers F, Wassie A, Sterck FJ, Bekele T, Teketay D (2006) Ecological restoration and church forests in northern Ethiopia. J Drylands 1(1), 35–44

Bewket W, Sterk G (2003) Assessment of soil erosion in cultivated fields using a survey methodology for rills in the Chemoga watershed, Ethiopia. Agric Ecosyst Environ 97:81–93. https://doi.org/10.1016/S0167-8809(03)00127-0

Cardelús CL, Scull P, Wassie Eshete A, Woods CL, Klepeis P, Kent E, Orlowska I (2017) Shadow conservation and the persistence of sacred church forests in Northern Ethiopia. Biotropica 49(5):726–733

Carrasco LR, Le Nghiem TP, Chen Z, Barbier EB (2017) Unsustainable development pathways caused by tropical deforestation. Sci Adv 3(7), e1602602.https://doi.org/10.1126/sciadv.1602602

Darbyshire I, Lamb H, Umer M (2003) Forest clearance and regrowth in northern Ethiopia during the last 3000 years. Holocene 13(4):537–546

Foley JA, Asner GP, Costa MH, Coe MT, DeFries R, Gibbs HK et al (2007) Amazonia revealed. Forest degradation and loss of ecosystem goods and services in the Amazon Basin. Front Ecol Environ 5(1):25–32. https://doi.org/10.1890/1540-9295(2007)5[25:ARFDAL]2.0.CO;2

Gashaw T, Tulu T, Argaw M, Worqlul AW, Tolessa T, Kindu M (2018) Estimating the impacts of land use/land cover changes on ecosystem service values: the case of the Andassa watershed in the Upper Blue Nile basin of Ethiopia.

Gebregeorgis EG, Robertson I, Koprowski M, Zhou LP, Gao P, Williams AP, Eshetu Z, Wils THG (2020) Historical droughts recorded in extended Juniperus procera ring-width chronologies from the Ethiopian Highlands. Int J Biometeorol 64(5):739–753

Gebrekirstos A et al (2014) Opportunities and applications of dendrochronology in Africa. Curr Opin Environ Sustain 6:48–53

Guariguata MR, Balvanera P (2009) Tropical forest service flows: improving our understanding of the biophysical dimension of ecosystem services. For Ecol Manage 258(9):1825–1829

IPCC (2014) Climate change 2014 synthesis report summary chapter for policymakers. https://doi.org/10.1017/CBO9781107415324

Kindu M, Schneider T, Teketay D, Knoke T (2013) Land use/land cover change analysis using object-based classification approach in Munessa-Shashemene landscape of the Ethiopian Highlands. Remote Sens 5(5):2411–2435

Kindu M, Schneider T, Teketay D, Knoke T (2016) Changes of ecosystem service values in response to land use/land cover dynamics in Munessa-Shashemene landscape of the Ethiopian highlands

Kindu M, Schneider T, Döllerer M, Teketay D, Knoke T (2018) Scenario modelling of land use/land cover changes in Munessa-Shashemene landscape of the Ethiopian highlands. Sci Total Environ 622–623:534–546

Knoke T, Paul C, Rammig A, Gosling E, Hildebrandt P, Härtl F, Peters T, Richter M, Diertl K-H, Castro LM, Calvas B, Ochoa S, Valle-Carrión LA, Hamer U, Tischer A, Potthast K, Windhorst D, Homeier J, Wilcke W, Velescu A, Gerique A, Pohle P, Adams J, Breuer L, Mosandl R, Beck E, Weber M, Stimm B, Silva B, Verburg PH, Bendix J (2020) Accounting for multiple ecosystem services in a simulation of land-use decisions: does it reduce tropical deforestation? Global Change Biol 26:2403–2420

Lemenih M, Teketay D (2004) Restoration of native forest flora in the degraded highlands of Ethiopia: constraints and opportunities. Ethiop J Sci 27:75–90

Lewis SL, Edwards DP, Galbraith D (2015) Increasing human dominance of tropical
Mokria M et al (2015) Tree dieback affects climate change mitigation potential of a dry Afromontane forest in northern Ethiopia. For Ecol Manage 344:73–83. https://doi.org/10.1016/j.foreco.2015.02.008
Nyssen J, Poesen J, Moeyersons J, Deckers J, Haile M, Lang A (2004) Human impact on the environment in the Ethiopian and Eritrean highlands—a state of the art. Earth Sci Rev 64:273–320
Paul C, Hanley N, Meyer ST, Fürst C, Weisser WW, Knoke T (2020) On the functional relationship between biodiversity and economic value. Sci Adv 6(5):eaax7712. https://doi.org/10.1126/sciadv.aax7712
Tilahun A (2015) The contribution of Ethiopian orthodox Tewahido church in forest management and its best practices to be scaled up in north Shewa Zone of Amhara Region, Ethiopia. Agric for Fish 4(3):123
Votrin V (2005) The orthodoxy and sustainable development a potential for broader involvement of the orthodox churches In Ethiopia and Russia. Environ Dev Sustain 7(1):9–21
Wassie A (2002) Opportunities, constraints, and prospects of the Ethiopian orthodox tewahido churches in conserving forest resources: the case of churches in South Gonder, Northern Ethiopia, MSc thesis, Uppsala, Sweden
Wassie A (2007) Ethiopian church forests: opportunities and challenges for restoration, PhD thesis, Wageningen, The Netherlands, Wageningen Univeristy
Wassie A, Sterck FJ, Bongers F (2010) Species and structural diversity of church forests in a fragmented Ethiopian Highland landscape. J Veg Sci 21(5):938–948
Wassie A, Sterck FJ, Teketay D, Bongers F (2009a) Effects of livestock exclusion on tree regeneration in church forests of Ethiopia. For Ecol Manage 257(3):765–772
Wassie A, Sterck FJ, Teketay D, Bongers F (2009b) Tree regeneration in church forests of Ethiopia: effects of microsites and management Biotropica 41(1):110–119
Wassie A, Teketay D, Powell N (2005) Church forests in North Gondar administrative zone, Northern Ethiopia. Forests, Trees Livelihood 15:349–373
Wassie A, Teketay D (2006) Soil seed banks in church forests of northern Ethiopia: implications for the conservation of woody plants. Flora Morphology Distrib Funct Ecol Plants 201(1):32–43
Willemen L, Barger NN, Bt B, Cantele M, Erasmus BFN, Fisher JL, Gardner T, Holland TG, Kohler F, Kotiaho JS, von Maltitz GP, Nangendo G, Pandit R, Parrotta JA, Potts MD, Prince SD, Sankaran M, Brainich A, Montanarella L, Scholes R (2020) How to halt the global decline of lands. Nat Sustain 3:164–166. https://doi.org/10.1038/s41893-020-0477-x
Wils THG, Robertson I, Eshetu Z, Touchan R, Sass-Klaassen U, Koprowski M (2011) Crossdating Juniperus procera from North Gondar, Ethiopia. Trees 25(1):71–82
Woldemedhin TT, Teketay D (2016) Forest conservation tradition of the Ethiopian Orthodox Tewahdo Church: a case study in West Gojjam Zone, north-western Ethiopia. Symb Bot Upsal 38:57–73
Yilma G, Derero A (2020) Carbon stock and woody species diversity patterns in church forests along church age gradient in Addis Ababa, Ethiopia. Urban Ecosyst 551:404

Part I
General Overview of Ethiopian Church Forests

Chapter 2
Understanding Land Use/Land Cover Dynamics in and Surrounding the Ethiopian Church Forests

Mengistie Kindu, Degefie Tibebe, Demeke Nigussie, Thomas Schneider, Martin Döllerer, Demel Teketay, and Thomas Knoke

Abstract A better understanding of the causes and consequences of complex changes of land use/land cover (LULC) in a given landscape call for global attention for scientific research. This chapter reviews the state-of-the-art of scientific knowledge on LULC dynamics within and in the surrounding areas of Ethiopian Church Forests (ECFs). We conducted the review using systematically selected articles on LULC changes associated with ECFs, with additional literature to contextualize the broad scope of our understanding of the changes. Our review highlights approaches used for change studies, overall LULC changes and underling reasons/drivers of changes. Our study demonstrated that remote sensing and ground inventory based approaches are essential tools for change detection within as well as in the surrounding areas of ECFs. We revealed a mix of forest cover changes within and the ECFs. For increasing trends in forest area, the existence of stone walls, exclosures, intense rehabilitation, and protection activities were mentioned as major drivers. Conversion to croplands, population pressure, livestock grazing, government policies, isolation, and small sizes were the most common drivers for

M. Kindu (✉) · T. Schneider · M. Döllerer · T. Knoke
Institute of Forest Management, TUM School of Life Sciences Weihenstephan, Technical University of Munich, Hans-Carl-von-Carlowitz-Platz 2, 85354 Freising, Germany
e-mail: mengistie@tum.de

T. Schneider
e-mail: tomi.schneider@tum.de

M. Döllerer
e-mail: doellerer@tum.de

T. Knoke
e-mail: knoke@tum.de

D. Tibebe · D. Nigussie
Ethiopian Institute of Agricultural Research, P.O. Box 2003, Addis Ababa, Ethiopia

D. Teketay
Department of Range and Forest Resources, Botswana University of Agriculture and Natural Resources, Private Bag 0027, Gaborone, Botswana
e-mail: dteketay@buan.ac.bw

M. Kindu et al. (eds.), *State of the Art in Ethiopian Church Forests and Restoration Options*, https://doi.org/10.1007/978-3-030-86626-6_2

declining trends of forest areas. Further research challenges, and opportunities within and surrounding the ECFs are outlined.

Keywords Satellite images · Remote sensing · Inventory · Changes · Drivers · Ethiopia

2.1 Introduction

Land use/land cover (LULC) change is a challenging key research theme in the field of land system science (Turner et al. 2020). The LULC changes can affect biodiversity (Dayamba et al. 2016), and other ecosystem services across local and regional levels (Vitousek 1997; Paz-Kagan et al. 2014; Doffana and Yildiz 2017). LULC change can also affect climate from local to global scales through its influence on surface energy budgets and biogeochemical cycling mechanisms (Pielke et al. 2002; Fairman et al. 2011). Hence, it has increasingly become a topic of paramount importance. There have been calls for global attention for continuous monitoring and understanding of the complex LULC changes that have important consequences for socio-economic and environmental dynamics in a given landscape.

Among the LULC types, forests are very dynamic, being extremely important to the acquisition of accurate and up-to-date data for proper monitoring (Vieilledent et al. 2018; Armenteras et al. 2019). Continuous monitoring is key to effectively identify and preserve important forests. In particular, tropical forests contain over half of Earth's species (Lewis et al. 2015) are among the terrestrial biomes with the largest flows of ecosystem services (Foley et al. 2007; Guariguata and Balvanera 2009; Kindu et al. 2016; Knoke et al. 2016; Gashaw et al. 2018; Knoke et al. 2020), on which billions of people depend (FAO 2011). Only very few patches of forest remain in the central and northern Ethiopian highlands, and these are almost entirely confined within the areas of the Ethiopian Orthodox Tewahido Churches (EOTC), monasteries, and other holy sites (Wassie 2002; Tilahun 2015; Aerts et al. 2016). They are relicts of the Afromontane forests (Wassie et al. 2010). For centuries, EOTC and followers of EOTC have conserved patches of native trees around churches (Wassie 2002; Woldemedhin and Teketay 2016). Each church operates largely autonomously to manage its forests, and management varies from strict protection—with some churches surrounded by walls and patrolled by paid forest guards—to lose protection with harvesting of wild fruits, honey, or fuelwood from dead woody species in church forests (Amare et al. 2016).

The church forests are usually surrounded by agricultural land and pasture (Woods et al 2017), settlements (Daye and Healey 2015; Tura et al. 2016) or other patches of woodlots (Gebresamuel et al. 2010; Daye and Healey 2015). Such church forests have certain spiritual and religious values to the church communities, and represent important cultural heritage sites, as an integral part of EOTC (Wassie 2007; Woldemedhin and Teketay 2016). Apart from this, nowadays, a growing number of studies are conducted on the contribution of Ethiopian Church Forests (ECFs) to the conservation of biodiversity and ecosystem services (Votrin 2005; Wassie and Teketay 2006; Wassie et al. 2009; Aynekulu et al. 2016; Woldemedhin and Teketay 2016; Cardelús et al. 2017; Yilma and Derero 2020). Specifically, the ECFs provide habitat as critical sanctuaries for many endangered and endemic plant and invertebrate taxa in Ethiopia (Bongers et al. 2006). For example, Wassie et al. (2010) documented 160 different species of indigenous trees in a survey of only 28 church forests, documenting the highest tree species richness in the study region. These show that ECFs are within complex social-ecological systems that have provided a variety of cultural, ecological, and economic benefits for churches and surrounding communities for centuries (Wassie et al. 2010; Berhane et al. 2013; Woldemedhin and Teketay 2016).

In Ethiopia, various studies revealed historical LULC changes and drivers behind the dynamics within or surrounding the ECFs (e.g., Tedla 1998; Zeleke and Hurni 2001; Awoke 2010; Demissie et al. 2017). The LULC change studies found different types and rates of changes in various parts of the country over different time periods (Zeleke and Hurni 2001; Tegene 2002; Hurni et al. 2005; Dessie and Kleman 2007; Tsegaye et al. 2010; Temesgen et al., 2013; Desalegn et al. 2014; Mekasha et al. 2014; Nigatu et al. 2014; Kassawmar et al. 2018; Berihun et al. 2019; Mohamed et al. 2019). This shows that the LULC changes in the Ethiopian landscapes are nonlinear and geographically heterogeneous (Bewket 2002; Tegene 2002; Kindu et al. 2013). For example, Zeleke & Hurni (2001) found that natural forest cover in the Dembecha area (Gojjam) declined from 27% in the 1950s to less than 1% in 1995, a decline of over 99% of the forest that existed in 1957. Tegene (2002) also reported dramatic decline in shrub land in the Derekolli Catchment, South Wello, from 16% in 1957 to 2.5% in 1986. The drivers of these changes maybe well-defined depending on their location. such drivers include population pressure, cropland explanation, livestock grazing, fuelwood collection (Bewket 2002; Tegene 2002; Gashaw et al. 2014; Kindu et al. 2015).

Although several studies have evaluated the LULC changes within or surrounding the ECFs, to date, there is no comprehensive review of the changes and their associated drivers. This information is, however, urgently needed to fully understand the ongoing changes and their drivers for comprehensive management of ECFs and the surrounding areas.

This chapter, thus, aims to review LULC dynamics in and surroundings the ECFs. Specifically, the following questions will be addressed: (1) What are the current states of studies about LULC dynamics in and surroundings the ECFs? (2) Which approaches are mainly used for the study of such changes? (3) What do the overall changes in and surrounding the church forests look like? And (4) What are the underlying reasons/drivers of the changes?

To answer these questions we conducted a systematic literature review described in Sect. 2.2. In the subsequent Sects. 2.3, 2.4, 2.5, and 2.6, we present a summary of systematically selected articles with discussions. Here, we will focus on a general overview of the selected articles, review approaches used for LULC change study, overall changes, and underlying reasons (drivers) for changes. In the last section, this review will draw some conclusions, which are intended to stimulate further research on LULC changes in and surrounding the ECFs.

2.2 Materials and Methods

We performed a systematic search, which was carried out using the following search engines and platforms: Web of Science (https://apps.webofknowledge.com), Research Gate (https://www.researchgate.net), Google Scholar (https://scholar.google.com), and Science Direct (https://www.sciencedirect.com) (visited between December 2019 and April 2020) to identify articles about LULC changes in and surrounding areas of ECFs. We searched for articles with titles combining a keyword for changes of LULCs in Ethiopia, a keyword indicating church forests, and a keyword indicating reasons/drivers of changes. Related books and PhD dissertations sourced from different university archives were also used for this review.

Our search retrieved more than 500 articles from different search engines. Among them, 51 were finally selected after reading the abstracts. Our emphasis of criteria for this further selection was to retain only articles dedicated to LULC changes in or surrounding areas of ECFs. We supplemented this list of articles with additional literature for the background and discussion part of the review.

From the systematically selected articles, we extracted key information such as: publication years, geographical locations of case study areas, data sources, analytical approaches used, overall LULC changes, and reasons/drivers for the observed changes.

2.3 General Overview of the Reviewed Articles

The number of articles on LULC dynamics in and surrounding areas of ECFs strongly increased in recent years (Fig. 2.1), as we can see that 42 articles (about 82%) were published after 2010. This shows that the topic has gained increasing scientific attention in recent years.

The studies in our selected articles were carried out, mainly, in the Ethiopian highlands (altitude > 1500 m), which represent about 96% of the studies analyzed (Fig. 2.2). East Tigray, South Tigray, and South Gonder Administrative Zones accounted for the largest share of studies in the selected articles (52%).

2.4 Approaches of Change Studies in and Surrounding ECFs

Among the 51 articles, about half used a remote sensing-based study approach, followed by 35% using a ground inventory-based approach. The remaining studies were based on questionnaire surveys and literature reviews. The trend of publication year corresponding to study approach showed that studies based on remote sensing

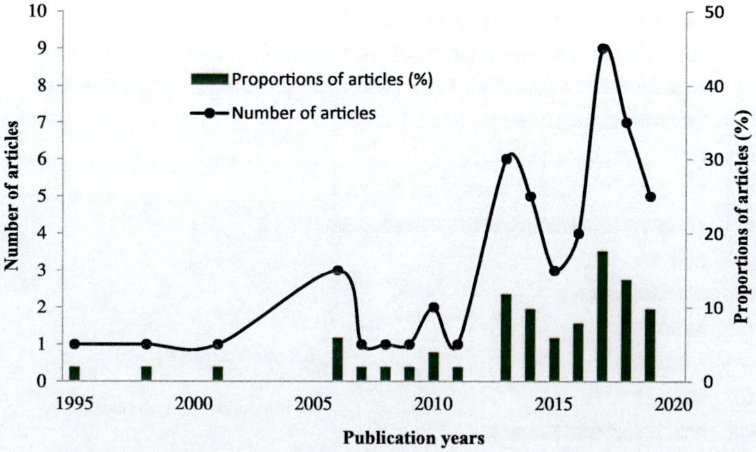

Fig. 2.1 Number of articles on LULC changes in and surrounding ECF per publication year (n = 51)

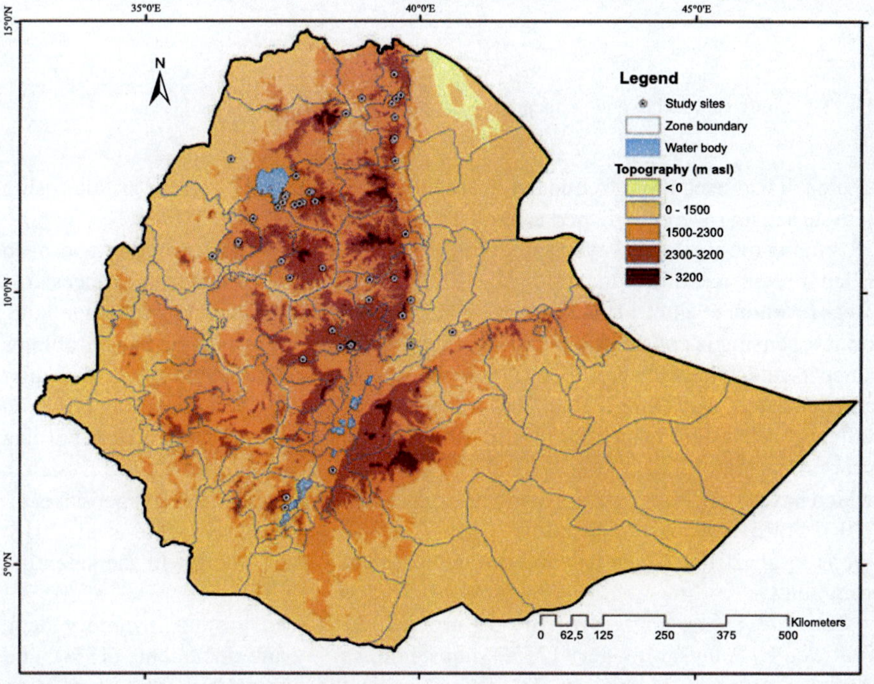

Fig. 2.2 Distribution of reviewed articles on LULC dynamics in and the surrounding areas of ECFs based on geographical location of their sites

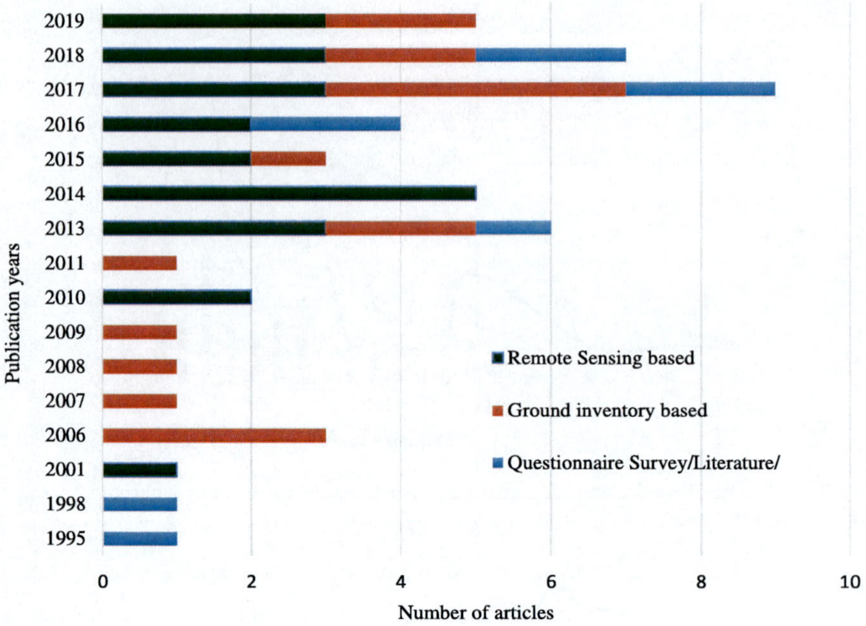

Fig. 2.3 Study approaches and publication years of reviewed articles ($n = 51$)

approach was used more frequently in recent years. Indeed, 87% of the articles using remote sensing were published after 2013 (Fig. 2.3).

Among the available ways of studying LULC changes, ground inventory approach is labor intensive, time consuming, and difficult for capturing data from inaccessible areas (Wondie et al. 2011; Kindu et al. 2013; Wallner et al. 2017). On the other hand, remote sensing is considered the most efficient technology to handle these problems since it can explicitly reveal LULC changes over a large geographic area in a regular and consistent way (Singh et al. 2012; Kindu et al. 2013). Remote sensing has been influenced by significant progress in several technologies, such as advanced data processing algorithms, Geographical Information Systems, and Satellite Systems, which have contributed to improving and expanding its applications (Nagendra et al. 2013; Politi et al. 2016; Agapiou 2016; Vihervaara et al. 2017; Côté et al. 2018; Wang et al. 2019). These advantages have attracted great interest in the scientific community.

Regarding data sources of the 51 articles, 35% used ground inventory data, followed by Landsat imagery (25%), household survey and documents (18%), and Aerial photographs (10%) while the rest utilized Google Earth and Very High resolution imagery (Ikonos, quick bird) (Fig. 2.4).

Overall, about half of the studies used imagery/aerial photographs as sources of datasets with higher frequency in recent years (Figs. 2.3 and 2.4). The rich archive and spectral resolution of satellite images are the most important reasons for their

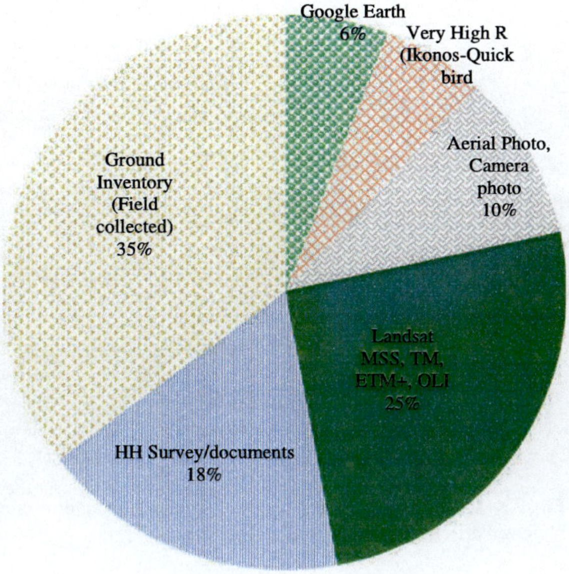

Fig. 2.4 Data sources of reviewed articles ($n = 51$). MSS = Multispectral Scanner System, TM = Thematic Mapper, ETM+ = Enhanced Thematic Mapper Plus, OLI = Operational Land Imager

use (Lillesand et al. 2015). Thus, imagery datasets have become major sources for LULC change because of repetitive coverage at short time intervals, which is useful for tracking changes over longer periods of time and at more varied temporal scales than what is typically done with field experiments or ground inventory (Fan et al. 2007; Kindu et al. 2013; Lillesand et al. 2015).

2.5 Changes of LULCs in and Surrounding the ECFs

Changes of LULC, particularly forest cover, in and surrounding the ECFs differed in the reviewed articles. Of the 51 articles, 13 (25%) showed increasing, 34 (67%) found declining, and only four (8%) identified no trends of change. Of these, 20 articles (39%) looked at the area inside the church forests, and 31 (61%) were focused on the surrounding areas of the church forests. Among those conducted in the church forests, 14 (27%) showed declining forest cover, followed by 4 (8%) no change and 2 (4%) increasing trends of forest cover. The articles conducted in the surrounding areas of church forest showed 21 (41%) declining, and 10 (20%) increasing trends (Fig. 2.5).

Mixed trends of LULC changes were also reported by previous studies in other landscapes covering different parts of the country, most of which revealed a declining trend of forest cover (Tegene 2002; Kindu et al. 2013, 2020; Temesgen et al. 2013; Meshesha et al. 2014; Hailemariam et al. 2016). For example, Dessie and Kleman (2007) reported more than 82% of high forest conversion to other LULC types in

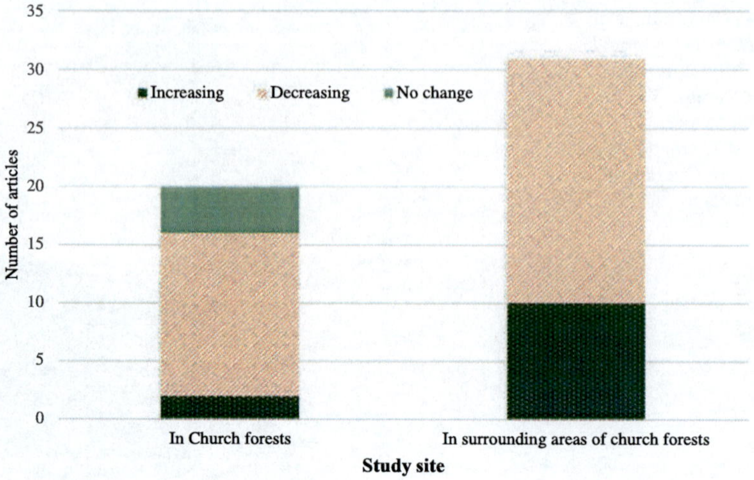

Fig. 2.5 Overall forest cover changes in and surrounding areas of church forests based on the reviewed articles ($n = 51$)

the south-central Rift Valley of Ethiopia within about 28 years (1972–2000). Kindu et al. (2013) also revealed that about 95% of woodlands and 59% of natural forests that existed in 1973 have been converted to other LULC types in the past four decades in the Munessa-Shashemene landscape of the Ethiopian highlands. Also, Hailemariam et al. (2016) pointed out the loss of forests and gain of farmlands at the same magnitude in Bale mountain of Ethiopia during 1985–2015. In Ghana, Campbell (2005) studied the sustainability of sacred groves or forests (forests of special religious importance to a particular culture) between 1968 and 1990 and found a greater decline in area extent and tree species than in non-sacred forests. However, Osuri et al. (2014) reported a decline of forest cover within sacred groves between 2000 and 2010 in a human modified landscape of India's Western Ghats. Contrary to these, an increase in forest cover was observed by Bewket (2002) in non-church forest areas of the Chemoga watershed within the Blue Nile, which was attributed to community afforestation programs.

2.6 Drivers of Change

Various drivers determine the trajectories of LULC dynamics (Campbell 2005; Kindu et al. 2015; Batunacun et al. 2019). Understanding such drivers is very important to enhance the ability of intervention strategies for sustainable land use systems (Mottet et al. 2006; Kindu et al. 2015; Kindu et al. 2018). We identified multiple drivers of LULC changes in and surrounding the ECFs from the systematically reviewed articles (Table 2.1).

Table 2.1 Summary of reasons/drivers of LULC changes in and the surrounding areas ECFs. Frequency refers to the number of studies mentioning a specific driver

Study sites	Overall change	Reasons/drivers for change	Frequency	Study area/zones
	Increasing	Stone walls	1	South Gondar
In church forests	Decreasing	Livestock grazing	4	South Gondar, North Shewa, Asebot (West Harerge)
		Isolation	3	
		Small sizes	3	
		Edge effect	2	
		Seedling mortality and regeneration decline	2	
		Shifts in values	2	
		Replacement by exotic species	1	
		Lack of laws and enforcement	1	
		Disturbance	1	
		Forest fire	1	
		Illegal wood loggers	1	
In surrounding areas of church forests	Increasing	Exclosures	7	South and East Tigray, Central Tigray
		Rehabilitation	2	
		Protection	2	
		Plantation	1	
		Land abandonment	1	
		Awareness	1	
	Decreasing	Conversion to cropland	11	South Gondar, East and West Gojjam, Central and South Tigray, West and Southwest Shewa Addis Ababa, Gamo Gofa
		Population pressure	8	
		Government policies	5	
		Livestock grazing	3	
		Settlement expansion	2	
		Construction and furniture	2	
		Infrastructure developments	2	
		Fire wood collection	2	

(continued)

Table 2.1 (continued)

Study sites	Overall change	Reasons/drivers for change	Frequency	Study area/zones
		Regime changes	1	
		Drought cycles	1	
		Inadequate exclosures	1	
		Erosion of cultural values	1	

In those studies that were carried out within church forests, we found one driver for increasing and 11 drivers for declining trends of changes. The use of stone walls as an effective conservation tool in the South Gondar Zone (Woods et al. 2017) was the only driver for the increasing trend of church forests. Stone walls are erected by communities around the perimeters of the church forests to demarcate the boundary and/or protect the interior of the forests. Woods et al. (2017) evaluated the effectiveness of these walls at protecting ecological conditions by examining tree and seedling communities among church forests with and without walls. They found that the density and species richness of seedlings was significantly higher in forests with a wall. They also revealed forests with walls had seedlings of many native tree species that were not found in forests without walls.

For the declining trends of changes within church forests, the top three drivers mentioned in three or more studies were livestock grazing (Wassie 2007; Wassie et al. 2009a, b; Amare et al. 2016; Reynolds et al. 2017), isolation (Cardelus et al. 2013, 2017; Scull et al. 2016) and small sizes (Aerts et al. 2016; Cardelus et al. 2013, 2017). Similarly, edge effects (Cardelus et al. 2017, 2019), increasing mortality rates of seedlings and declining regeneration (Tilahun 2015; Cardelus et al. 2019) and shifts in the values of church forests and emphasis on built structures within them (Klepeis et al. 2016; Orlowska and Klepeis 2018) were each indicated in two studies each as drivers of declining trends of church forests. Remaining drivers, such as disturbance (Aerts et al. 2016), replacement by exotic species (Eucalypts) (Cardelus et al. 2017), forest fire (Yadav and Mekonnen 2013) and illegal wood loggers (Yadav and Mekonnen 2013) were mentioned in one study each as responsible for the declining trends. The set of studies in church forests that showed drivers of all the decreasing trends were conducted in the South Gondar, North Shewa, and Asebot (West Harerge) Zones of Ethiopia.

The most mentioned influential factor affecting church forests was livestock grazing. Livestock grazing has been reported to cause tree damage through trampling and selectively grazing particular species that result in loss of species richness and diversity (Hiernux 1998; Mayer et al. 2006; Vandenberghe et al. 2007; Giday et al. 2018). In addition, isolation, small sizes, and edge effects were the other drivers negatively influencing forest health. In managing forest fragments, Hill and Curran (2003) suggested the importance of quantifying the effects of forest area, shape, and isolation on tree species diversity, and comparing their impacts with those of other environmental variables. Laurance et al. (2002) reported that small forests are more susceptible to edge effects and their fragment isolation disconnects them from seed sources. Haddad et al. (2015) also pointed out that the loss of area, increase in isolation, and greater exposure to human land uses along fragment edges initiate long-term changes to the structure and function of the remaining fragments and lead to altered patterns of selection and trait evolution.

We also identified six drivers for increasing change and 11 for declining change in forest cover in those studies conducted in the surrounding areas of ECFs. Of these, the role of establishment of exclosures (Descheemaeker et al. 2006a, b; Descheemaeker et al. 2008; Mekuria and Veldkamp 2012; Mekuria and Yami 2013; Meire et al. 2013; Birhane et al. 2017) was the main driver indicated in seven studies responsible for the increasing trends of forest cover. Similarly, intense rehabilitation activities (Descheemaeker et al. 2006a; Nyssen et al. 2014) and protection activities (de Mûelenaere et al. 2014; Zeratsion and Tejada Moral 2019) were mentioned as positive drivers in two studies each. Additionally, continuous plantation (Zeratsion and Tejada Moral 2019), abandonment of marginal farm and rangeland because of food aid and import of food from other regions (Belay et al. 2013), and growing awareness of local communities (Yeshaneh et al. 2013) were mentioned in one study each as positive drivers. Those sets of studies carried out in the surrounding areas of church forests show all the drivers for the increasing trend were conducted in the South, East, and Central Tigray Zones of Ethiopia.

Exclosures are areas closed-off or protected from human and animal interference by a physical boundary or a social fence (that means restriction to access) to allow regeneration of native vegetation and reduce further degradation (Aerts et al. 2009; Birhane et al. 2017; Mekuria et al. 2017). The role of such approaches as drivers or causes of improving forest cover was also reported in different studies in other areas of the country or elsewhere (e.g., Habrova and Pavlis 2017; Qasim et al. 2017; Gebregziabher and Soltani 2019; Ebabu et al. 2020). Success of intense rehabilitation and protection activities as drivers of improving vegetation cover were also reported (Jong 2010; Pancel 2016; Duncan et al. 2016; Bremer et al. 2019; Kinoti and Mwende 2019; Poudel et al. 2020). Similarly, expansion of plantations was reported a driver for increasing forest cover (Pirard et al. 2017; Pliscoff et al. 2020).

As main drivers responsible for the declining trends of forest area, we found conversion to croplands in 11 studies (Zeleke and Hurni 2001; Aynekulu et al. 2006; Awoke 2010; Assefa and Bork 2014; Assefa et al. 2017; Amsalu et al. 2018; Gebremicael et al. 2018; Tekalign et al. 2018; Wondie 2018; Angessa et al. 2019; Berihun et al. 2019), population pressure in eight studies (Pankhurst 1995; Selassie and Ayanna 2013; Yeshaneh et al. 2013; Demissie et al. 2017; Tekalign et al. 2018; Berihun et al. 2019; Deribew and Dalacho 2019), government policies in five studies (Zeleke and Hurni 2001; Yeshaneh et al. 2013; Mekonnen et al. 2016; Tekalign et al. 2018; Deribew and Dalacho 2019) and conversion to grazing lands in three studies (Aynekulu et al. 2006; Assefa et al. 2017; Amsalu et al. 2018). We also found the drivers, i.e., infrastructure developments (Gebremicael et al. 2018; Angessa et al. 2019), fire wood collection (Assefa and Bork 2014; Amsalu et al. 2018), harvesting of wood for construction and furniture (Amsalu et al. 2018; Angessa et al. 2019), and settlement expansion (Pankhurst 1995; Angessa et al. 2019) in two studies each. The failure of exclosures to compensate for the lost vegetation (Gebresamuel et al. 2010) and erosion of traditional cultural values (Daye and Healey 2015) were also found as drivers in one study each. All of the studies, which were carried out in the surrounding areas of church forests that show all the drivers for declining trends, were conducted in South Gondar, East & West Gojjam, Central and South Tigray, West and Southwest Shewa, and Gamo Gofa Zones as well as Addis Ababa, Ethiopia.

Many studies in other areas of the country or elsewhere have also demonstrated conversions of forests for cropland expansion as being drivers or causes for LULC changes (e.g., Bewket 2002; Hurni et al. 2005; Dessie and Kleman 2007; Kidane et al. 2012; Woodhouse 2012; Messerli et al. 2013; Gashaw et al. 2014; Kindu et al. 2015). Similarly, population pressure was also reported as the other factor responsible for observed LULC changes (e.g., Garedew et al. 2012; Pellikka et al. 2018; Mucova et al. 2018; Alturk and Konukcu 2019; Adenle et al. 2020; Liu et al. 2020; Elagouz et al. 2020; Tang et al. 2020). The change, failure, or absence of government policies have also played a considerable role as drivers for forest cover loss. Similar effects were observed in other countries, including Australia (Simmons et al. 2018), China (Batunacun et al. 2019), Kenya (Schürmann et al. 2020) and Brazil (Azevedo-Ramos et al, 2020).

Like factors within church forests, the other contributory influential driver leading to changes in the surrounding areas of ECFs was livestock grazing. Free grazing occurs in the remaining forests due to the low carrying capacity of the available existing grazing lands, which could induce forest degradation through hindering natural regeneration of the remaining forests (Teketay et al. 2010). Similar effects were also reported in other studies (Giday et al. 2018; Muhati et al. 2018; Souther et al. 2019; Dominguez Lozano et al. 2020).

2.7 Conclusions

Based on our review of a systematically selected set of studies and the wider scholarly literature related to LULC changes in and surrounding areas of ECFs, we conclude the following.

(1) The importance of considering the underlying causes of changes within and surrounding areas of ECFs in order to develop effective management plans.
(2) The systematic review showed a mix of changes within and surrounding areas of ECFs, which reveal the location dependencies.
(3) Conversion to croplands, population pressure, livestock grazing, government policies, isolation, and small sizes were reported as the most common drivers for the declining trends of forest area within and surrounding the ECFs among published articles. The top three identified drivers or causes of changes are consistent with studies conducted elsewhere and need urgent action to stop

further degradation. The survival to date of many ECFs is a testament to the historical and contemporary importance of the cultural, spiritual, and religious values with which they are associated. This review also revealed that the changes in such values put additional pressure on the ECFs that also need attention. On the other hand, for the increasing trends of forest area, existence of stone walls, exclosures, intense rehabilitation, and protection activities were mentioned as major drivers.

(4) Remote Sensing and ground inventory based approaches are and will remain essential tools for change detection studies for church forests within and surrounding ECFs.

(5) There are exciting research opportunities for future studies on LULC changes in and surrounding ECFs, especially studies that investigate future patterns of changes, as well as their consequences in case studies based on high resolution images and field, collected data. While facing changes within and surrounding areas of ECFs, there are, remote sensing based methodological challenges in highly structured church forests for generating and delivering high quality information. Thus studies are urgently needed to answer the question of if and how to effectively monitor these changes.

Acknowledgments This work was made possible through the church forest project entitled "Exploring Church Forests as ecosystem baseline for restoration concepts in the Ethiopian highlands" by the financial support of Bavarian State Chancellery (Bayerische Staatskanzlei), which is gratefully acknowledged.

References

Adenle AA, Eckert S, Adedeji OI, Ellison D, Speranza CI (2020) Human-induced land degradation dominance in the Nigerian Guinea Savannah between 2003–2018. Remote Sen Appl Soc Environ 19:100360

Aerts R, Nyssen J, Haile M (2009) On the difference between "exclosures" and "enclosures" in ecology and the environment. J Arid Environ 73(8):762–763

Aerts R, van Overtveld K, November E, Wassie A, Abiyu A, Demissew S, Daye DD, Giday K, Haile M, TewoldeBerhan S, Teketay D, Teklehaimanot Z, Binggeli P, Deckers J, Friis I, Gratzer G, Hermy M, Heyn M, Honnay O, Paris M, Sterck FJ, Muys B, Bongers F, Healey JR (2016) Conservation of the Ethiopian church forests: threats, opportunities and implications for their management. Sci Total Environ 551–552:404–414

Agapiou A (2016) Remote sensing heritage in a petabyte-scale: satellite data and heritage Earth Engine© applications. Int J Digit Earth 10(1):85–102

Alturk B, Konukcu F (2019) Modeling land use/land cover change and mapping morphological fragmentation of agricultural lands in Thrace Region/Turkey. Environ Dev Sustain 5(3):75

Amare D, Mekuria W, T/wold T, Belay B, Teshome A, Yitaferu B, Tessema T, Tegegn B (2016) Perception of local community and the willingness to pay to restore church forests: the case of Dera district, northwestern Ethiopia. For Trees Livelihoods 25(3):173–186

Amsalu N, Bezie Y, Fentahun M, Alemayehu A, Amsalu G (2018) Use and conservation of medicinal plants by indigenous people of Gozamin Wereda, East Gojjam Zone of Amhara Region, Ethiopia: an Ethnobotanical approach. Evid Based Complement Altern Med 1:1–23

Angessa AT, Lemma B, Yeshitela K (2019) Land-use and land-cover dynamics and their drivers in the central highlands of Ethiopia with special reference to the Lake Wanchi watershed. GeoJournal 13(4):267

Armenteras D, Murcia U, González TM, Barón OJ, Arias JE (2019) Scenarios of land use and land cover change for NW Amazonia: Impact on forest intactness. Global Ecol Conserv 17:e00567

Assefa D, Rewald B, Sandén H, Rosinger C, Abiyu A, Yitaferu B, Godbold DL (2017) Deforestation and land use strongly effect soil organic carbon and nitrogen stock in Northwest Ethiopia. CATENA 153:89–99

Assefa E, Bork H-R (2014) Deforestation and forest management in Southern Ethiopia: investigations in the Chencha and Arbaminch Areas. Environ Manage 53(2):284–299

Awoke BG (2010) The role of geoinformation technology for predicting and mapping of forest cover spatio-temporal variability: Dendi District case study, Ethiopia. J Sustain Dev Africa 12(6)

Aynekulu E, Birhane E, Wubneh W, Begashaw N (2006) Monitoring and evaluating land use/land cover change using participatory geographic information system (PGIS) tools: a case study of Begasheka Watershed, Tigray, Ethiopia. Electron J Inf Syst Dev Countries 3(25):1–10

Aynekulu E, Aerts R, Denich M, Negussie A, Friis I, Demissew S, Boehmer HJ (2016) Plant diversity and regeneration in a disturbed isolated dry Afromontane forest in northern Ethiopia. Folia Geobot 51(2):115–127

Azevedo-Ramos C, Moutinho P, Arruda VLdS, Stabile MCC, Alencar A, Castro I, Ribeiro JP (2020) Lawless land in no man's land: the undesignated public forests in the Brazilian Amazon. Land Use Policy 99:104863

Batunacun WR, Lakes T, Yunfeng H, Nendel C (2019) Identifying drivers of land degradation in Xilingol, China, between 1975 and 2015. Land Use Policy 83:543–559

Belay B, Pötzelsberger E, Sisay K, Assefa D, Hasenauer H (2018) The carbon dynamics of dry tropical afromontane forest ecosystems in the Amhara Region of Ethiopia. Forests 9(1):18

Belay KT, van Rompaey A, Poesen J, van Bruyssel S, Deckers J, Amare K (2013) Spatial analysis of land cover changes in Eastern Tigray (Ethiopia) from 1965 to 2007: are there signs of a forest transition? Land Degrad Dev 26(7):680–689

Berhane A, Totland Ø, Moe SR (2013) Woody plant assemblages in isolated forest patches in a semiarid agricultural matrix. Biodivers Conserv 22(11):2519–2535

Berihun ML, Tsunekawa A, Haregeweyn N, Meshesha DT, Adgo E, Tsubo M, Masunaga T, Fenta AA, Sultan D, Yibeltal M (2019) Exploring land use/land cover changes, drivers and their implications in contrasting agro-ecological environments of Ethiopia. Land Use Policy 87:104052

Bewket W (2002) Land cover dynamics since the 1950s in Chemoga Watershed, Blue Nile Basin, Ethiopia. Mountain Res Dev 22(3):263–269

Birhane, E.; Mengistu, T., Seyoum, Y., Hagazi, N., Putzel, L., Mekonnen Rannestad, M., Kassa, H. (2017) Exclosures as forest and landscape restoration tools: lessons from Tigray Region, Ethiopia. Int. For. Rev, vol. 19, no. 4.

Bongers F, Wassie A, Sterck JF, Bekele T, Teketay D (2006) Ecological restoration and church forests in northern Ethiopia. J Drylands 1(1):35–44

Bremer LL, Wada CA, Medoff S, Page J, Falinski K, Burnett KM (2019) Contributions of native forest protection to local water supplies in East Maui. Sci Total Environ 688:1422–1432

Campbell MO'N (2005) Sacred groves for forest conservation in Ghana's coastal Savannas: assessing ecological and social dimensions. Singap J Trop Geogr 26(2):151–169

Cardelús C, Scull P, Hair J, Baimas-George M, Lowman M, Eshete A (2013) A preliminary assessment of Ethiopian sacred grove status at the landscape and ecosystem scales. Diversity 5(2):320–334

Cardelús CL, Scull P, Wassie Eshete A, Woods CL, Klepeis P, Kent E, Orlowska I (2017) Shadow conservation and the persistence of sacred church forests in northern Ethiopia. Biotropica 49(5):726–733

Cardelús CL, Woods CL, Bitew Mekonnen A, Dexter S, Scull P, Tsegay BA, Bond-Lamberty B (2019) Human disturbance impacts the integrity of sacred church forests, Ethiopia. Plos One 14(3):e0212430

Côté J-F, Fournier RA, Luther JE, van Lier OR (2018) Fine-scale three-dimensional modeling of boreal forest plots to improve forest characterization with remote sensing. Remote Sens Environ 219:99–114

Dayamba SD, Djoudi H, Zida M, Sawadogo L, Verchot L (2016) Biodiversity and carbon stocks in different land use types in the Sudanian Zone of Burkina Faso, West Africa. Agr Ecosyst Environ 216:61–72

Daye DD, Healey JR (2015) Impacts of land-use change on sacred forests at the landscape scale. Global Ecol Conserv 3:349–358

de Jong W (2010) Forest rehabilitation and its implication for forest transition theory. Biotropica 42(1):3–9

de Mûelenaere S, Frankl A, Haile M, Poesen J, Deckers J, Munro N, Veraverbeke S, Nyssen J (2014) Historical landscape photographs for calibration of Landsat land use/cover in the northern Ethiopian highlands. Land Degrad Dev 25(4):319–335

Demissie F, Yeshitila K, Kindu M, Schneider T (2017) Land use/Land cover changes and their causes in Libokemkem District of South Gonder, Ethiopia. Remote Sen Appl Soc Environ 8:224–230

Deribew KT, Dalacho DW (2019) Land use and forest cover dynamics in the North-eastern Addis Ababa, central highlands of Ethiopia. Environ Syst Res 8(1):13

Desalegn T, Cruz F, Kindu M, Turrión MB, Gonzalo J (2014) Land-use/land-cover (LULC) change and socioeconomic conditions of local community in the central highlands of Ethiopia. Int J Sust Dev World 21(5):406–413

Descheemaeker K, Nyssen J, Poesen J, Raes D, Haile M, Muys B, Deckers S (2006) Runoff on slopes with restoring vegetation: a case study from the Tigray highlands, Ethiopia. J Hydrol 331(1–2):219–241

Descheemaeker K, Nyssen J, Rossi J, Poesen J, Haile M, Raes D, Muys B, Moeyersons J, Deckers S (2006) 'Sediment deposition and pedogenesis in exclosures in the Tigray highlands, Ethiopia. Geoderma 132(3–4):291–314

Descheemaeker K, Poesen J, Borselli L, Nyssen J, Raes D, Haile M, Muys B, Deckers J (2008) Runoff curve numbers for steep hillslopes with natural vegetation in semi-arid tropical highlands, northern Ethiopia. Hydrol Process 22(20):4097–4105

Dessie G, Kleman J (2007) Pattern and magnitude of deforestation in the South Central Rift Valley Region of Ethiopia. Mt Res Dev 27(2):162–168

Doffana ZD, Yildiz F (2017) Sacred natural sites, herbal medicine, medicinal plants and their conservation in Sidama, Ethiopia. Cogent Food Agric 3(1):61

Domínguez Lozano F, Zurdo Jorda J, Sánchez de Dios R (2020) The role of demography and grazing in the patterns of endangerment of threatened plants. Global Ecol Conserv 23:e01151

Duncan C, Primavera JH, Pettorelli N, Thompson JR, Loma RJA, Koldewey HJ (2016) Rehabilitating mangrove ecosystem services: a case study on the relative benefits of abandoned pond reversion from Panay Island, Philippines. Mar Pollut Bull 109(2):772–782

Ebabu K, Tsunekawa A, Haregeweyn N, Adgo E, Meshesha DT, Aklog D, Masunaga T, Tsubo M, Sultan D, Fenta AA, Yibeltal M (2020) Exploring the variability of soil properties as influenced by land use and management practices: a case study in the Upper Blue Nile basin, Ethiopia. Soil Tillage Res 200:104614

Elagouz MH, Abou-Shleel SM, Belal AA, El-Mohandes MAO (2020) Detection of land use/cover change in Egyptian Nile Delta using remote sensing. Egypt J Remote Sens Space Sci 23(1):57–62

Fairman JG, Nair US, Christopher SA, Mölg T (2011) Land use change impacts on regional climate over Kilimanjaro. J Geophys Res 116(D3):32141

Fan F, Weng Q, Wang Y (2007) Land use and land cover change in guangzhou, china, from 1998 to 2003, based on landsat TM /ETM+ imagery. Sensors 7(7):1323–1342

FAO (2011) Assessing forest degradation—towards the development of globally applicable guidelines. Italy, Rome

Foley JA, Asner GP, Costa MH, Coe MT, DeFries R, Gibbs HK, Howard EA, Olson S, Patz J, Ramankutty N, Snyder P (2007) Amazonia revealed: forest degradation and loss of ecosystem goods and services in the Amazon Basin. Front Ecol Environ 5:25–32

Garedew E, Sandewall M, Soderberg U (2012) A dynamic simulation model of land-use, population, and rural livelihoods in the Central Rift Valley of Ethiopia. Environ Manage 49(1):151–162
Gashaw T, Tulu T, Argaw M, Worqlul AW, Tolessa T, Kindu M (2018) Estimating the impacts of land use/land cover changes on ecosystem service values: the case of the Andassa watershed in the Upper Blue Nile basin of Ethiopia. Ecosyst Serv 219–228
Gashaw T, Bantider A, Mahari A (2014) Population dynamics and land use/land cover changes in Dera District, Ethiopia. Global J Biol Agric Health Sci 2:137–140
Gebregziabher D, Soltani A (2019) Exclosures in people's minds: perceptions and attitudes in the Tigray region, Ethiopia. Forest Policy Econ 101:1–14
Gebremicael TG, Mohamed YA, van der Zaag P, Hagos EY (2018) Quantifying longitudinal land use change from land degradation to rehabilitation in the headwaters of Tekeze-Atbara Basin, Ethiopia. Sci Total Environ 622–623:1581–1589
Gebresamuel G, Singh BR, Dick Ø (2010) Land-use changes and their impacts on soil degradation and surface runoff of two catchments of Northern Ethiopia. Acta Agriculturae Scandinavica, Section B—Plant Soil Sci 60(3):211–226
Giday K, Humnessa B, Muys B, Taheri F, Azadi H (2018) Effects of livestock grazing on key vegetation attributes of a remnant forest reserve: the case of Desa'a Forest in northern Ethiopia. Global Ecol Conserv 14:e00395
Guariguata MR, Balvanera P (2009) Tropical forest service flows: Improving our understanding of the biophysical dimension of ecosystem services. For Ecol Manage 258(9):1825–1829
Habrova H, Pavlis J (2017) Dynamic response of woody vegetation on fencing protection in semi-arid areas; case study: pilot exclosure on the Firmihin Plateau, Socotra Island. Saudi J Biol Sci 24(2):338–346
Haddad NM, Brudvig LA, Clobert J, Davies KF, Gonzalez A, Holt RD, Lovejoy TE, Sexton JO, Austin MP, Collins CD, Cook WM, Damschen EI, Ewers RM, Foster BL, Jenkins CN, King AJ, Laurance WF, Levey DJ, Margules CR, Melbourne BA, Nicholls AO, Orrock JL, Song DX, Townshend JR (2015) Habitat fragmentation and its lasting impact on Earth's ecosystems. Sci Adv 1(2):e1500052
Hailemariam S, Soromessa T, Teketay D (2016) Land use and land cover change in the bale mountain eco-region of Ethiopia during 1985 to 2015. Land 5(4):41
Hiernaux P (1998) Effects of grazing on plant species composition and spatial distribution in rangelands of the Sahel. Plant Ecol 138:191–202
Hill JL, Curran PJ (2003) Area, shape and isolation of tropical forest fragments: effects on tree species diversity and implications for conservation. J Biogeogr 30:1391–1403
Hurni H, Tato K, Zeleke G (2005) The implications of changes in population, land use, and land management for surface runoff in the upper Nile basin area of Ethiopia. Mt Res Dev 25(2):147–154
Kassawmar T, Zeleke G, Bantider A, Gessesse GD, Abraha L (2018) A synoptic land change assessment of Ethiopia's Rainfed agricultural area for evidence-based agricultural ecosystem management. Heliyon 4(11):e00914
Kidane Y, Stahlmann R, Beierkuhnlein C (2012) Vegetation dynamics, and land use and land cover change in the Bale Mountains, Ethiopia. Environ Monitor Assess 184(12):7473–7489
Kindu M, Schneider T, Teketay D, Knoke T (2013) Land use/land cover change analysis using object-based classification approach in Munessa-Shashemene landscape of the Ethiopian highlands. Remote Sens 5(5):2411–2435
Kindu M, Schneider T, Teketay D, Knoke T (2015) Drivers of land use/land cover changes in Munessa-Shashemene landscape of the south-central highlands of Ethiopia. Environ Monit Assess 187(7):60
Kindu M, Schneider T, Teketay D, Knoke T (2016) Changes of ecosystem service values in response to land use/land cover dynamics in Munessa-Shashemene landscape of the Ethiopian highlands
Kindu M, Schneider T, Döllerer M, Teketay D, Knoke T (2018) Scenario modelling of land use/land cover changes in Munessa-Shashemene landscape of the Ethiopian highlands. Sci Total Environ 622–623:534–546

Kindu M, Angelova D, Schneider T, Döllerer M, Teketay D, Knoke T (2020) Monitoring of urban growth patterns in rapidly growing Bahir Dar City of Northwest Ethiopia with 30 year landsat imagery record. ISPRS Int J Geo Inf 9(9):1–19

Kinoti KD, Mwende KM (2019) Spatial-temporal assessment of forest rehabilitation along Mt. Kenya east forest buffer zone using remote sensing and GIS. Int J Environ Plann Dev 5(1)

Klepeis P, Orlowska IA, Kent EF, Cardelús CL, Scull P, Wassie Eshete A, Woods C (2016) Ethiopian church forests: a hybrid model of protection. Hum Ecol 44(6):715–730

Knoke T, Paul C, Hildebrandt P, Calvas B, Castro LM, Härtl F, Döllerer M, Hamer U, Windhorst D, Wiersma YF, Curatola Fernández GF, Obermeier WA, Adams J, Breuer L, Mosandl R, Beck E, Weber M, Stimm B, Haber W, Fürst C, Bendix J (2016) Compositional diversity of rehabilitated tropical lands supports multiple ecosystem services and buffers uncertainties. Nat Commun 7:11877

Knoke T, Paul C, Rammig A, Gosling E, Hildebrandt P, Härtl F, Peters T, Richter M, Diertl K-H, Castro LM, Calvas B, Ochoa S, Valle-Carrión LA, Hamer U, Tischer A, Potthast K, Windhorst D, Homeier J, Wilcke W, Velescu A, Gerique A, Pohle P, Adams J, Breuer L, Mosandl R, Beck E, Weber M, Stimm B, Silva B, Verburg PH, Bendix J (2020) Accounting for multiple ecosystem services in a simulation of land-use decisions: does it reduce tropical deforestation? Global Change Biol 26:2403–2420

Laurance WF, Lovejoy TE, Vasconcelos HL, Bruna EM, Didham RK, Stouffer PC, Gascon C, Bierregaard RO, Laurance SG, Sampaio E (2002) Ecosystem decay of Amazonian forest fragments: a 22-year investigation. Conserv Biol 16(3):605–618

Lewis SL, Edwards DP, Galbraith D (2015) Increasing human dominance of tropical forests

Lillesand TM, Kiefer RW, Chipman JW (2015) Remote sensing and image interpretation. Wiley, Hoboken

Liu SA, Li X, Chen D, Duan Y, Ji H, Zhang L, Chai Q, Hu X (2020) Understanding land use/land cover dynamics and impacts of human activities in the Mekong delta over the last 40 years. Global Ecol Conserv 22:e00991

Mayer AC, Stöckli V, Konold W, Kreuzer M (2006) Influence of cattle stocking rate on browsing of norway spruce in subalpine wood pastures. Agrofor Syst 66(2):143–149

Meire E, Frankl A, de Wulf A, Haile M, Deckers J, Nyssen J (2013) Land use and cover dynamics in Africa since the nineteenth century: warped terrestrial photographs of North Ethiopia. Reg Environ Change 13(3):717–737

Mekasha A, Gerard B, Tesfaye K, Nigatu L, Duncan AJ (2014) Inter-connection between land use/land cover change and herders'/farmers' livestock feed resource management strategies: a case study from three Ethiopian eco-environments. Agric Ecosyst Environ 188:150–162

Mekonnen M, Sewunet T, Gebeyehu M, Azene B, Melesse MA (2016) GIS and remote sensing-based forest resource assessment, quantification, and mapping in Amhara Region, Ethiopia. In: Melesse AM, Abtew W (eds) Landscape dynamics, soils and hydrological processes in varied climates [Online]. Springer International Publishing, Cham. Available at https://doi.org/10.1007/978-3-319-18787-7

Mekuria W, Barron J, Dessalegn M, Adimassu Z, Amare T, Wondie M (2017) Exclosures for ecosystem restoration and economic benefits in Ethiopia: a catalogue of management options

Mekuria W, Veldkamp E (2012) Restoration of native vegetation following exclosure establishment on communal grazing lands in Tigray, Ethiopia. Appl Veg Sci 15(1):71–83

Mekuria W, Yami M (2013) Changes in woody species composition following establishing exclosures on grazing lands in the lowlands of Northern Ethiopia. Afr J Environ Sci Technol 7(1):30–40

Meshesha DT, Tsunekawa A, Tsubo M, Ali SA, Haregeweyn N (2014) Land-use change and its socio-environmental impact in Eastern Ethiopia's highland. Reg Environ Change 14(2):757–768

Messerli P, Heinimann A, Giger M, Breu T, Schönweger O (2013) From 'land grabbing' to sustainable investments in land: Potential contributions by land change science. Curr Opin Environ Sustain 5(5):528–534

Mohamed A, Worku H, Kindu M (2019) Quantification and mapping of the spatial landscape pattern and its planning and management implications a case study in Addis Ababa and the surrounding area, Ethiopia. Geol Ecol Landscapes 99(3):1–12

Mottet A, Ladet S, Coqué N, Gibon A (2006) Agricultural land-use change and its drivers in mountain landscapes: a case study in the Pyrenees. Agr Ecosyst Environ 114(2–4):296–310

Mucova SAR, Filho WL, Azeiteiro UM, Pereira MJ (2018) Assessment of land use and land cover changes from 1979 to 2017 and biodiversity & land management approach in Quirimbas National Park, Northern Mozambique, Africa. Global Ecol Conserv 16:e00447

Muhati GL, Olago D, Olaka L (2018) Land use and land cover changes in a sub-humid Montane forest in an arid setting: a case study of the Marsabit forest reserve in northern Kenya. Global Ecol Conserv 16:e00512

Nagendra H, Lucas R, Honrado JP, Jongman RHG, Tarantino C, Adamo M, Mairota P (2013) Remote sensing for conservation monitoring: assessing protected areas, habitat extent, habitat condition, species diversity, and threats. Ecol Ind 33:45–59

Nigatu W, Dick ØB, Tveite H (2014) GIS based mapping of land cover changes utilizing multi-temporal remotely sensed image data in Lake Hawassa Watershed, Ethiopia. Environ Monitor Assess 186(3):1765–1780

Nyssen J, Frankl A, Haile M, Hurni H, Descheemaeker K, Crummey D, Ritler A, Portner B, Nievergelt B, Moeyersons J, Munro N, Deckers J, Billi P, Poesen J (2014) Environmental conditions and human drivers for changes to north Ethiopian mountain landscapes over 145 years. Sci Total Environ 485–486:164–179

Orlowska I, Klepeis P (2018) Ethiopian church forests: a socio-religious conservation model under change. J East Afr Stud 12(4):674–695

Osuri AM, Madhusudan MD, Kumar VS, Chengappa SK, Kushalappa CG, Sankaran M (2014) Spatio-temporal variation in forest cover and biomass across sacred groves in a human-modified landscape of India's Western Ghats. Biol Cons 178:193–199

Pancel L (2016) Forest restoration and rehabilitation in the tropics. In Pancel L, Köhl M (eds) Tropical forestry handbook. Springer, Berlin

Pankhurst R (1995) The history of deforestation and afforestation in ethiopia prior to world war I. Northeast Afr Stud 2(1):119–133

Paz-Kagan T, Shachak M, Zaady E, Karnieli A (2014) Evaluation of ecosystem responses to land-use change using soil quality and primary productivity in a semi-arid area, Israel. Agr Ecosyst Environ 193:9–24

Pellikka PKE, Heikinheimo V, Hietanen J, Schäfer E, Siljander M, Heiskanen J (2018) Impact of land cover change on aboveground carbon stocks in Afromontane landscape in Kenya. Appl Geogr 94:178–189

Pielke RA, Marland G, Betts RA, Chase TN, Eastman JL, Niles JO, Niyogi DDS, Running SW (2002) The influence of land-use change and landscape dynamics on the climate system: Relevance to climate-change policy beyond the radiative effect of greenhouse gases. Philos Trans Ser A Math Phys Eng Sci 360(1797):1705–1719

Pirard R, Petit H, Baral H (2017) Local impacts of industrial tree plantations: an empirical analysis in Indonesia across plantation types. Land Use Policy 60:242–253

Pliscoff P, Simonetti JA, Grez AA, Vergara PM, Barahona-Segovia RM (2020) Defining corridors for movement of multiple species in a forest-plantation landscape. Global Ecol Conserv 23:e01108

Politi E, Rowan JS, Cutler MEJ (2016) Assessing the utility of geospatial technologies to investigate environmental change within lake systems. Sci Total Environ 543:791–806

Poudel A, Sasaki N, Abe I (2020) Assessment of carbon stocks in oak forests along the altitudinal gradient: a case study in the Panchase Conservation Area in Nepal. Global Ecol Conserv 23:e01171

Qasim S, Gul S, Shah MH, Hussain F, Ahmad S, Islam M, Rehman G, Yaqoob M, Shah SQ (2017) Influence of grazing exclosure on vegetation biomass and soil quality. Int Soil Water Conserv Res 5(1):62–68

Reynolds TW, Stave KA, Sisay TS, Eshete AW (2017) Changes in community perspectives on the roles and rules of church forests in northern Ethiopia: Evidence from a panel survey of four Ethiopian Orthodox communities. Int J Commons 11(1):355

Schürmann A, Kleemann J, Fürst C, Teucher M (2020) Assessing the relationship between land tenure issues and land cover changes around the Arabuko Sokoke Forest in Kenya. Land Use Policy 95:104625

Scull P, Cardelús CL, Klepeis P, Woods CL, Frankl A, Nyssen J (2016) The resilience of Ethiopian church forests: interpreting aerial photographs, 1938–2015. Land Degrad Dev 28(2):450–458

Selassie GY, Ayanna G (2013) Effects of different land use systems on selected physico-chemical properties of soils in Northwestern Ethiopia. J Agric Sci 5(4)

Simmons BA, Law EA, Marcos-Martinez R, Bryan BA, McAlpine C, Wilson KA (2018) Spatial and temporal patterns of land clearing during policy change. Land Use Policy 75:399–410

Singh P, Thakur JK, Kumar S, Singh UC (2012) Assessment of Land Use/Land Cover Using Geospatial Techniques in a Semi-arid Region of Madhya Pradesh, India. In: Thakur JK, (ed) Geospatial Techniques for Managing Environmental Resources. Springer, Dordrecht, the Netherland/New York, NY, USA, pp 152–163

Souther S, Loeser M, Crews TE, Sisk T (2019) Complex response of vegetation to grazing suggests need for coordinated, landscape-level approaches to grazing management. Global Ecol Conserv 20:e00770

Tang J, Li Y, Cui S, Xu L, Ding S, Nie W (2020) Linking land-use change, landscape patterns, and ecosystem services in a coastal watershed of southeastern China. Global Ecol Conserv 23:e01177

Tedla S (1998) Forest/vegetation cover in some parts of Ethiopia: a story to tell', Walia

Tegene B (2002) Land-cover/land-use changes in the Derekolli Catchment of the South Welo Zone of Amhara Region, Ethiopia. East Afr Soc Sci Res Rev 18(1):1–20

Teketay D, Lemenih M, Bekele T, Yemshaw Y, Feleke S, Tadesse W, Moges Y, Hunde T, Nigussie D (2010) Forest Resources and Challenges of Sustainable Forest Management and Conservation in Ethiopia. In: Bongers F, Tennigkeit T (eds) Degraded forests in Eastern Africa. Management and restoration. Earthscan, London, Washington, DC, pp 19–64

Tekalign M, Flasse C, Frankl A, van Rompaey A, Poesen J, Nyssen J, Muys B (2018) Forest cover loss and recovery in an East African remnant forest area: Understanding its context and drivers for conservation and sustainable ecosystem service provision. Appl Geogr 98:133–142

Temesgen H, Nyssen J, Zenebe A, Haregeweyn N, Kindu M, Lemenih M, Haile M (2013) Ecological succession and land use changes in a lake retreat area (Main Ethiopian Rift Valley). J Arid Environ 91:53–60

Tilahun A (2015) The contribution of Ethiopian orthodox Tewahido church in forest management and its best practices to be scaled up in north Shewa Zone of Amhara Region, Ethiopia. Agric for Fish 4(3):123

Tsegaye D, Moe SR, Vedeld P, Aynekulu E (2010) Land-use/cover dynamics in Northern Afar rangelands, Ethiopia. Agric Ecosyst Environ 139(1–2):174–180

Tura TT, Soromessa T, Leta S, Argew M, Eshetu Z (2016) Urban church forests for local temperature regulation: implications the role of managing and incorporating urban green space in urban planning. Taiwania 61(1):305–313

Turner BL, Meyfroidt P, Kuemmerle T, Müller D, Roy Chowdhury R (2020) Framing the search for a theory of land use. J Land Use Sci 15:489–508

Vandenberghe C, Freléchoux F, Moravie M-A, Gadallah F, Buttler A (2007) Short-term effects of cattle browsing on tree sapling growth in mountain wooded pastures. Plant Ecol 188(2):253–264

Vieilledent G, Grinand C, Rakotomalala FA, Ranaivosoa R, Rakotoarijaona J-R, Allnutt TF, Achard F (2018) Combining global tree cover loss data with historical national forest cover maps to look at six decades of deforestation and forest fragmentation in Madagascar. Biol Cons 222:189–197

Vihervaara P, Auvinen A-P, Mononen L, Törmä M, Ahlroth P, Anttila S, Böttcher K, Forsius M, Heino J, Heliölä J, Koskelainen M, Kuussaari M, Meissner K, Ojala O, Tuominen S, Viitasalo M, Virkkala R (2017) How Essential Biodiversity Variables and remote sensing can help national biodiversity monitoring. Global Ecol Conserv 10:43–59

Vitousek PM (1997) Human domination of Earth's ecosystems. Science 277(5325):494–499
Votrin V (2005) The orthodoxy and sustainable development a potential for broader involvement of the orthodox churches in Ethiopia and Russia. Environ Dev Sustain 7(1):9–21
Wallner A, Elatawneh A, Schneider T, Kindu M, Ossig B, Knoke T (2017) Remotely sensed data controlled forest inventory concept. Euro J Remote Sens 51(1):75–87
Wang L, Jia M, Yin D, Tian J (2019) A review of remote sensing for mangrove forests: 1956–2018. Remote Sens Environ 231:111223
Wassie A (2002) Opportunities, constraints, and prospects of the Ethiopian orthodox tewahido churches in conserving forest resources: the case of churches in south Gonder, northern Ethiopia, MSc thesis, Uppsala, Sweden
Wassie A (2007) Ethiopian church forests: opportunities and challenges for restoration, PhD thesis, Wageningen, The Netherlands, Wageningen Univeristy
Wassie A, Sterck FJ, Bongers F (2010) Species and structural diversity of church forests in a fragmented Ethiopian Highland landscape. J Veg Sci 21(5):938–948
Wassie A, Sterck FJ, Teketay D, Bongers F (2009) Effects of livestock exclusion on tree regeneration in church forests of Ethiopia. For Ecol Manage 257(3):765–772
Wassie A, Sterck FJ, Teketay D, Bongers F (2009) Tree regeneration in church forests of ethiopia: effects of microsites and management. Biotropica 41(1):110–119
Wassie A, Teketay D (2006) Soil seed banks in church forests of northern Ethiopia: implications for the conservation of woody plants. Flora Morpholog, Distrib Funct Ecol Plants 201(1):32–43
Woldemedhin TT, Teketay D (2016) Forest conservation tradition of the Ethiopian Orthodox Tewahdo Church: a case study in West Gojjam Zone, north-western Ethiopia. Symb Bot Upsal 38:57–73
Wondie A (2018) Ecological conditions and ecosystem services of wetlands in the Lake Tana Area, Ethiopia. Ecohydrol Hydrobiol 18(2):231–244
Wondie M, Schneider W, Melesse AM, Teketay D (2011) Spatial and temporal land cover changes in the Simen mountains national park, a world heritage site in Northwestern Ethiopia. Remote Sens 3(4):752–766
Woodhouse P (2012) New investment, old challenges. Land deals and the water constraint in African agriculture. J Peasant Stud 39(3–4):777–794
Woods CL, Cardelús CL, Scull P, Wassie A, Baez M, Klepeis P (2017) Stone walls and sacred forest conservation in Ethiopia. Biodivers Conserv 26(1):209–221
Yadav RH, Mekonnen E (2013) Degradation of preserved and isolated church forests in Asebot, Ethiopia. Euro Acad Res 1(4)
Yeshaneh E, Wagner W, Exner-Kittridge M, Legesse D, Blöschl G (2013) Identifying land use/cover dynamics in the Koga catchment, Ethiopia, from multi-scale data, and implications for environmental change. ISPRS Int J Geo Inf 2(2):302–323
Yilma G, Derero A (2020) Carbon stock and woody species diversity patterns in church forests along church age gradient in Addis Ababa, Ethiopia. Urban Ecosyst 551:404
Zeleke G, Hurni H (2001) Implications of land use and land cover dynamics for mountain resource degradation in the Northwestern Ethiopian Highlands. Mt Res Dev 21(2):184–191
Zeratsion BT, Tejada Moral M (2019) Urban forest change detection in Endayesus area, Tigray, Ethiopia. Cogent Food Agric 5(1):263

Chapter 3
Land Use Land Cover Changes and Forest Fragmentation on the Surrounding of Selected Church Forests in Ethiopia

Nesibu Yahya, Getachew Kebede, Melaku Getachew, and Abeje Eshete

Abstract Church forest (CF) is the most important biodiversity hotspot in Ethiopia. Land use land cover change (LULCC) is among the threats of CF. However, little is known about the quantitative trends of LULCC and forest fragmentation around CF. The aim of this study was to analyze LULCC and forest fragmentation of four selected CF and their surrounding (within 3 km buffer). The selected churches were: 1. Kedest Arsema Monastery (KAM) from Tigray region, 2. Woji Abune Aregawi Church (WAA) from Amhara region, 3. Abune Tekele Hayimanot Monastery of Bole Bulbula (AT-B) from Addis Ababa 4. Abune Tekele Haymanot Monastery of Wolayta (AT-W) from Southern Nations, Nationalities and People's Regional State (SNNPRS). Satellite images of Landsat TM (1984/86/87, 1995&1999) and OLI_TIRS (2017/18) were used. After preprocessing activities conducted, the images were classified using pixel-based supervised classification technique. By using the classified LULC maps, the landscape metrics of fragmented patches was calculated. We found that farmland and forest/shrub land are the most dominant LULC classes of the study areas. Reduction of forestland was recorded within the boundary of AT-W (from 28.2 to 17.6%), WAA (from 2.4 to 1.4%) and AT-B (from 7.1 to 3.5%) between 1986/87 and 2017. Forest area increased from 7.3 to 26.8% inside the boundary of KAM between 1986 and 2018, while shrub land decreased from 92.7 to 73.2% during the same period. The study areas except KAM shows a decreasing trend of forest cover and shrub land inside their boundary over the study periods. Forest fragmentations and attrition of natural vegetation were also observed from WAA, AT-B, and AT-W along the study periods. Thus, effective conservation activities are urgent in and around CF to conserve the biological resources of CF.

Keywords Land use land cover · Forest fragmentation · Patch analysis

---

N. Yahya (✉) · G. Kebede · M. Getachew · A. Eshete
Ethiopian Environment and Forest Research Institute, PO Box: 24536, 1000 Addis Ababa, Ethiopia

M. Kindu et al. (eds.), *State of the Art in Ethiopian Church Forests and Restoration Options*, https://doi.org/10.1007/978-3-030-86626-6_3

3.1 Introduction

Land use land cover change (LULCC) is a concern of global environmental challenges as it is one of a major factor to the global climate change (IPCC 2014). It is also one of the primary causes of environmental challenges such as soil erosion, habitat loss, and hydrological imbalance (Costa et al. 2003). The most significant pattern of LULCC is an expansion of crop and pastor land at the expense of natural forest ecosystems (Lambin and Meyfroidt 2011). The global forest cover converted to other ecosystems at an average rate of 0.8% over the past four to five decades, and it is the major driver for loss of ecosystem functions and services (Millennium Ecosystem Assessment 2005). According to modeling information, the global cropland increased from 2.7 million km^2 to 15 million km^2, while the global pastureland increased from 5.2 million km^2 to 34 million km^2 for the last three hundred years (Goldewijk 2001; Goldewijk et al. 2011). In tropical biome, more than 55% of the new agricultural land expanded at the expense of intact forests, and another 28% derived from disturbed forests between the 1980–2000 period (Gibbs et al. 2010). The urgency of sustainability in the scarce land resources, call global research concern, and priority about LULCC. In some parts of the world, forest patches conserved by local communities in the form of sacred forests for religious and cultural purposes. These practices are mostly frequent in Asia (Wadley and Colfer 2004) and Africa (Alohou et al. 2017), particularly in India, Japan, and Ethiopia (Bhagwat et al. 2014). Besides the cultural and religious benefit of sacred forest, sacred forests provide many other ecosystem services such as provisioning of fuel wood, honey, animal feed, clean water, and medicinal plant (Tura et al. 2017). The conservation of biological diversity is one of the most intrinsic roles of sacred forests because of its species composition usually comprises many rare and endemic flora and fauna. Many scientific evidences show that church forests (i.e., a type of sacred forest) in Ethiopia hold up the highest diversity of flora (Aerts et al. 2006; Wassie et al. 2010).

Ethiopian church forests are typically found as isolated patches on the mosaic landscape, and there is a high anthropogenic pressure through encroachment and LULCC. The increasing rate of LULCC over time triggers deforestation and forest degradation in the central and northern highlands of Ethiopia, where many church forests are present (Nyssen et al. 2004). In addition, church forests are threatened by many anthropogenic disturbances either directly or indirectly; cutting, grazing, drought, and fires are among them (Bongers et al. 2006; Cardelús et al. 2017).

Understanding of the past trends of LULCC is fundamental information in formulating appropriate biodiversity conservation policy and strategy both at national and local scale. Analyzing the pattern of deforestation and forest fragmentation is helpful to implement appropriate land use planning. Such knowledge is crucial to implement sustainable management of forest. Understanding the LULCC pattern around church forest gives a better hint about the degree of anthropogenic pressure from surrounding. The spatial distribution and patch characteristics within church compound and their surrounding are crucial because the existence of a species within a patch could be explained by various patch factors (such as size of forest patch, perimeter of forest patch… etc.). Associating LULCC with forest fragmentation at different temporal scale is basic information to plan appropriate action to conserve church forests. The objectives of this study are therefore to: (1) quantify the LULCC of church compound and the surrounding buffer zone (3 km) of selected church forests using multi-temporal image analysis for the year (1984/86/87, 1995/99 and 2017/18) and (2) characterize the changes associated with forest fragmentation occurring at class scale.

3.2 Methodology

3.2.1 Description of the Study Area

This study was conducted on the surrounding buffer zone of selected church forests in Ethiopia. Four church forests (two from the northern, one from central and one from southern highland of Ethiopia) were selected (Fig. 3.1). The detailed description of the study areas is summarized in Table 3.1.

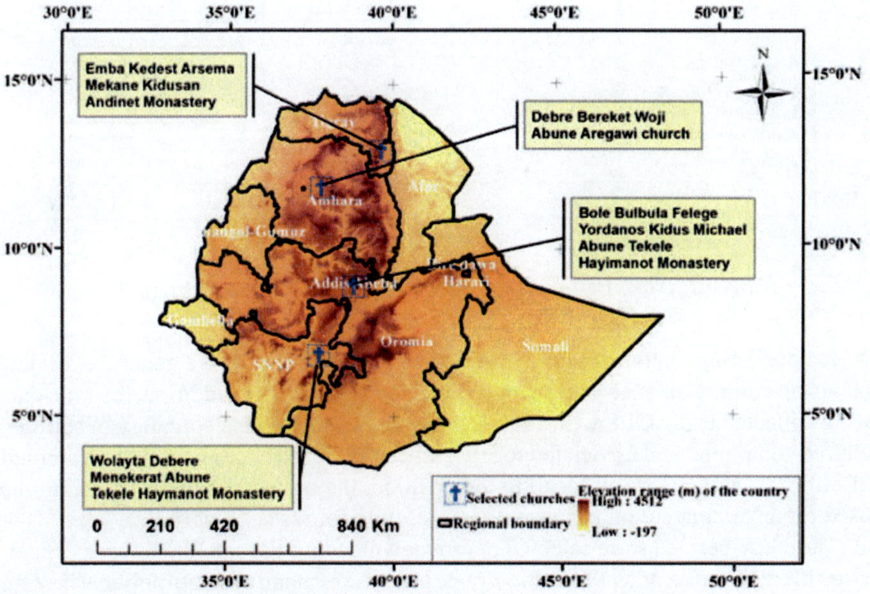

Fig. 3.1 Location map of the selected church forests

Table 3.1 Location description and extent of the study areas

No	Name of the church	Longitude	Latitude	Altitude (m a.s.l)	Total area within church compound (ha)	Total area within 3 km buffer zone (ha)
1	Debre Bereket Woji Abune Aregawi church (WAA)	37° 48′ 54.1″	11° 55′ 44.2″	1993	35.3	4278
2	Emba Kedest Arsema Mekane Kidusan Andinet Monastery (KAM)	39° 38′ 59.7″	13° 01′ 26.7″	2419	41	4153
3	Bole Bulbula FelegeYordanos Kidus Michael AbuneTekele Hayimanot Monastery in Addis Ababa (AT- B)	38° 48′ 19.9″	08° 56′ 54.8″	2212	21	5044
4	Wolayta Debere Menekerat Abune Tekele Haymanot Monastery (AT-W)	37° 45′ 35.5″	06° 52′ 37.7″	2122	25	4261
Total					**122.3**	**17,736**

3.2.2 *Remote Sensing Data Collection and Classification*

At the beginning, reconnaissance survey was performed to have a general overview about the current land use land cover types of the study area. In addition, location data were collected using GPS with church communities to map the boundary of selected church compound and associated forests. Since the GPS devices have certain amount of error (at least ±3 m) in locating the position, the boundary shape file cannot be used for any demarcation purpose. We use it only for research purposes.

The study periods were selected before downloading the satellite images considering the 1991 government transition since LULCC respond to socio-political driving forces particularly in Ethiopian context. Moreover, the availability of cloud free satellite image (except for AT-W where getting cloud free image was difficult especially on the peak of the mountain) determines the study periods. Then, multi-dated Landsat imageries were downloaded from the United States Geological Survey (USGS) archive (Table 3.2). The images were geometrically and radio metrically

Table 3.2 Summary of remote sensing images used in this study

Image type	Scene (Path/Row)	Processed spatial resolution (m)	Acquisition date
Landsat 5_TM	169/055	30 × 30	22/11/1984
Landsat 5_TM	168/051	30 × 30	05/01/1986
Landsat 5_TM	169/052	30 × 30	22/11/1987
	167/168/054	60 × 60	22/11/1987
Landsat 5_TM	169/055	30 × 30	21/01/1995
Landsat 5_TM	168/051	30 × 30	25/01/1999
Landsat 8_OLI_TIRS	169/055	30 × 30	02/02/2017
	169/052	30 × 30	05/11/2017
	167/168/054	30 × 30	05/11/2017
Landsat 8_OLI_TIRS	168/051	30 × 30	22/02/2018

(i.e., Top of Atmosphere) corrected. Image preprocessing techniques such as subsetting, layer stacking, and image enhancement were conducted for the downloaded images. The pixel resampling technique was implemented to align images that have unequal spatial resolution. The spectral bands of the satellite images covering the blue, green, red, near, and shortwave infrared spectrum (L5 TM: bands 1–5 and 7/L8 OLI: bands 2–7) were selected for analysis (Table 3.2).

The team of expertise independently conducted the image analyses of the selected churches. After satellite data prepossessing, a minimum of 30 and a maximum of 50 plots per class were evenly distributed over the images (for all study images). During classification, random forest algorithm (for the case of AT-W) (Breiman Leo 2001) and maximum likelihood algorithm of supervised classification technique were implemented. The classification was conducted bearing in mind the definition of each type of LULC class (Table 3.3). The classification of target classes was assisted by calculating spectral indices such as normalized difference vegetation index (NDVI).

Meanwhile, the collected samples (GPS waypoints) from the ground were used to compute accuracy assessment of the classified images. From each land use land cover classes, a minimum of 16 and a maximum of 30 random sample points (i.e., independent points from those used for training sample) were generated for accuracy assessment. For accuracy assessment and training samples, raw satellite image and Google earth image have been used as source of data. Accuracy assessment was conducted for the most recent classified image. Kappa statistics and accuracy parameters were calculated to evaluate the accuracy of classified images (van Vliet et al. 2011).

Table 3.3 Descriptions of land use land cover classes used in the analysis

Land use land cover types	Descriptions
Bare land	Land covered with vascular plants, composed of exposed rock, sand, and soil surface
Grass land	Land covered with open grassland. Mostly found in flat areas that uses as grazing
Settlement	Land covered by residences, road networks, buildings and small industrial areas in both rural and urban areas
Shrub land	Land with shrubs/bushes canopy cover $\leq$10%. Shrubs and bushes are woody perennial plants, 2 m in height at maturity in situ
Farmland	Part of land prepared mainly for growing agricultural crops
Forest land	Land spanning at least 0.5 ha covered by trees attaining a height of at least 2 m and a canopy cover of at least 20%
Water bodies	Any type of surface water such as lakes including other intermittent ponds

3.2.3 Patch Dynamics Analysis

By using the classified LULCC map, the landscape metrics of forest on the surrounding church forests were estimated. For each fragment at landscape and class level, various patch parameters that were expected to have relation with biodiversity were calculated. Among various patch matrices, we calculate number of patches and mean patch size (Aerts et al. 2016).

The preprocessing and classification activities were carried out on QGIS and R software program (R Core Team 2015). All other spatial analyses including patch dynamics and mapping were performed on ARC GIS and patch analyst extension of ARC GIS (ESRI 2014).

3.3 Results and Discussions

3.3.1 Classification Accuracies

The overall accuracies of the four classified satellite images range from 77 to 90% with their kappa statistics ranging from 0.69 to 0.85. (Landis and Koch 1977) categorized the value of kappa statistics into slight agreement (Kappa range = 0.0–0.2), fair agreement (Kappa range = 0.21–0.40), moderate agreement (Kappa range = 0.41–0.60), substantial agreement (Kappa range = 0.61–0.80), and almost perfect agreement (Kappa range = 0.81–1.00). Hence, the results of kappa statistics of the classified images suggest substantial to almost perfect agreement (Table 3.4).

Table 3.4 Confusion matrix result of the classified images for 2017 and 2018

1. Kedest Arsema Monastery (KAM) for the 2018 classification

		Reference						
		Bare land	Farmland	Forest land	Settlement	Shrub land	Water body	Total
Classified	Bare land	**24**	0	0	0	0	6	**30**
	Farmland	2	**28**	0	0	0	0	**30**
	Forest land	0	0	**25**	0	5	0	**30**
	Settlement	0	0	0	**27**	0	3	**30**
	Shrub land	0	6	0	0	**24**	0	**30**
	Water body	0	1	0	0	0	**29**	**30**
Total		26	35	25	27	29	**38**	**180**

Overall accuracy: 87%
Kappa statistics: 0.85

2. Woji Abune Aregawi church (WAA) for the 2017 classification

		Reference						
		Forest land	Farmland	Shrub land	Settlement			Total
Classified	Forest land	18	0	0	0	–	–	18
	Farm land	2	20	2	4	–	–	28
	Shrub land	0	0	18	0	–	–	18
	Settlement	0	0	0	16	–	–	16
Total		20	20	20	**20**			80

Overall accuracy: 90% Kappa statistics: 0.83

3. Abune Tekele Hayimanot Monastery of Bole Bulbula (AT-B) for the 2017 classification

		Reference						
		Forest land	Farmland	Settlement				Total
Classified	Forest land	17	2	2	–	–	–	21
	Farm land	2	18	3	–	–	–	23
	Settlement	1	0	15	-	-	-	16
Total		20	20	**20**				60

Overall accuracy: 83%
Kappa statistics: 0.75

4. Abune Tekele Haymanot Monastery of Wolayta (AT-W) for the 2017 classification

		Reference						
		Farmland	Forest land	Grassland	Settlement			Total
Classified	Farm land	21	7	2	0	–	–	30

(continued)

Table 3.4 (continued)

1. Kedest Arsema Monastery (KAM) for the 2018 classification

		Reference						
		Bare land	Farmland	Forest land	Settlement	Shrub land	Water body	Total
	Forest land	2	26	2	0	–	–	30
	Grass land	10	1	19	0	–	–	30
	Settlement	1	1	2	26	–	–	30
Total		34	35	25	**26**	–	–	120

Overall accuracy: 77%
Kappa statistics: 0.69

3.3.2 *State of Land Use Land Cover Changes at Spatial Scale of Within 3 km Buffer Zone of Church Forests*

Within the 3 km buffer zone of KAM, we identified six possible land use land cover classes, namely bare land, farmland, forest, settlement, shrub land, and water body (Table 3.5, Fig. 3.2). Within this buffer zone, farmland, forest, and settlement show an increasing trend constantly, while shrub land shows a constant declining trend throughout the study periods (1986–2018). In contrary, the trend of bare land shows variation among study periods. There is a declining trend of bare lands between 1986 and 1999 and shows increasing trend between 1999 and 2018. The water body within the buffer of KAM has not been observed from the 1986 and 1999 image, since it is an artificial pond that constructed in 2014 for irrigation purpose.

Within the buffer zone of WAA, farmland, forest, settlement, and shrub land were the types of land use land cover (Table 3.5, Fig. 3.3). In this buffer zone, the proportion of settlement and farmland increased between 1987 and 2017, while forest and shrub land declined in the same period.

We identified three types of land use land cover inside and the surrounding landscape of AT-B. Within this buffer zone, the land use land cover classes include settlement, farmland, and forest. As this church compound located inside the capital Addis Ababa, settlement was the second dominant class after farmland. (Table 3.5, Fig. 3.4). We found that forest and farmland were decreasing rapidly within the buffer zone of AT-B during 1987 and 2017 mainly due the fast growing pattern of settlement.

The LULC result of AT-W showed that the area within the buffer zone comprises four land use land cover classes, namely farmland, forest, grassland, and settlement (Table 3.5, Fig. 3.5). The trend of forest and grassland was continuously declining throughout study periods (1984–2017). However, the trend of other land cover classes varied among the two study periods. Hence, farmland has shown increasing trend between 1984 and 1995, while it slightly declined between 1995 and 2017. The trend of settlement, within this buffer zone, has been almost constant between 1984 and

Table 3.5 Land use land cover change results of the four selected church forests within 3 km buffer zone

1. Kedest Arsema Monastery (KAM)								
Land use land cover classes	1986		1999		2018		1986–1999	1999–2018
	Area (ha)	Proportion (%)	Area (ha)	Proportion (%)	Area (ha)	Proportion (%)	Rate of change (ha/yr)	Rate of change (ha/yr)
Bare land	43	1.04	38	0.92	52	1.25	−0.38	0.74
Farmland	1398	33.66	1568	37.76	2545	61.28	13.08	51.42
Forest	19	0.46	30	0.72	61	1.47	0.85	1.63
Settlement	0	0.00	3	0.07	11	0.26	0.23	0.42
Shrub land	2693	64.84	2514	60.53	1477	35.56	−13.77	−54.58
Water body	0	0.00	0	0.00	7	0.17	0.00	0.37
Total	**4153**	**100**	**4153**	**100**	**4153**	**100**		
2. Woji Abune Aregawi church (WAA)								
Land use land cover classes	1987		2017		1987–2017			
	Area (ha)	Proportion (%)	Area (ha)	Proportion (%)	Rate of change (ha/yr)			
Settlement	0	0.00	39	0.90	1.3			
Farmland	3024	70.70	4002	93.60	32.6			
Forest	102	2.40	60	1.40	−1.4			
Shrub land	1151	26.90	177	4.10	−32.5			
Total	**4278**	**100**	**4278**	**100**				
3. Abune Tekele Hayimanot Monastery of Bole Bulbula (AT-B)								
Land use land cover classes	1987		2017		1987–2017			
	Area (ha)	Proportion (%)	Area (ha)	Proportion (%)	Rate of change (ha/yr)			
Farmland	3942	78.20	3217	63.80	−24.17			
Forest	358	7.10	176	3.50	−6.07			
Settlement	744	14.70	1651	32.70	30.23			
Total	**5044**	**100**	**5044**	**100**				
4. Abune Tekele Haymanot Monastery of Wolayta (AT-W)								
Land use land cover classes	1984		1995		2017		1984–1995	1995–2017
	Area (ha)	Proportion (%)	Area (ha)	Proportion (%)	Area (ha)	Proportion (%)	Rate of change (ha/yr)	Rate of change (ha/yr)
Farmland	2860	67.12	3335	78.27	3024	70.97	43.18	−14.14
Forest	1202	28.21	761	17.86	752	17.65	−40.09	−0.41

(continued)

Table 3.5 (continued)

1. Kedest Arsema Monastery (KAM)								
Land use land cover classes	1986		1999		2018		1986–1999	1999–2018
	Area (ha)	Proportion (%)	Area (ha)	Proportion (%)	Area (ha)	Proportion (%)	Rate of change (ha/yr)	Rate of change (ha/yr)
Grassland	155	3.64	134	3.14	83	1.95	−1.91	−2.32
No data	12	0.28	0	0.00	0	0.00	−1.09	0.00
Settlement	32	0.75	31	0.73	402	9.43	−0.09	16.86
Total	**4261**	**100**	**4261**	**100**	**4261**	**100**		

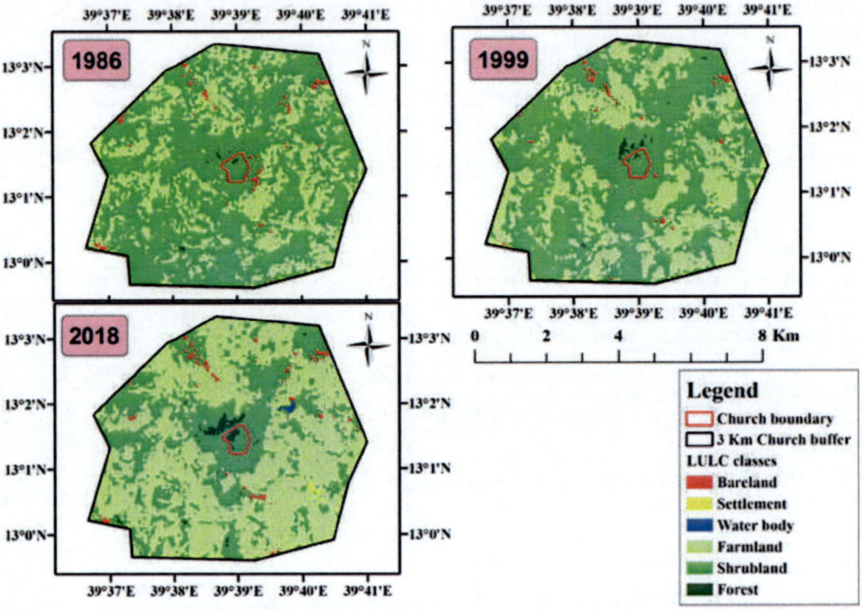

Fig. 3.2 Land use land cover map of Kedest Arsema Monastery (KAM) within 3 km buffer zone

1995 and then rapidly increased between 1995 and 2017. Changes in demographic and socio-economic issues reflected by population growth and rural–urban migration are among the major drivers of rapid settlement growth. According to CSA (2013), the population of Wolayta Sodo town increased by 17.4% between 2014 and 2017, and it is among densely populated (>300 persons/km^2 (CSA 2011). This resulted a similar increasing of demand for additional settlement. More rapid rate of urbanization was recorded from Addis Ababa city that shows a steady growth from 22 to 40% between 1999 and 2011 (Bisrat et al. 2018).

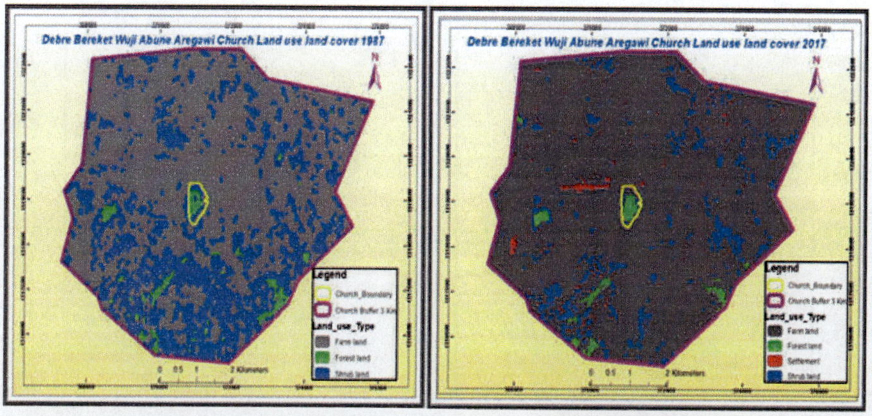

Fig. 3.3 Land use land cover map of Woji Abune Aregawi church (WAA) within 3 km buffer zone

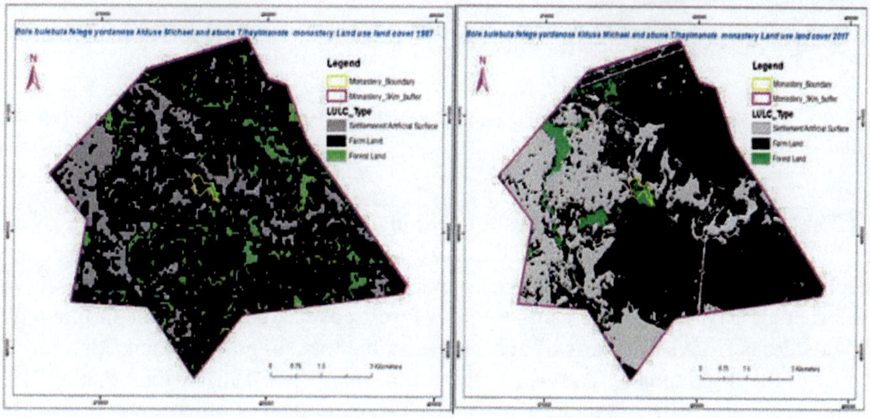

Fig. 3.4 Land use land cover map of Abune Tekele Hayimanot Monastery of Bole Bulbula (AT-B) within 3 km buffer zone

The high floristic and faunal diversity of church forests can play a significant role for biodiversity conservation (Wassie et al. 2005, 2010). However, the reduction of forest, shrub land, and grassland around church forests may have influence on church forest via exerting an extra pressure. As the livelihood of the study areas characterized by mixed farming, free grazing is a common practice on communal grassland as well as ranching to remaining natural forests. Uncertainties the boundary and lack of land ownership (Aerts et al. 2016) together with the aforementioned concern may exacerbate the threat of land—use change on the church forests.

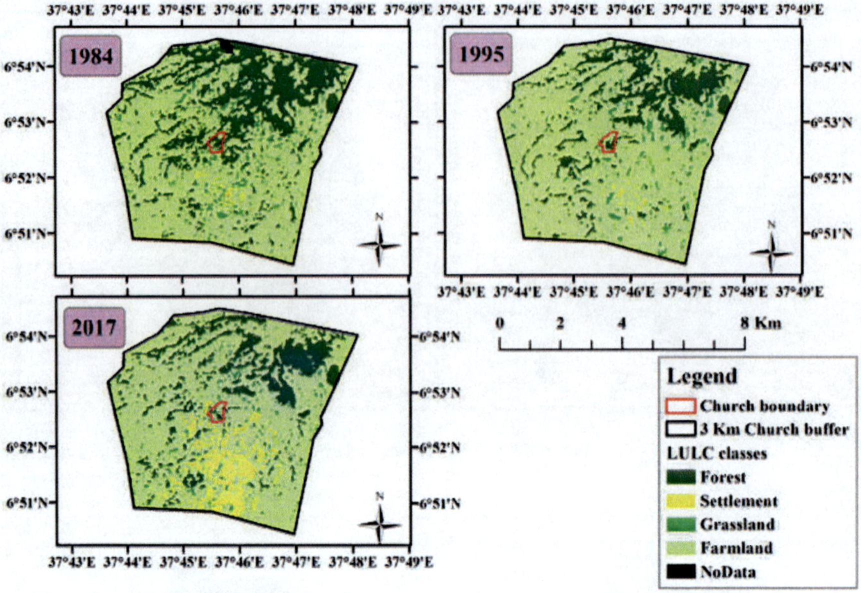

Fig. 3.5 Land use land cover map of Abune Tekele Haymanot Monastery of Wolayta (AT-W) within 3 km buffer zone

The general forest cover shows a declining trend over time for the case of all churches except KAM. Most of land use land cover change studies depict the declining trend of forest cover in many parts of Ethiopia (Gebrekidan et al. 2014; Teferi et al. 2013). The increasing trend of forest cover around KAM is in line with other studies from some parts of northern Ethiopia that suggest the increasing forest cover in the highlands of Tigray, Northern Ethiopia, since 1961 (Meire et al. 2013; Mûelenaere et al. 2014; Wondie et al. 2011). The increment of forest cover was likely from the successional conversion of shrub land into forest, which in turn implies the declining of shrub land.

The trend of farmland expansion was observed in most of the study periods at 3 km buffer zone of the churches. Similar reports were obtained that shows alarming expansion of farmland over time in the southern Ethiopia (Assefa and Bork 2014; Worku et al. 2014), in south central highland (Kindu et al. 2013), in Northwestern highland (Garede and Minale 2014; Zeleke and Hurni 2001) in Eastern highland (Meshesha et al. 2014). However, very limited studies indicated the decline of farmland area over time in northern highland due to the expansion of plantation woodlot on the subsistence farmland (Mûelenaere et al. 2014; Wondie et al. 2011).

3.3.3 State of Land Use Land Cover Changes at Spatial Scale of Church Forest Boundary

Focusing at the scale of church compound, forest was identified as a common type of land use land cover for all studied church sites (Table 3.6, Fig. 3.6). Shrub land was common for KAM and WAA at different proportion, while farmland is only found in WAA, AT-B and AT-W. Grassland and settlement were found only in AT-W and AT-B, respectively.

The spatial resolution (30 m) of Landsat image was inadequate to detect classes of LULC unless its area coverage is wide enough. All church compounds comprises grasslands, and it was detected only from AT-W due to limited spatial resolution. Inside the boundary of studied churches, the trend of LULCC along study periods was varied. Forest shows a continuous decreasing trend inside the boundary of WAA and AT-W, and it showed a continuous increasing trend inside KAM and AT-B throughout the study periods. The continuous increment of forest inside the boundary of KAM and AT-B was at the expense of shrub land and farmland, respectively. Similarly, throughout the study period, shrub land has been showing a continuous declining trend inside the boundary of KAM and WAA. In contrary to shrub land, a continuous expansion of farm land has been observed from the boundary of WAA and AT-W, while it decline inside AT-B that converted to forest and settlement. Settlement has been expanded inside AT-B between 1987 and 2017. Grassland has declined continuously throughout the study period inside the boundary of AT-W. This is due to the expansion of farmland inside the church compound of AT-W.

Considering the spatial and temporal trend of forest at the two study scales (3 km buffer and inside church boundary), the studied churches could be categorized into three groups.

The first group was a church that exhibited increment of a forest cover at both spatial scales. KAM belongs to this group, because the forest has been continuously increasing throughout the study periods and at both scales.

The second group was represented by the churches that exhibited the declining of forest cover at 3 km buffer zone scale, while the forest shows increasing trend at church boundary scale. AT-B belongs to this group since the forest was continuously declined at 3 km buffer zone while the forest increased inside the church boundary.

The third group exhibited the declining of forest cover at both spatial scales. WAA and AT-W could be categorized under this category, since the forest cover declined continuously throughout study period at both scale.

3.3.4 The State of Patch Dynamics at a Class Level

Within 3 km buffer zone of the churches, fragmentation was also considered at the scale of individual land use land cover classes (Table 3.7). Focusing on farmland, the number of patches is significantly decreasing, with increasing the mean patch

Table 3.6 Land use land cover change of the four selected church forests at the scale of church boundary

1. Kedest Arsema Monastery (KAM)								
Land use land cover classes	1986		1999		2018		1986–1999	1999–2018
	Area (ha)	Proportion (%)	Area (ha)	Proportion (%)	Area (ha)	Proportion (%)	Rate of change (ha/yr)	Rate of change (ha/yr)
Forest	3	7.32	4	9.76	11	26.83	0.08	0.37
Shrub land	38	92.68	37	90.24	30	73.17	−0.08	−0.37
Total	**41**	**100**	**41**	**100**	**41**	**100**		
2. Woji Abune Aregawi church (WAA)								
Land use land cover classes	1987		2017		1987–2017			
	Area (ha)	Proportion (%)	Area (ha)	Proportion (%)	Rate of change (ha/yr)			
Farmland	9	27.0	18	53.0	0.30			
Forest	18	50.0	14	39.0	−0.13			
Shrub land	8	23.0	3	8.0	−0.17			
Total	**35**	**100**	**35**	**100**				
3. Abune Tekele Hayimanot Monastery of Bole Bulbula (AT-B)								
Land use land cover classes	1987		2017		1987–2017			
	Area (ha)	Proportion (%)	Area (ha)	Proportion (%)	Rate of change (ha/yr)			
Farmland	16	73.9	10	47.7	−0.20			
Forest	5	25.7	11	50.0	0.20			
Settlement	0.1	0.4	0.48	2.3	0.01			
Total	**21.1**	**100**	**21.48**	**100**				

(continued)

Table 3.6 (continued)

1. Kedest Arsema Monastery (KAM)								
Land use land cover classes	1986		1999		2018		1986–1999	1999–2018
	Area (ha)	Proportion (%)	Area (ha)	Proportion (%)	Area (ha)	Proportion (%)	Rate of change (ha/yr)	Rate of change (ha/yr)
4. Abune Tekele Haymanot Monastery of Wolayta (AT-W)								
Land use land cover classes	1984		1995		2017			
	Area (ha)	Proportion (%)	Area (ha)	Proportion (%)	Area (ha)	Proportion (%)	Rate of change (ha/yr)	Rate of change (ha/yr)
Farmland	9	36.0	16	64.0	18	72.0	0.64	0.09
Forest	15	60.0	9	36.0	7	28.0	−0.55	−0.09
Grassland	1	4.0	0	0.0	0	0.0	−0.09	0.00
Total	**25**	**100**	**25**	**100**	**25**	**100**		

Fig. 3.6 Photo showing the forest of WAA **a**, AT-B **b** and AT-W **c**

size. This result clearly provides supportive information on the dramatic expansion of farmland (less fragmentation trend on farmland).

Within 3 km buffer zone of the churches, the result of patch dynamics shows two spatial pattern of forest loss (for the case of WAA, AT-B and AT-W). The first pattern was removing or loss of patches (attrition) which is indicated by the decreasing trend of both mean patch size and number of patches (i.e., the case of WAA and AT-B). The second spatial pattern was breaking apart (fragmentation) that indicated by the decreasing patch size while increasing number of patches (i.e., the case of AT-W). In the case of KAM, increasing of mean patch sizes was detected while increasing forest patches at the same time. Shrub land from both KAM and WAA shows the declining trend of mean patch size, while the number of patch was increasing which is likely to indicate the process of fragmentation over the study period.

Dynamic of patches and fragmentation could be assessed by using several metrics such as number of patches, patch density, and mean patch size (Castillo et al. 2015). Particularly, number of patch and size of the patch are related with forest fragmentation. Thus, the resulted forest attrition and fragmentation likely to have negative impact on biodiversity via altering population size, genetic diversity (Shafer 1981), pollination, and dispersal (Aizen and Feinsinger 1994). Together these changes are likely to have influence on species composition and stand structure (Forman and Godron 1981). This finding agrees with report by (Daye and Healey 2015) who found

Table 3.7 Fragmentation metrics of the four selected churches over the study periods at a class level

1. Kedest Arsema Monastery (KAM)					
Study years	LULC classes	Mean patch size	Number of patches	Total landscape area (ha)	Class area (ha)
1986	Bare land	0.38	115	4153	43
1999	Bare land	0.30	125	4153	38
2018	Bare land	0.37	145	4153	52
1986	Farmland	1.36	1029	4153	1398
1999	Farmland	2.69	584	4153	1568
2018	Farmland	8.34	305	4153	2545
1986	Forest	0.19	98	4153	19
1999	Forest	0.34	89	4153	30
2018	Forest	0.59	103	4153	61
1986	Settlement	–	–	–	–
1999	Settlement	0.14	24	4153	3
2018	Settlement	0.53	20	4153	11
1986	Shrub land	5.14	524	4153	2693
1999	Shrub land	5.66	444	4153	2514
2018	Shrub land	2.05	721	4153	1477
1986	Water body	–	–	–	
1999	Water body	–	–	–	–
2018	Water body	3.51	2	4153	7
2. Woji Abune Aregawi church (WAA)					
Study years	LULC classes	Mean patch size	Number of patches	Total landscape area (ha)	Class area (ha)
1987	Settlement	–	–	–	–
2017	Settlement	0	233	4277.5	31
1987	Farm land	31	98	4277.5	3051
2017	Farm land	86	47	4277.5	4030
1987	Forest	3	35	4277.5	100
2017	Forest	2	29	4277.5	58
1987	Shrub land	4	280	4277.5	1121
2017	Shrub land	0	509	4277.5	156
3. Abune Tekele Hayimanot Monastery of Bole Bulbula (AT-B)					
Study years	LULC classes	Mean patch size	Number of patches	Total landscape area (ha)	Class area (ha)
1987	Farm land	72.92	55.00	5043.9	4010.3
2017	Farm land	6.83	471.00	5043.9	3215.64
1987	Forest land	1.39	225.00	5043.9	311.8

(continued)

Table 3.7 (continued)

Study years	LULC classes	Mean patch size	Number of patches	Total landscape area (ha)	Class area (ha)
1. Kedest Arsema Monastery (KAM)					
2017	Forest land	1.25	139.00	5043.9	174.41
1987	Settlement	3.80	188.00	5043.9	714.3
2017	Settlement	4.83	342.00	5043.9	1652.32
4. Abune Tekele Haymanot Monastery of Wolayta (AT-W)					
Study years	LULC classes	Mean patch size	Number of patches	Total landscape area (ha)	Class area (ha)
1984	Farmland	5.76	469	4261	2860
1995	Farmland	23.61	135	4261	3335
2017	Farmland	9.42	305	4261	3024
1984	Forest	1.45	872	4261	1202
1995	Forest	1.20	695	4261	761
2017	Forest	0.92	919	4261	752
1984	Grass land	0.25	949	4261	155
1995	Grass land	0.35	550	4261	134
2017	Grass land	0.27	471	4261	83
1984	No data	12.42	1	4261	12
1995	No data	–	–	–	–
2017	No data	–	–	–	–
1984	Settlement	0.30	144	4261	32
1995	Settlement	0.28	169	4261	31
2017	Settlement	1.34	307	4261	402

extensive conversion from natural to human modified land-use on Gamo highland that trigger fragmentation of remaining forest patches.

3.4 Conclusion and Recommendations

Our result clearly shows the rapid expansion of mainly farmland and settlement around the studied church forests. In addition, the expansion occurred inside church compound. Over time, it influences the natural habitat. The reductions in both size and number of forest patches around church forest negatively affect species richness and composition of church forest. Since most church forests in Ethiopia existed in the form of isolated patch on the degraded highland landscape, forest fragmentation could be a threat on the conservation potential of church forests. Encouraging improvement of forest cover was obtained from KAM at both spatial scales that need to be scale-up to other church forests.

Therefore, we forward the following recommendations based on our findings:

- Constructing stone-wall or fence around boundary of church forests could be a solution since it is effective conservation mechanism. The local communities are more likely willing to participate in restoring church forests as the case of Dera district, North Western Ethiopia (Dagninet et al. 2016).
- Detail study is important to identify the socio-economic and ecological determinant factors to forest cover improvement in KAM.
- Further study is needed on the possibilities establishing forest corridors and buffer zone plantation to reduce the magnitude of forest fragmentation around church forest.

References

Aerts R, Van OK, Haile M et al (2006) Species composition and diversity of small Afromontane forest fragments in northern Ethiopia. Plant Ecol 187:127–142

Aerts R, Van OK, November E et al (2016) Conservation of the Ethiopian church forests: threats, opportunities and implications for their management. Sci Total Environ 551:404–414. https://doi.org/10.1016/j.scitotenv.2016.02.034

Aizen MA, Feinsinger P (1994) Forest fragmentation, pollination, and plant reproduction in a chaco dry forest, Argentina. Ecology 75:330–351

Alohou EC, Gbemavo DSJC, Mensah S et al (2017) Fragmentation of forest ecosystems and connectivity between sacred groves and forest reserves in southeastern Benin, West Africa. Trop Conserv Sci https://doi.org/10.1177/1940082917731730.

Assefa E, Bork HR (2014) Deforestation and forest management in southern Ethiopia: investigations in the Chencha and Arbaminch areas. Environ Manage 53:284–299. https://doi.org/10.1007/s00267-013-0182-x

Bhagwat SA, Nogue S, Willis KJ (2014) Cultural drivers of reforestation in tropical forest groves of the Western Ghats of India. For Ecol Manag 329:393–400

Bisrat A, Gizaw T, Gerrit S et al (2018) Influence of urbanization-driven land use/cover change on climate: The case of Addis Ababa, Ethiopia. Phys Chem Earth 105:212–223

Bongers F, Wassie A, Sterck FJ et al (2006) Ecological restoration and church forests in northern Ethiopia. Journal of the Drylands 1:35–44

Breiman Leo (2001) Random Forests. Machine Learning 45:5–32. https://www.stat.berkeley.edu/~breiman/randomforest2001.pdf, checked on 7/19/2015.

Cardelús CL, Scull P, Wassie EA et al (2017) Shadow conservation and the persistence of sacred church forests in northern Ethiopia. Biotropica 49:726–733. https://doi.org/10.1111/btp.12431

Castillo EM, Martin AG, Aladrén LAL et al (2015) Evaluation of forest cover change using remote sensing techniques and landscape metrics in Moncayo natural park (Spain). Appl Geogr 62:247–255

Costa MH, Botta A, Cardille JA (2003) Effects of large-scale changes in land cover on the discharge of the Tocantins river, Southeastern Amazonia. J Hydrol 283:206–217. https://doi.org/10.1016/S0022-1694(03)00267-1

CSA (2011) Atlas of agricultural statistics 2006/07–2010/11. Central Statistical Agency (Addis Ababa). Addis Ababa

CSA (2013) Population projection of Ethiopia for all regions at wereda level from 2014–2017. Central statistical agency (Addis Ababa). Addis Ababa. http://www.csa.gov.et/images/general/news/pop_pro_wer_2014-2017_final

Dagninet A, Wolde M, Tilaye T et al (2016) Perception of local community and the willingness to pay to restore church forests: the case of Dera district, northwestern Ethiopia. Forests, Trees Livelihood. https://doi.org/10.1080/14728028.2015.1133330

Daye D, Healey JR (2015) Impacts of land-use change on sacred forests at the landscape scale. Global Ecology Conserv 3:349–358. https://doi.org/10.1016/j.gecco.2014.12.009

ESRI (2014) ArcMap 10.2.2. Redlands, California

Forman RT, Godron M (1981) Patches and structural components for a landscape ecology. Bioscience 31:733–740

Garede N, Minale A (2014) Land use/cover dynamics in Ribb Watershed, North Western Ethiopia. J Nat Sci Res 4:16–16. http://www.iiste.org/Journals/index.php/JNSR/article/view/15029/15444

Gebrekidan W, Bantider A, Temesgen H (2014) Land use and land cover change in Ameleke watershed, South Ethiopia. Nat Sci Res 4:42–47

Gibbs HK, Ruesch AS, Achard F et al (2010) Tropical forests were the primary sources of new agricultural land in the 1980s and 1990s. In: Proceedings of the national academy of sciences of the United States of America. https://doi.org/10.1073/pnas.0910275107

Goldewijk KK (2001) Estimating global land use change over the past 300 years: the HYDE database. Global Biogeochem 15:417–433. https://doi.org/10.1029/1999GB001232

IPCC (2014) Climate change 2014: synthesis report. In: Pachauri RK, Meyer LA (eds) Contribution of working groups I, II and III to the fifth assessment report of the intergovernmental panel on climate change. Edited by core writing team, IPCC. Geneva, Switzerland p 151

Kindu M, Schneider T, Teketay D et al (2013) Land use/land cover change analysis using object-based classification approach in munessa-shashemene landscape of the Ethiopian highlands. Remote Sens 5:2411–2435. https://doi.org/10.3390/rs5052411

Lambin EF, Meyfroidt P (2011) Global land use change, economic globalization, and the looming land scarcity. In: Proceedings of the national academy of sciences of the United States of America https://doi.org/10.1073/pnas.1100480108

Landis J, Koch G (1977) The measurement of observer agreement for categorical data. Biometrics 33:159–174. https://doi.org/10.2307/2529310

Meire E, Frankl A, Wulf A et al (2013) Land use and cover dynamics in Africa since the nineteenth century, warped terrestrial photographs of North Ethiopia. Reg Environ Change 113:717–737

Meshesha DT, Tsunekawa A, Tsubo M et al (2014) Land-use change and its socio-environmental impact in Eastern Ethiopia's highland. Reg Environ Change 14:757–768. https://doi.org/10.1007/s10113-013-0535-2

Millennium Ecosystem Assessment (2005) Ecosystems and human well-being. Synthesis. Island Press, Washington DC

Mûelenaere S, Frankl A, Haile M et al (2014) Historical landscape photographs for calibration of landsat land use/cover in the northern Ethiopian highlands. Land Degrad Develop 25:319–335. https://doi.org/10.1002/ldr.2142

Nyssen J, Poesen J, Moeyersons J et al (2004) Human impact on the environment in the Ethiopian and Eritrean highlands—a state of the art. Earth Sci Rev 64:273–320

R Core Team (2015) R: a language and environment for statistical computing. http://www.R-project.org/

Shafer ML (1981) Minimum population sizes for species conservation. Bioscience 31:131–134

Teferi E, Bewket W, Uhlenbrook S et al (2013) Understanding recent land use and land cover dynamics in the source region of the upper blue Nile, Ethiopia: spatially explicit statistical modeling of systematic transitions. Agr Ecosyst Environ 165:98–117. https://doi.org/10.1016/j.agee.2012.11.007

Tura T, Soromess T, Argaw M et al (2017) Holy hills role in biodiversity conservation and mountain management: implication of multi-purpose Ethiopian orthodox tewahido church in sustainable

mountain conservation. J Biodivers Endanger Species. https://doi.org/10.4172/2332-2543.1000179
Van J, Bregt A, Hagen-Zanker A (2011) Revisiting kappa to account for change in the accuracy assessment of land-use change models. Ecol Model 222:1367–1375. https://doi.org/10.1016/j.ecolmodel.2011.01.017
Wadley RL, Colfer CJP (2004) Sacred forest, hunting, and conservation in West Kalimantan, Indonesia. Hum Ecol 32:313–338. https://doi.org/10.1023/B:HUEC.0000028084.30742.d0
Wassie A, Teketay D, Powell N (2005) Church forests in North Gondar administrative zone, Northern Ethiopia. Forests, Trees Livelihood 15:349–373. https://doi.org/10.1080/14728028.2005.9752536
Wassie A, Sterck FJ, Bongers F (2010) Species and structural diversity of church forests in a fragmented Ethiopian highland landscape. J Veg Sci 21:938–948
Wondie M, Schneider W, Melesse A et al (2011) Spatial and temporal land cover changes in the simen mountains national park, a world heritage site in Northwestern Ethiopia. Remote Sens 3:752–766. https://doi.org/10.3390/rs3040752
Worku G, Bantider A, Temesgen H (2014) Land use and land cover change in Ameleke watershed, South Ethiopia. J Nat Sci Res 4:42–47
Zeleke G, Hurni H (2001) Implications of land use and land cover dynamics for mountain resource degradation in the Northwestern Ethiopian highlands. Mt Res Dev 21:184–191. https://doi.org/10.1659/0276-4741(2001)021[0184:IOLUAL]2.0.CO;2

Chapter 4
Sacred Texts and Environmental Ethics: Lessons in Sustainability from Ethiopia

David K. Goodin

Abstract This chapter examines the cultural importance of Ethiopian church forests with respect to environmental ethics and eco-theology, both within Tewahedo tradition for restoration projects in Ethiopia and for the potential to inform and enrich Christian traditions throughout the developed and developing world. Sustainability solutions are largely dependent on local acceptance for successful implementation, a dynamic shaped by culture and religion: Sensitivity to local traditions can be crucial for project success. This essay presents specific recommendations for improving project success by identifying how the sacred forests of Ethiopia are protected by Tewahedo theology, and how this cultural protection can be increased or lost in the eyes of the people. One factor is the protective presence of wandering monastics in these forests; restoration project managers can help ensure that these forests will continue to be safeguarded by providing the forest species and other protections needed for their subsistence lifeways. Another factor is the level of cultural memory among the people regarding the religious significance of these forests. Accordingly, this essay brings forward the ancient and sacred Tewahedo text, *Śənä Fəṭrät* (ሥነ ፍጥረት). Here we find a theocentric cosmology that uniquely empowers humankind to protect the natural world. This textual tradition is reflected in the present-day practice of encompassing churches and monasteries with Edenic forests that serve as habitat for many threatened and endangered endemic species. It is a religious narrative that can further restoration efforts throughout Ethiopia by recovering cultural memory and investing it with deep historical context and theological meaning.

Keywords Chruch forests · Eco-theology · Historical theology · Traditional ecological practice · Public policy acceptance · Edenic forests

---

D. K. Goodin (✉)
McGill School of Religious Studies, Montreal, Canada
e-mail: david.goodin@mail.mcgill.ca

M. Kindu et al. (eds.), *State of the Art in Ethiopian Church Forests and Restoration Options*, https://doi.org/10.1007/978-3-030-86626-6_4

4.1 Introduction

The remnants of native Afromontane forests exist in Ethiopia today either in inaccessible regions or as small islands of biodiversity around rural churches, monasteries, and other ecclesial lands. It is estimated that between 35,000 and 55,000 of these forest fragments remain throughout Ethiopia, with most in the northern highlands. This represents less than 4% of the forest cover of what used to exist and a marked decrease from the estimated 35% of native forests that remained only century ago. Alarmingly, the native forests of Ethiopia continue to decline through both natural and anthropogenic factors, all of which necessitates immediate invention to persevere these biodiversity hot spots of endemic flora and fauna. The forest remnants have been rightly described as a genetic "Noah's Ark" for the hopes of re-greening the degraded lands of Ethiopia with native, drought-tolerant species, and for restoring critical ecosystem services needed in traditional subsistence agriculture. Put simply, preserving these forests and reforesting denuded landscapes promote both ecosystem resilience and food security in this drought-prone nation.

These forest fragments have survived into the twenty-first century due to their status as "scared lands" protected by the Ethiopian Orthodox Tewahedo Church (EOTC). However, this protection exists as cultural mores and traditions rather than through any kind of legal authority; the people refrain from harvesting or otherwise degrading these forests out of deference to their religious significance in the EOTC faith. Those who desecrate or degrade these forests face social sanctions, community shunning, and/or risk condemnation by those priests, monks, and nuns who they would otherwise look to as their intercessors before God. This has proven to be a powerful deterrent, so much so that the church forests have survived by the sheer force of these religious prohibitions alone. But as already mentioned, these remnant forests are under threat from a number of anthropogenic factors including climate change, unrestricted livestock grazing, and developmental pressure. This has placed the forests at the focal point of a false dilemma which has sustainable agriculture on one side and Afromontane biodiversity on the other. As discussed elsewhere (Goodin et al. 2019), it is not necessary to choose one over the other, since the key to food security in Ethiopia are the ecosystem services provided by these church forests.

Even so, there is another issue that must be discussed. Sustainability solutions are largely dependent on local acceptance for successful implementation, a dynamic shaped by culture and religion: *Sensitivity* to local traditions can be crucial for project success. It is crucial that reforestation scientists and project managers to fully understand exactly how these forests are being protected by the EOTC, and how reforestation plans can be developed that respect and highlight the sacred place these forests have in the EOTC. But the status of these forests as "sacred" cannot be taken for granted, and it cannot be simply assumed that these forests enjoy the same level of religious deference. The sacred status of the forests can, in fact, deteriorate in the eyes of the local people. At the same time, it is also possible to accentuate the sacred status of these forests through careful project design. The key determinants here, as will be argued below, are the size and quality of the forest, the origin and types of flora within it, and the presence (or lack) of the wandering monastics known as the "invisible saints" by the local people. Detailing exactly how the status of these forests can change in EOTC valuation will be the first topic discussed in the sections that follow.

Yet it is not just the scientists that can benefit from this discussion. A key variable in any restoration project is the level of religious knowledge of the local people regarding the "place" of the forests in EOTC theology. The same can be true of the clergy and monastics too. Taking an anecdote from my own experience with the Orthodox Church in America (OCA), because of a pressing need, priests and deacons have sometimes been ordained without having any formal training at seminary—instead, it is their Godly devotion to the faith, reputation within the community, and their knowledge of the liturgy, rituals, hymns, and other ecclesial matters that qualifies them for ministry. This is sometimes also the case in Ethiopia, especially in rural areas. Interviews conducted by Orlowska and Klepeis (2018) found that only one out of twenty-nine priests in the South Gondar Administrative Zone in northern Ethiopia could identify where in the sacred texts of the EOTC, the protection of church forests, is prescribed (p. 7f.). The consequence is that familiarity with the deep and rich history of EOTC is sometimes lacking among the clergy, which in turn can lead to a corresponding lack of awareness by the faithful in their communities. This can all result in an underappreciation of the church forests with respect to their true place in EOTC theology.

Bringing that theology forward is another aim of this article, beginning with an overview of the religious history of the EOTC, and then finally moving on to a detailed discussion of one of its ancient and underappreciated holy texts. This will serve three purposes. The first is to awaken a deeper awareness of the textual history of EOTC theology within the faith itself. The second is to bring this history to the attention of scientists and project managers so that they may engage in their work with all due respect and reverence. And finally, it is hoped that this discussion will inspire Christians outside Ethiopia. In this age of global environmental crises, Ethiopia stands uniquely qualified to speak with particular theological authority to worldwide Christian traditions: Orthodox, Catholic, and Protestant alike. The lessons for sustainability to be learned from the EOTC are a message not limited to Ethiopia, but for the entire world to hear.

4.2 Diversity and Resilience

The overall quality and health of a forest are typically described by conservation scientists as its resilience. Put simply, this is the measure of a forest to withstand environmental shocks due to its biodiversity and the interdependence of its constituent species that preserves the integrity of the whole. By way of analogy, the same can be said of EOTC theology. It too is strong, resilient, and capable of surviving throughout time, whole and uncompromised, due to the diverse heritage of its traditions—and because of this, its theological insights are capable of imparting its resilience to other Christian eco-theologies. And so, with this in mind, let me continue building this analogy, and in the process, detail for the reader the unique history of the EOTC.

Ethiopia was, and still is, at the crossroads of civilization itself. This has allowed for a unique, homegrown synthesis of religious thought from across the known world. Ethiopia's religious history begins with Jewish missionaries who converted people in the Amhara and Tigray regions of ancient Abyssinia. According to church tradition, these settlers arrived with Emperor Menilek I, who was the son of King Solomon with the Queen of Sheba (Makeda), sometimes in the late 900s BCE. They would become Beta Israel (House of Israel) and forever establish the bloodlines and languages of modern Ethiopia as a Semitic people and more importantly as heirs to the rich theological tradition of the Hebrew Bible. Because of this, the history of Ethiopia is recorded in the later books of the Hebrew Bible, beginning with the Messianic announcement in the Book of Isaiah that, "all those from [the Queen of] Sheba shall come; they shall bring gold and incense,[1] and they shall proclaim the praises of the Lord" (Isaiah 60:6, NKJV). This prophecy is also reflected in the Book of Psalms, which proclaims, "Envoys will come out of Egypt; Ethiopia will quickly stretch out her hands to [worship] God" (Psalm 68:31, NKJV). All this would be fulfilled with the visitation of the infant Jesus by the three wise men from the east, as told in the Gospel of Matthew (2:9–11), bearing gifts of gold and frankincense. Christian tradition has even given a name to one of these wise men, Saint Gaspar of Ethiopia (Nederveen 1992, p. 26).

The story of Ethiopia continues with the New Testament and beyond, beginning with an unnamed Ethiopian official studying the aforementioned Book of Isaiah, a man who was baptized by St. Philip the Evangelist (Acts 8). It is most conspicuous that Saint Luke made special note of this encounter in his testimony about the birth of the Christian church; it reveals the special place for Ethiopia in Christian history. Saint Luke also did this to unite New Testament with the Hebrew Bible by reminding the reader of an unnamed Ethiopian official (referred to as *Ebed-melech*, which mean "servant to the king") who saved the Prophet Jeremiah from death (Jeremiah 38: 7–13). It is Jeremiah who preached that God would make a New Covenant with the people (31:31–34), a prophecy fulfilled by Christ in the New Testament. The place of the Ethiopian officials at both events testifies to the fulfillment of Messianic prophecy and to the centrality of Ethiopia in the salvation history of the world.

The conversion of Ethiopia into a Christian kingdom would take place, famously, through the evangelization by Saint Frumentius, who is also known in Ethiopia as Abbā Salāmā Kaśātē Berhān (Grillmeier and Hainthaler 1996, p. 295ff.). A Greek by birth, Frumentius was taken to Ethiopia as young man along with his brother, Adesyos (known in Ethiopia as Sidrakos), by his father Mērpyos, who was a merchant. Bandits attacked the family, killing the father and taking the

[1] The Hebrew word in this verse, הָנוֹבָל, means frankincense, a plant native to Ethiopia.

two brothers as hostage. They were ransomed by King Ella Allādā (also known as Ella 'Améda), ruler of the Axumite kingdom, who brought them into his household to help educate the young prince and heir, 'Ēzānā. In time, the brothers would distinguish themselves, and when the king died, Adesyos was appointed the "master of the household" and Frumentius as "guardian of laws and scribe of Axum" until the young prince was of age (p. 299). In addition to a Greek education in philosophy and science, the two brothers also instructed the prince in the Christian faith and built a house of prayer for worship. When the prince, 'Ēzānā, became king of Axum (ruling c. 320 s–c. 360 AD), the two brothers were released from service. Adesyos returned to their family home in tire, but Frumentius journeyed to Alexandria to present himself to the Coptic Patriarch, who was none other than Saint Athanasius the Great (also known as Abbā Atnātyos) in 328 AD. The Patriarch immediately ordained Frumentius a Bishop for the region of Ag 'azi (meaning Ethiopia),[2] where he returned to preach throughout the land with the support of King 'Ēzānā.[3]

A second wave of evangelization came in the late fifth century through the efforts of a group of Syrian monks, who are known today as the Nine Saints. Their names are Abbā Aftse, Abbā Alef, Abbā Aragawi, Abbā Garima (Isaac, or Yeshaq), Abbā Guba, Abbā Liqanos, Abbā Pantelewon, Abbā Sehma, and Abbā Yem'ata. Through them, the wealth of Syriac spirituality was added to the Ethiopian theological heritage, including the celebrated mystical writings of Saints Aphrahat (fourth century), Ephrem (fourth century), and Yaʿqub (Jacob) of Serugh (sixth century). Such was the strength of the relationship that there was even a Syrian appointed as the head of the Ethiopian Church during the reign of Yekunno Amlāk from 1270 to 1285 (Tamrat 1972, p. 69ff).

The Tewahedo tradition brings all this worldwide heritage together in a synthesis that makes it purely Ethiopian. It also embraces certain aspects of traditional African spirituality through popular religion, which has seamlessly syncretized with Orthodox faith. This has resulted in a holistic and communal view of the land of one's birth, owing in part to the animistic "spirits" who inhabit that land. Berhane-Selassie (1994) documented one such popular narrative concerning these spirits. In the story, God is said to have given Adam and Eve thirty children after Eden. God would visit them frequently, counting each of her children, one by one, each time. Eve, embarrassed by this meticulous counting of her children, decided to

---

[2] It is said that Frumentius found that the people either followed the Torah or were unconverted pagans worshiping "the dragon" (Grillmeier and Hainthaler, p. 295–6). It is a curious detail in his hagiography. The reference to the dragon (evidently meaning pagan deities) is perhaps part of the reason why Tewahedo Orthodoxy has embraced iconography of St. George slaying a dragon, which is very popular today throughout Ethiopia. St. George was a Cappadocian of the early fourth century, though some sources place him in Libya when he reportedly slew a dragon and converted the pagans to Orthodoxy—a story which parallels the life of St. Frumentius. These great saints are, of course, universal to worldwide Christian heritage and venerated by Catholics and Orthodox of all confessions.

[3] King 'Ēzānā is recognized as a Tewahedo saint today.

hide fifteen of them. God, at his next visit, counted only fifteen, asked her if this was all the children she had. She said yes. God then announced, "All right, then, let the others remain invisible." Her invisible children would become the ancestors of the spirits that still live alongside the people today, most often in elder trees known as the *adbar* (p. 158f.). Because of this, an *adbar* tree becomes an auspicious communal place for prayer, rituals, and important meetings—also owing in part to the natural beauty and shade, these ancient trees provide. Veneration of the Virgin Mary (*Theotokos*) may even be celebrated under ancient *adbar* trees, with some of the same people at another time venerating a female fertility deity from traditional religion known as *Atete* (p. 161). As argued by Berhane-Selassie, "Ethiopian Orthodox Christians fill the space around them with spirits [which inform] how they relate to life on earth and define their environment and country in terms of their concepts of sin and God's covenant of mercy" (p. 155).[4] All this has become integral with Tewahedo eco-theology.

In summary, the resilience of the church forests is due, in significant part, to the strength of the Ethiopian eco-theology which sustains them. This theology represents a homegrown synthesis of a rich religious diversity from Hebrew, Coptic, Greek, Syrian, and traditional African sources, all of which represents a treasure for worldwide Christianity, and something that comes together as uniquely Tewahedo. Acknowledging and appreciating Ethiopian theology is the essential first step in protecting the church forests. But this takes us to the next topic. How exactly are the forests considered sacred?

4.3 The Sacred Forests

The term sacred can be misleading since it typically means "to be set apart" and "untouched" in certain western paradigms, especially when contrasted with the opposing term, the profane. But in EOTC theology, there is not a strong demarcation between the sacred and the profane since the very confession of faith, Tewahedo (ተዋሕዶ, *Täwaḥedo*), is a reference to God's redemption of fallen human nature through Christ's own holy flesh through the Incarnation and Crucifixion.[5] What would have been otherwise alienated from God was brought

[4] Spirits are even said to inhabit certain forests by some Muslim communities, an animism that transcends religious boundaries, at least with respect to a deference to sacred spaces in nature (Berhane-Selassie, p. 162).

[5] The EOTC, together with the Syriac Orthodox Church of Antioch, the Coptic Orthodox Church of Alexandria, the Armenian Apostolic Church, the Eritrean Orthodox Church, and the Malankara Orthodox Church of India, comprise what is commonly known as the Oriental Orthodox Christians —an expression of Christendom that once extended as far east as China's Yellow Sea, but today only exist geographically as these six churches. In terms of theological tradition, these churches share the first three ecumenical councils with the Eastern Orthodox and Catholic Christians. However, disputes over the Incarnation of Christ caused, in large part, the Oriental Orthodox to form their own Miaphysite confession of faith.

together with God's own healing essence "in one united nature," which is how the Amharic term Tewahedo is translated. Now, only sin sets one apart from God. There is no profane reality any longer, properly speaking, since God's overshadowing grace has healed humanity. The great Church Father, Saint Gregory of Nazianzus (reposed 390 AD), in his *Letter to Cledonius the Priest (Against Apollinarius)*, emphasized this very point: "For which He [Christ] has not assumed [that is, human nature through the Incarnation] He has not healed; but that which is united to His Godhead is also saved." He stated it this way to *insist* that Christ took on all of humanity into the Incarnation and how as our High Priest opened the way for all of us to find peace with God (Hebrews 4:14–16). This was the very purpose for the word of God becoming flesh and dwelling among us, as it is declared in the Johannine prolog (John 1:14).

This is the foundation of salvation history, and it is not limited to people alone. The entire world is also to be redeemed. This point is found in the words of the aforementioned Saint Athanasius of Alexandria, who wrote that "no part of Creation had ever been without Him [Christ] Who, while ever abiding in union with the Father, yet fills all things that are [through indwelling grace]" (*On the Incarnation* 2.8; cf. 7.42). Put simply, the created world is included in God's redemptive grace because it will be necessary for resurrected humanity to have an earthly abode. This is confirmed in the New Testament, where the future rebirth of Eden is announced: It is referred to as "the times of restoration [in Greek, ἀποκαταστάσεως] of all things, which God has spoken by the mouth of all His holy prophets since the world began" (Acts 3:21, NKJV). St. Paul also spoke of the rebirth of Eden when he wrote: "For the earnest expectation of the creation eagerly waits for the revealing of the sons of God. For the creation was subjected to futility [after Eden], not willingly, but because of Him who subjected it in hope; because the creation itself also will be delivered from the bondage of corruption into the glorious liberty of the children of God. For we know that the whole creation groans and labors with birth pangs together until now" (Romans 8:19–22, NKJV). The rebirth of Eden is even preached by the very first systematic theologian in the history of the Church, St. Ireneaus (reposed 202 AD), who wrote of the end times where there "shall be the new heaven and the new earth, in which the new man shall remain, always holding fresh converse with God" (*AH* 5.36.1). All of this to say, the church forests of Ethiopia have survived and exist today as a testimony of God's promises, as well as to inspire the faithful by its beauty (Psalm 104). These are the deeper scriptural and theological traditions that testify to the sanctity of church forests.

But even so, and as mentioned before, the degree religious-based protection enjoyed by the church forests cannot assume to be permanent and unchanging. It can be increased or degraded in the eyes of the people. This is an important subject and is discussed next.

4.4 Degraded Forests

The average parishioner, and perhaps even a few of the clergy, will not be very familiar with the historical and theological foundations for the church forests discussed above. What they will be more familiar with is the reverential journey through the forest to a church that is typically atop a mountain. It is an act that mirrors the ancient pilgrimages of Jews to Jerusalem, who sang the Psalms of Ascent (Psalms 120–134) while processing up the slopes of Mount Zion to the Temple of Solomon. There, they would be in very presence of God within the Holy of Holies. For the faithful of the EOTC, this is the *tabot* (ታቦት) representing the Ark of the Covenant. The *tabot* is the symbol of consecration for the church, and the forest which houses it, as holy ground. The faithful also know that holy bodies of the priests, monks, and nuns who have reposed over the centuries are buried in the forest; the trees serve as living shrines for their remembrance. All this adds to the sense of holiness the people feel in the church forests.

But perhaps the most important trait that makes the forests sacred in the eyes of the people is the presence of wandering monastics who lead lives of seclusion and prayer in the forests. The forests, in very real effect, are the *home* for these holy men and women, and the people greatly revere them for their lives of holiness. They have rejected worldly concerns to draw closer to God, as captured in the expression associated with these monastics, *Alem beqagn* (አለም በቃኝ), and because of this, they have been given the power to become intercessors on behalf of the people (Orlowska and Klepeis 2018, p. 8).[6] Here it is important to note that these secluded saints subsist from the forest resources for their food and shelter, at least in part. "The *menagn*, as the priests refer to them, are said to feed themselves in church forests, infused with sacred energy radiating from the *tabot*" (p. 9). Their presence in a church forest greatly adds to the sense of reverence the people feel for the trees, wildlife, and other natural features of the landscape; correspondingly, the

[6] Technically speaking, there are two pathways to this kind of monasticism: those who, like those described here, have rejected the secular world out of love for God, and then there are those who embraced monasticism without having lived secular lives beforehand, being inspired to embrace this holy life from youth. This latter group can be described is *Alem lemenie* (አለም ለምኔ). In both cases, it must be stressed that this is a movement toward God and not escapism from secular concerns. I am grateful to the reviewers of this manuscript for highlighting these distinctions.

absence of these wandering monastics diminishes the value of the church forest in the eyes of the people, a conclusion also reached by Orlowska and Klepeis (p. 9).[7]

The essential question here for researchers is what are the flora species needed by these hermits? It is vital that reforestation projects include these subsistence species in their plans so that the sacred presence that these monastics impart to the forest can be safeguarded for project success. It is also important to know the minimum viable size of a church forest to support such a community of wandering monastics and well as the types of landscapes and tree density they find most useful. Unfortunately, the answers to these questions are not yet known, and until such research is done, it will be best for restoration managers to provide the greatest size and variety possible. Most critically, the research of Orlowska and Klepeis has also found that the sacred qualities of these forests can become lost or degraded in the eyes of local people when restoration scientists undertake mass replanting programs (p. 8):

> Important for external actors [that is, conservation scientists] wanting to engage local communities in church forest protection is the fact that priests and older people tend to embrace the idea through expressions such as *dur woled* and *wef zerash* (translated respectively as "born from the forest" and wild, lit. "planted by birds"), which perceives church forest as the product of natural planting. In other words, church forest is seen to consist of "natural", "wild" trees, which are God's creation. [...] *Wef zerash* (i.e. wild or native trees) are considered quite distinct from trees planted from seedlings. The implication here is that native forest cannot be planted and the prestige and respect attached to them [that is, naturally regenerated forests] is greater than those planted by people.

Their research has found that bringing in mass-produced seedlings from outside the community has led to a devaluation of the forest; these plantations are seen as being less special and merely a collection of easily replaced objects. Put simply, a tree plantation is not seen as holy, but just another consumer good to be used and discarded. It is for this reason that, whenever possible, church forests should be protected and improved through enclosures (e.g., perimeter walls built by the community) that allows natural regrowth and expansion into the enclosed areas. The subsequent forest growth is more likely to be seen as *dur woled* and *wef zerash* and an expression of God's protective grace.

[7] The forests sometimes serve as a refuge for those who have become landless and then turn to a monastic life: "By becoming nuns these individuals are able to maintain respectability since monks and nuns are well regarded by society. Their extreme poverty and living through alms becomes a virtue and a way of getting closer to the divine. Living in the church forests as a nun or monk offers a respectful place in society, despite one's destitution. Monks and nuns are usually invited to the feasts that take place in church forests, which is a respectful way to be fed by others" (Orlowska and Klepeis, p. 6). The loss of these forests, or the loss of the forest resources needed to support this life choice, can force the landless to become beggars in urban centers, at great risk of losing both culture and faith.

4.5 Sacred Text and Sacred Trees

The ancient and sacred Tewahedo text, *Śənä Fəṭrät* (ሥነ ፍጥረት), was composed in the liturgical language of *Gəʿəz* at some unknown date before the seventeenth century and was later translated into Amharic.[8] It may in fact be very ancient since it is found in numerous Ethiopic manuscripts, among them commentaries on the first chapters of the Book of Genesis, the *Gädlä Addam* (the Acts of Adam and Eve), in *Tä'ammerä Iyäsus* (the Miracles of Jesus), and the *Aksimaros* (the Ethiopic *Hexameron*). The *Śənä Fəṭrät* describes the world as an anticipation of the eschatological Kingdom of Heaven; it prefigures the life of the saints in a reborn Eden at the end of time. Here, Adam is revealed as the king of creation, and also its priest, a holy prophet, and a child of God through a gift of grace—this last declaration echoing Psalm 82, itself a commentary on Genesis, where it is said of humanity that, "You are gods, and all of you are children of the Most High" (verse 6, NKJV).

What is noteworthy here of the *Śənä Fəṭrät* is twofold. The first is a contrast to the Western conception of stewardship, which when viewed by some in light of the "good steward" passages of the New Testament (e.g., the parable of the talents in Matthew 25:14–30), suggests (wrongfully) that it is God's desire for humankind to misuse the earth's resources in an exclusionary and selfish capitalistic ethos (see the critique by Metropolitan John of Pergamon 2003). This can, and indeed has, further the exploitation of nature, all in the name of development, as documented by Lynn White Jr. in (1967). His analysis argued that Western Christianity is in large part responsible for facilitating modern environmental crises by desacralizing nature, inaugurating a theological dualism between humanity and the world. Here it should be remembered that the true biblical view of good stewardship is that the wealthy should be generous to the poor and for a person to be "rich" only in good deeds done for others because the true reward is in heaven (1 Timothy 6:17–21). But more than that, the actual message of the creation narrative is not stewardship, but something else.

Tewahedo presents a "priestly" understanding of the human place in creation, a view shared with other Orthodox Christians (see, for example, Lossky 1976, p. 133). In the priestly paradigm, humanity serves as the intermediary between heaven and earth and can bring the earth into sacred relation with God. This is exemplified in Holy Communion where people take wheat and grapes, transform

---

[8] The Amharic version of the text reveals a *Qeb'at* influence in a single alteration to paragraph 23, indicating a translation date of around 1620 AD. This change points to the Ewostathian monks of Gojjam, who advocated the Christological formula, "Through Unction Christ the Son was consubstantial with the Father," for which they were known as the *Qeb'at* ("Unction") faction (Pankhurst 1992, p. 254). This belief was based on a misreading of Acts 10: 38 that the two natures of Christ were fully united by the Unction (or *Qeb'at*) of the Holy Spirit (Crummey 1972, p. 20). It must be emphasized that the original *Gəʿəz* version of the text does not contain this influence, and that no other parts of the Amharic translation reveal heterodox changes. Therefore, with this one caveat in mind, the text can be said to be *Tewahedo*.

them into bread and wine, and consecrate them to God to become the flesh and blood of Christ for the faithful. Here, it must be recalled that all Christians are part of the "royal priesthood" according to the Order of Melchizedek (1 Peter 2:9 and Hebrews 7:13–17). So too all of humanity, through our collective activity here on earth, are called upon to bring the entire created order into the presence of God, a priestly *anaphora* (consecration) made manifest through wise and virtuous use of nature as its priest. The key difference here with the stewardship model is that this understanding is communitarian, not capitalist—meaning, the priestly relationship to nature is to be expressed liturgically through religious feasts where the people share their livestock and agricultural goods in displays of magnanimous generosity, all of which serves to bring the entire biotic community together in Eucharistic sharing realized throughout village life (for further details, see Goodin et al. 2019).

While *Śənä Fəṭrät* offers a theological model for an ennobling human relationship to nature, it cannot be helped but to notice that the commentary is framed in anthropocentric terms. True, for example, another part of the text describes creation in terms like, "grass that is mown with a sickle and trees that are chopped with an axe" (paragraph 10). This is because people are not apart *from* the world, but a part *of* the world. Instead, we find a theocentric anthropocentrism where Adam and Eve's rulership is called upon to represent the love, justice, and the wisdom of God, in part by taking instruction from the wisdom and harmony of created world. All of this echoes the words of Psalm 104, "O Lord, how manifold are Your works! In wisdom You have made them all" (verse 24, NKJV). For this reason, the *Śənä Fəṭrät* advises that, "The sun is an example of righteousness. It is (always) complete; there is no waning fullness in its light as (with) the moon" (paragraph 15). The moral lesson here for the faithful is a call to never waiver in virtue and righteousness—that they are like the light of Christ radiating into the world, just as the sun illuminates the earth with its warmth. The text continues: "Furthermore, the moon has the stars (as its) helpers" (paragraph 15). The lesson is that a person is never truly independent, but lives in a community with others, and is responsible to help one another in Christian love. *Śənä Fəṭrät* goes on to describe those who embrace these lessons of wisdom from God as having become complete in their knowledge of righteousness, such that, now, "Their hearts contain what their mouths utter (and) their mouths utter what their hearts contain" (paragraph 15). And so, rather than *just* an environmental ethic, what we find here is a virtue ethic where humanity becomes fully actualized as complete human beings, ethical and wise, from a harmonious relationship to nature and the community, both.

With respect to this knowledge from wisdom and harmony of creation, the *Śənä Fəṭrät* continues with an admonition, one that reveals a shared heritage with the older Syriac and Orthodox understandings of unknowable realities—which in the academic literature is called the apophatic. The *Śənä Fəṭrät* speaks of a comprehensible and incomprehensible creation (paragraph 4). God is also said to have four attributes: omnipotence, justice, wealth (meaning, superabundance in glory), and to be both comprehensible and incomprehensible at the same time (paragraph 4 in the *Gə'əz* text). This follows the theology of Saint Clement of Alexandria (reposed 215 AD) concerning the limits of human knowledge, writing that: "For both is it a difficult task to discover the Father and Maker of this universe; and having found Him, it is impossible to declare Him to all. For this is by no means capable of expression, like the other subjects of instruction" (*Stromata* 5.11). The ultimate incomprehensibility of both God and creation is a call to epistemological humility. Stated another way, while we can contemplate the wisdom of harmony of creation (Psalm 104), we can never impiously begin to think we can know and master natural processes better than God. In more familiar terms, this is called the precautionary principle in conservation science: It is better to try to conserve, protect, and expand the remaining church forests through natural regrowth than believing it is possible to re-green Ethiopia through large-scale tree plantings alone.

But more than that, the theological message is that we are not to make an "idol" of the human intellect, to borrow now the words of the Cappadocian Saint, Gregory of Nyssa (who reposed 394 AD, before the Chalcedonian schism) (*On the Life of Moses* §165, p. 96). This contrasts with Western understanding of the human intellect as the image of God, most notably in the writings of René Descartes (1596–1650). Because of him, Western Christianity tends to see humanity as rendering itself the masters and possessors of creation through intellectual prowess and technological mastery. This invariably led to the modern environmental crises as documented by Lynn White, Jr. Tewahedo tradition instead offers another perspective for humanity to embrace, that of humility. Dominion is not ownership, nor permission to do as we wish for our own self-satisfaction. It is a call to personify the virtues of communitarian righteousness.

4.6 Closing Words

Biotic diversity gives resilience to a forest, yet it is also true that the church forests of Ethiopia derive their resilience, in large part, from the religious history that comes together in the Tewahedo eco-theology. For restoration scientists and project managers, it is not enough to just look at the first and more familiar definition of resilience. The second should also be a foremost concern because the sacredness of the church forests can be diminished or increased through several factors. The first of which is the presence or absence of the wandering monastics who make these forests their home. It is essential to restore those biotic communities and landscape features most needed for their subsistence lifeways; it is also important to note that these monastics are a vital part of the local communities, both through serving as intercessors before God through prayer, and as a welcoming vocation for the landless to have respect and community support (instead of the communityless life in urban centers they would otherwise face). Interviews with local priests, who can speak to these monastics on another's behalf, will be the essential first step to achieving this end. In addition to preserve, or in some cases, and to recreate this protective sense of sacredness, it is vital for restoration projects to allow natural regrowth of the forests through enclosure walls, rather than mass plantings of seedlings that are not seen as sacred in the eyes of the people.

Another foundation for successful church forest restoration is highlighting the protection afforded to these lands through Tewahedo theology. The fullness and depth of this theology may not be known by the laity and perhaps even by some of the clergy. This can lead to an underappreciation of the true theological significance of these forests. This short essay is just one small step toward awakening a desire in the reader to know more and to invite others to contribute to this recovery of historical memory and to make it accessible to a wider readership. That is one hope of this essay—another is to awaken a new ecological awareness in other Christian denominations. Lynn White Jr. (1967) had written that Western Christianity needed to rethink its own religion if it wishes to be part of the answer to the many and varied environmental crises of the modern world, instead of part of the problem. Perhaps what it needs to do is to rediscover its deeper theological heritage that can be found in Tewahedo Orthodoxy.

Lastly, I want to conclude with a statement that the research presented here is not a handbook or manual; it is a call for a public participation element to reforestation projects that includes faith leaders from the surrounding community, to ask the right questions, and to look for ways to embed these projects in the cultural and religious landscape. To this end, I have sought to show where a theocentric cosmology exists in Tewahedo tradition, one that uniquely empowers humankind to protect the natural world as an eschatological anticipation for a reborn Eden. It is a religious narrative that, once brought to the fore, can further preservation and restoration efforts throughout Ethiopia with deep historical context and theological meaning.

References

Athanasius of Alexandra (2012) On the incarnation. Vladimir's Seminary Press, Yonkers, St

Berhane-Selassie T (1994) Ecology and ethiopian orthodox theology. In: Hallman DG (ed) Ecotheology: voices from the South and North. Geneva, World Council of Churches

Clement of Alexandria (1885) The Stromata. In: Roberts A, Donaldson J, Coxe AC (eds) Ante-Nicene Fathers, vol 2. Buffalo, Christian Literature Publishing Co. http://www.newadvent.org/fathers/02105.htm. Accessed on 5 Mar 2020

Crummey D (1972) Priests and politicians: protestant and catholic missions in Orthodox Ethiopia, 1830–1868. Oxford University Press, London

Goodin DK, Wassie A, Lowman M (2019) The Ethiopian orthodox Tewahedo Church forests and economic development: the case of traditional ecological management. J Relig Soc 21:1–23

Gregory of Nyssa (1978) On the life of Moses. Paulist Press, New York

Grillmeier SJA, Hainthaler T (1996) Christ in christian tradition—volume two: from the council of Chalcedon (451) to gregory the great (590–604), part four: the church of Alexandria with Nubia and Ethiopia after 451. Louisville, Westminster John Knox Press

Getatchew Haile G, Amare M (trans.) (1991) Beauty of Creation (Śənä Fəṭrät). J Semitic Stud, Monograph 16, University of Manchester, 1991

Lossky V (1976) The mystical theology of the eastern church. Vladimir's Seminary Press, Crestwood, St

Metropolitan John of Pergamon (2003) "Proprietors or priests of creation?" Keynote address, fifth symposium of religion, science, and the environment, June 2, 2003, https://www.orthodoxytoday.org/articles2/MetJohnCreation.php. Accessed on 5 Mar 2020

Nazianzus G (1894) To Cledonius the priest against Apollinarius. In: Schaff P, Wace H (eds) Nicene and post-Nicene fathers, Second Series 7. Buffalo, Christian Literature Publishing Co. http://www.newadvent.org/fathers/3103a.htm. Accessed on 5 Mar 2020

Nederveen J (1992) White on black: images of Africa and blacks in western popular culture. Yale University Press, Pieterse

Orlowska IA, Klepeis P (2018) Ethiopian church forests: a socio-religious conservation model under change. J Eastern Afr Stud (19 Sep 2018) https://doi.org/10.1080/17531055.2018.1519659.

Richard Pankhurst R (1992) Review of beauty of the creation (Śənä Fəṭrät) by Getatchew Haile and Misrak Amare. J Royal Asiatic Soc, Third Ser 2(2):253–255

Tamrat T (1972) Church and state in Ethiopia: 1270–1527. Tsehai Publishers, Los Angeles

White L Jr (1967) The historical roots of our ecologic crisis. Science 155:1203–1207

Part II
Present Role and Future Challenges of Ethiopian Church Forests

Chapter 5
Soil Carbon Stocks and Dynamics of Church Forests in Northern Ethiopian

Dessie Assefa, Abrham Abiyu, Boris Rewald, Hans Sandén, and Douglas Godbold

Abstract Rising atmospheric CO_2 concentration is a major cause of climate change. Release of soil carbon is one of the major causes. In the Northern part of Ethiopia, church forests are major stocks of above- and below-ground carbon. However, the significance of church forest ecosystem in soil carbon storage is not well documented. Three church forests along a climatic gradient were selected to examine soil carbon storage of church forest in comparison with adjacent land use systems (eucalyptus plantation, grazing land and cropland). Inputs from above- and below-ground biomass were quantified. Litter trap and in-growth core methods were used. Results showed that conversion of church forest to cropland or grazing land reduced SOC stocks by 58–69% in <50 years. Restoring the area through reforestation with eucalyptus species led to an increase in SOC; however, the rate of increase (ca. 0.3 kg m^{-2} $year^{-1}$) was lower than losses (ca. 0.4 kg m^{-2} $year^{-1}$). The annual carbon input through litterfall is nearly the same as below-ground input in the church forest. On the other hand, the above-ground inputs in eucalyptus, cropland, and grazing land were absent/minor due to litter raking, complete crop residue harvest, and overgrazing. The annual fine root production in the church forest was 723 g m^{-2} whereas in grazing land and cropland it was in the range of 50–60 g m^{-2} illustrating a reduction of carbon input into the soil by >90% if church forests are converted to these land uses. This calls for urgent attention to restore essential ecosystem functions and soil carbon sequestration.

Keywords Biomass · Fine roots · Land use conversion · Litterfall · Strontium to calcium ratio

D. Assefa (✉)
Department of Natural Resources Management, Bahir Dar University, Bahir Dar, Ethiopia

A. Abiyu
World Agroforestry (ICRAF), Addis Ababa, Ethiopia

B. Rewald · H. Sandén · D. Godbold
Institute of Forest Ecology, University of Natural Resources and Life Science, Vienna, Austria

M. Kindu et al. (eds.), *State of the Art in Ethiopian Church Forests and Restoration Options*, https://doi.org/10.1007/978-3-030-86626-6_5

5.1 Introduction

The world has experienced a dramatic conversion of natural forests to agriculture, grazing land or plantations during the past few decades (Guo and Gifford 2002). Geographically, the sub-Saharan African region recorded the fastest conversion of forest land to agriculture in the past 20 years (Nkonya et al. 2013). Although accurate records on forest cover and reliable data on the rate of deforestation are limited in Ethiopia, about 20 thousand hectares of forest are cleared annually in the Amhara region alone for the expansion of smallholder farmlands and biomass fuel (Meseret 2016). Today, remnants of the original vegetation are found mainly around churches, monasteries and very inaccessible mountainsides and gorges. The numerous monasteries and churches are playing a critical role in the conservation of forests and their biodiversity in the Region (Aerts et al. 2016).

The conversion of forest to other land use plays a major role in determining carbon (C) storage in ecosystems (Le Quéré et al. 2016). Global estimates showed that conversion of forest to cropland declines soil organic carbon (SOC) by 30–35% within the first 30 years in the top 10 cm soil (Oertel et al. 2016). In Ethiopia, conversion of forest to cropland reduced SOC by 62% within 10 years, and by 75% in 53 years after deforestation (Lemenih et al. 2006). Similarly, conversion of forest to open grazing land reduced SOC stock by 15–48% in Ethiopia (Girmay et al. 2008). These losses are also due to a substantial amount of soil is removed from topsoil by erosion (Hurni et al. 2010). A study from Ethiopian highland showed that the estimated rate of soil loss from cultivated fields is up to 79 Mg ha^{-1} $year^{-1}$ which by all measures exceed the rate of soil formation (Bewket and Sterk 2003). To gain an insight into changes in soils, strontium (Sr) to calcium (Ca) and barium (Ba) to calcium ratios have been widely used as markers to reconstruct environmental history (Bullen et al. 2005; Tabouret et al. 2010). Although Sr^{2+} is a non-essential element, it is strongly associated with and chemically very similar to Ca (Capo et al. 1998) with a similar ionic radius and charge. Strontium/ Ca and Ba/Ca ratios can be a powerful tool to trace the relative contributions of atmospheric dust input, mineral weathering to soil genesis (Bullen et al. 2005; Derry and Chadwick 2007; Li et al. 2016) and losses due to erosion (Assefa et al. 2017).

Besides soil erosion, litter inputs from above ground are absent in the cropland and grazing lands and potential inputs are mainly from below ground (Leppälammi-Kujansuu et al. 2014; Soleimani et al. 2019). This is because the traditional practice of free- and overgrazing in the region lead to virtually devoid of vegetation cover which leaves the area bare ground in most communal grazing lands. Complete removal of crop residue for cattle feeding at home restricts little input of carbon to the soil through below-ground biomass only. In addition, after forest clearance, rates of decay of soil C were 10 times higher in cultivated soil than in forest soil, and that soil C in some factions had become unprotected (Balesdent et al. 1998; Malhi et al. 1999).

On the other hand, conversion of cropland to plantation increases soil carbon stock by 29% (Don et al. 2011) or to secondary forest by 53% worldwide (Guo and Gifford 2002). This illustrates a great potential of recovering C stock through ecological restoration. However, soil C loss or gain estimates following deforestation and conversion to different land uses are highly variable and full of uncertainties (Houghton and Goodale 2004; Pan et al. 2011).

Even though the number of studies on soil carbon dynamics has increased in recent decades (Solomon et al. 2002; Lemenih et al. 2005, 2006; Nyssen et al. 2008), information about the role of protecting the native church forests in carbon stock conservation is limited. Therefore, the objective of this study was to investigate the role of church forests on soil carbon storage and how this process is affected by land use change in the northern highlands of Ethiopia. We investigated soil carbon stocks of four land use systems that include church forest, eucalyptus plantation, cropland, and grazing land, which may improve our mechanistic understanding of carbon dynamics. In this analysis, Sr/Ca and Ba/Ca ratios were used as putative markers of soil loss. However, accurate estimates of the amount of plant litter production (litter inputs from above and below ground) and changes in response to land use conversion are an important indicator of the overall ecosystem C fluxes and soil carbon budgets. Therefore, we determined the above- and below-ground litter production to estimate the amount of carbon fluxes into the soil. In addition, we conducted biomarker analysis with gas chromatograph-mass spectrometer (GC–MS) at molecular level after base hydrolysis to determine the relative input of carbon into the soil by roots and leaves. The results of this work can be used as a tool for policymakers at regional level setting up appropriate measures to combat key environmental issues such as climate change in relation to soil carbon storage.

5.2 Materials and Methods

5.2.1 Research Sites

Three research sites were selected in the Northwestern part of Amhara region that represents major climatic and edaphic conditions (Table 5.1). At each site, up to four land use types, namely church forest, eucalyptus plantation, cropland and grazing land, were identified adjacent to each on a similar topography, soil and climate conditions to ensure comparability among the land uses. However, not all land use types were available at each site. Geographical location, climate, soil types, and vegetation characteristics of the sites as well as the available land use types are presented in Table 5.1.

Forests at Gelawdios and Tara Gedam are dry Afromontane church forests composed of indigenous tree species. These forests are traditionally protected by churches and monasteries (Aerts et al. 2016). The dominant tree or shrub species for the Gelawdios and Tara Gedam forests are *Albizia schimperiana*, *Apodytes dimidiata*, *Calpurnia aurea*, *Carissa edulis*, *Croton macrostachyus*, *Ekebergia capensis*, *Maytenus arbutifolia*, *Olea europaea* and *Schefflera abyssinica*. The extensive lowland semiarid forest around Mahibere-Selassie monastery is a savannah

Table 5.1 Site characteristics of a range of sites in the highlands or lowland of the Amhara region of Northwestern Ethiopia

Site characteristics	Gelawdios	Tara Gedam	Mahibere-Selassie
Location	11° 38′ 25″ N 37° 48′ 55″ E	12° 8′ 47″ N 37° 44′ 45″ E	12° 32′ 20″ N 36° 35′ 9″ E
Altitude (m a.s.l.)	2500	2230	850
Air temperature (°C)*	19[a]	21[c]	27[e]
Precipitation* (mm $year^{-1}$)	1220[a]	1100[c]	965[d]
Soil type	Cambisols	Cambisols	Fluvisols
Land use types	F, EP, C, G	F, EP, G	F, C
Forest area	100	875	19,000
Basal area ** (m^2 ha^{-1})			
• Church forest	34.1	19.8	3.4
• Eucalyptus	19.0	na	Na
Tree density (N ha^{-1})**			
Chruch forest	6334	4170	205
• Eucalyptus	3031	na	Na

F forest, *EP* eucalyptus plantation, *G* grazing land, *C* cropland, *na* not available
*Climate data were obtained from previous studies ([b]Assaye et al. 2013; [d]Moges and Kindu 2006; [a]Wassie et al. 2009; [c]Workneh and Glatzel 2008)
**Inventory data was obtained from Gebrehana (2015). Basal area was measured at 1.3 m above the ground from a 10 to 15 m radius circular plot depending on forest size. Tree density includes saplings with height >1.5 m

woodland dominated by grasses with some scattered trees. This site, which represents the typical lowland forest type in the Region, is subject to frequent man-caused fires. Characteristic tree species in the lowland are *Acacia polyacantha, Balanites aegyptica, Boswellia papyrifera, Diospyros abbyssinica, Ficus sycomorus, Pterocarpus lucens, Sterculea setigera, Oxytenanthra abyssinica and Ziziphus spinachrist*. The *Eucalyptus globules* plantations at Gelawdios was planted on grazing land around 1985 and was thus ca. 30 years old at the time of sampling. In Tara Gedam, *Eucalyptus* camaldulensis was planted on former cropland and has been cut 3 times since planting. The date of planting is unknown but as in this part of Ethiopia, the coppicing intervals are between 5 to 10 years, the site must have been planted about 40 years ago. All the studied croplands and grazing lands were converted from church forest within the last 50 years but the exact date is unknown. The estimated number of years that the cropland was continuously cultivated after conversion of the forest was obtained from local knowledge. Croplands are ox-ploughed to a depth of approximately 30 cm. Owing to the nature of the climate, crops such as *Eragrostis tef*, *Eleusine coracana*,

Sorghum bicolour, *Zea mays*, *Triticum aestivum* and *Vicia faba* are grown during the wet season. Between the years, the types of crops grown are rotated.

5.2.2 Soil Sampling

Soil samples were taken in two rounds, one in March 2014 (end of the dry season) and another during the wet season in June 2015. In the church forests, 10 sampling points were marked every 100 m along a transect line. Since the forest area is large compared to other land use types, two soil samples were taken at a distance of about 2 m apart at each sampling point to include the variability. In eucalyptus plantations, croplands and grazing lands, one soil sample was taken at 10 sampling points. The distance between sampling points along transects in these land use types was 20 or 50 m according to the size of the land uses. After removing the litter layer (if present), a soil corer (6.6 cm internal diameter) was driven to a maximum of 50 cm depth or until the bedrock was reached. Soil samples were extracted into a Styrofoam tray and morphologically described, using the terminology of World Reference Base for soil resources (WRB 2014). Soil cores were divided into depth classes of 0–10, 10–20, 20–30 and 30–50 cm. For the samples from the church forest, a composite sample for each depth was made from the two cores taken at each sampling point. Each soil sample was sieved using a 2 mm sieve, homogenized and placed in a plastic bag. The samples were then transported to Vienna for laboratory analysis.

5.2.3 Bulk Density Determination

Bulk soil density samples were taken from five sampling points, i.e. from every second sampling point along the transect line. The samples were taken during the onset of the wet season when the soil had no cracks. At the edge of each sampling plot, a profile was dug to 60 cm to make the work easy. Soil samples of known volume were taken from the sides of the profile centred at 5, 15, 25 and 40 cm depth using a stainless steel bulk density ring. Using a trowel, the ring was removed from the horizon and the soil trimmed to the tops and bottoms of the ring using a sharp knife. Any stones (>2 mm) were sieved out and weighed separately. The volume of stones was quantified by displacement in a water bath. Bulk density (soil particle <2 mm) was determined after oven drying at 105 °C as stone-free dry weights according to Don et al. (2007).

5.2.4 Total C and N Stock Determination

Before analysis, the bulk soil samples from each horizon were homogenized again and a subsample of approximately 3–5 g was dried at 105 °C for 48 h, then crushed to a fine powder and remixed. From each subsample, a single large aliquot of about 200 mg was taken and total carbon and nitrogen concentration were determined using a CN elemental analyser (Truspec CNS LECO, St. Joseph, USA). The analyser was initially calibrated using the manufacturer's standard material (Part No. 502–309 purchased from LECO) and the calibration was controlled using the same standard every 30 samples. Baseline correction was carried out every 90 samples using 5–10 empty cells. Using the manufacturer's standard material, the instrument's typical precision was found to be ±0.005% for carbon and ±0.001% for N.

Carbon and nitrogen stocks were determined on weight to area basis (kg of C or N per m^2 of soil) per soil depth class and down to 50 cm. Total carbon stock was calculated by summing the carbon stock of each soil horizon determined. The carbon concentration was corrected for bulk density and soil volume and interpolated to an area basis as shown in the following equation:

$$\text{TC} = \sum_{i}^{n} C\%_i * \text{BD}_i * v_i * 0.001\ \text{kg} \tag{5.1}$$

TC is total carbon (kg C m^{-2}); $C\%_i$ is concentrations of carbon in percentage at depth *i*; BD_i (g cm^{-3}) is the bulk density at depth *i*; and V_i (cm^3) is the volume of soil at each horizon. Total nitrogen (TN) was calculated in the same way.

Due to difficulties in attaining all measurements in time and limited budget from all sites, elemental analysis, above and below-ground input measurements and their chemical analysis were done at one site only (Gelawdios).

5.2.5 Strontium, Calcium and Barium Elemental Analysis

For analysis of strontium, calcium and barium from the bulk soil sample of Gelawdios forest and cropland, a further 20 g subsample was taken from each sample, dried for 3 days at 50 °C and remixed. Before analysis, residual moisture was removed by drying at 105 °C for 30 min. From each sample, a 500–600 mg aliquot of dried soil was added into a 75 ml fusion tube with 20 ml of Aqua regia (3:1 HCl and HNO_3). To prevent the solution from foaming, 10 μl of Octanol was added. The soil samples were shaken for 4.5 h and then digested at 125 °C for 2 h. The cooled samples were then made up to a volume of 75 ml with deionized water and filtered with Whatman Grade 589/1 paper. The concentrations of total strontium (Sr), calcium (Ca) and barium (Ba) were determined by inductively coupled plasma optical emission spectroscopy

(Perkin Elmer Optima 8300; ICP-OES Spectrometer, Waltham, USA) using external calibration in line with ÖNORM L 1085. Two samples of standard soil were used as internal standards (ICP Multi-element Standard solution XIV, CertiPUR, purchased from Merck KGaA, Germany).

5.2.6 Litterfall Collection and Fine Root Biomass Determination

Litterfall was collected weekly from July 2014 to June 2015 from Gelawdios church forest. Ten litter traps were systematically placed at 100 m distance between each other along a transect line. Samples were collected on a 0.5 m × 0.5 m wooden frame furnished with a 1 mm polyamide mesh (Franz Eckert GmbH, Waldkirch, Germany); positioned 50 cm above the ground. At the end of each month, the weekly collected litter were combined and sorted into leaves, branches and twigs, reproductive organs (flowers, fruits and seed) and miscellaneous material (unidentified plant parts, mosses). The collected litter was dried at 70 °C to constant weight before weighing to the nearest 0.01 g.

For assessment of C inputs from below ground, fine root (<2 mm diameter) production was estimated using in-growth core method (Leppälammi-Kujansuu et al. 2014). In the church forest and eucalyptus plantation, five in-growth cores per plot (100 m^2 plot size) were established (50 cores per land use in total) at the end of June 2014. One in-growth core was positioned at the centre of the plot and the four others 2 m from the centre in N, E, W and S directions. Soil cores were extracted using a soil corer (6.6 cm internal diameter*40 cm length) and the holes were lined with a 2 mm mesh size polyethylene net and filled with the root free soil taken from the cores. An effort was made to restore the original soil horizons and compact the soil back to approximately the original soil bulk density. In the grazing land and cropland plots, three soil cores per plot (30 in total for each land use) were established as described above considering minor growth after crop harvest or in the dry periods. One in-growth core was positioned at the centre of the plot and two others 2 m apart from the centre. One in-growth core per plot was removed from the forest and eucalyptus plantation at the end of July, August, September 2014 and February and June 2015 corresponding to 1, 2, 3, 8 and 12 months in growth. In cropland and grazing land one core per plot was retrieved in July, August and September 2014 corresponding to 1, 2 and 3 months. Each time an in-growth core was removed, a new in-growth core per plot (ten in total per land use) was established in a new position to estimate root production and mortality between each sampling date. In total, the sampling dates correspond to 1, 2, 3, 4, 6, 8 and 12-month time intervals from the beginning of the measurement. After removal, the in-growth cores were carefully divided into 0–10, 10–20, 20–30 and 30–40 cm depths. Soil from the cores was spread on a plastic sheet, and fine roots were handpicked from the soil.

The root samples were placed in separate plastic bags and transported to Vienna (Austria) for further analysis. Samples were stored at 4 °C until processed. In Vienna, fine roots were placed in a 1 mm sieve and soil was washed away under running tap water and then the roots were handpicked with forceps from the sieve. All root samples were oven dried to constant weight at 70 °C, weighed to the nearest 0.01 g and converted to g m^{-2}.

The fine root was later divided into live fine roots (biomass) and dead fine roots (necromass). Fine root production (P) was estimated using a simple 'balancing model' (Santantonio and Grace 1987; Osawa and Aizawa 2012; Li et al. 2013) with some modifications. In the in-growth cores method, the last harvest at the end of the year is usually considered as direct root production (Vogt et al. 1998; Lukac 2012). However, in a one-year in-growth core method, biomass is increasing until it colonizes the whole space and reaches equilibrium. Nevertheless, fine root growth, death and decomposition are occurring simultaneously even between the sampling dates. Therefore, the interim root production and decomposition between sampling dates are missed. In addition, production, mortality and decomposition may vary in the growing season due to seasonal moisture and temperature variations. To overcome this problem, the additional in-growth cores were used to estimate root production between harvests. Here, it was assumed that root production favoured by soil disturbance during the installation of in-growth cores is compensated by the lag time of growth of severed roots. Therefore, we used a mass balance model from the time of insertion to any given time (t) in the growing season to estimate fine root production (P), mortality (M) and decomposition (D). The 'mass balance model' is the same as those of Santantonio and Grace (1987), Osawa and Aizawa (2012) and Li et al. (2013) except we considered additional growth between harvests. The interim production of biomass (b_i) between each harvest was calculated as:

$$b_i = b_t - b_0 = b_t, \quad \text{since} \quad b_0 = 0 \tag{5.2}$$

Similarly, the interim production of necromass (n_i) between each harvest was calculated as:

$$n_i = n_t - n_0 = n_t, \quad \text{since} \quad n_0 = 0 \tag{5.3}$$

Since there are no fine roots in the in-growth cores at the start of the installation, both b_0 and n_0 are always zero. Then, the actual mortality for in-growth cores and in-growth nets was calculated as:

$$M_t = b_i - B_i \quad \text{where} \quad B_i = B_{t+1} - B_t \tag{5.4}$$

This means that root production between interim periods from the short-term cores and nets should be equal to zero. If production with the short term within this time was higher than the calculated value from the initial installation (long-term), then the differences were assumed to be root mortality in the long-term cores. This mortality was then included to either necromass or partly to decomposition depending on the mass change of necromass due to this mortality.

$$M_j = N_{t+1} - N_t + D_j \tag{5.5}$$

Decomposition was therefore calculated from a decrease of necromass, or a decrease in fine root biomass that was not compensated for by an increase in necromass, with the following equation.

$$D_j = M_j + n_i + N_t - N_{t+1} = b_i + B_t - B_{t+1} + n_t + N_t - N_{t+1} \tag{5.6}$$

where small case letters b_i and n_i are root biomass and necromass from the interim periods.

$$\text{Annual mortality} = \sum_{j=1}^{n} M_j \tag{5.7}$$

$$\text{Annual decomposition} = \sum_{j=1}^{n} D_j \tag{5.8}$$

To calculate the annual root production, all production values obtained from all changes in biomass, necromass and decomposition from interim periods of consecutive sampling dates were summed from the start of sampling until the same time point in the following year (same as Eq. 5.2).

The annual fluxes of C into the soil (expressed in g m^{-2} $year^{-1}$) from litterfall and fine roots were estimated using litter/fine root production and litter/root C concentration in the roots according to Xia et al. (2015) as:

$$I_a = P * C\% \tag{5.9}$$

I_a, annual input of C into the soil (g m^{-2} $year^{-1}$): P, annual litter/fine root production (g m^{-2}); $C\%$, concentration of C in the fine roots (%).

5.2.7 *Determination of Suberin and Cutin Markers Using Base Hydrolysis*

Base hydrolysis was conducted on soil samples to determine total bound lipids (suberin and cutin) (Otto et al. 2005; Feng et al. 2008). Suberin and cutin are used to trace below- and above-ground plant origin, respectively. Suberin and cutin biomarkers were summarized and calculated according to their occurrence only in suberin or cutin based on parameters developed by Otto and Simpson (2006). Long-chain ω-hydroxyalkanoic acid ($\geq C_{20}$) and α,ω-alkanoic acids (C_{16}–C_{24}) are typical biomarkers for suberin (ΣS) whereas C_{16} and C_{18} ω-hydroxyalkanoic acids are used as cutin (ΣC) biomarker signatures. Biomarkers derived from both suberin or cutin (ΣSvC) are ω-Hydroxyalkanoic acids C_{16}, C_{18}, C_{18} di- and trihydroxy acids and α, ω-alkanedioic acids C_{16}, C_{18}.

5.2.8 *Data Analysis*

One-way analysis of variance was used to test the significance of mean differences in total carbon and nitrogen as dependent variables and land use systems and soil depths as factors. Assumptions of normality and homogeneous variance were examined by Shapiro–Wilk's and Levene's tests, respectively. When the assumptions of normality were not met, the data were $\log_{10}$-transformed to normalize the distribution. To determine differences between groups Scheffe post hoc tests were used. Means and standard errors were calculated using the IBM SPSS analytical software package (version 21) and graphs were prepared by using SigmaPlot (version 13). Significant level determined at $\alpha = 0.05$. Throughout the paper, error bars to the mean are ±SE.

5.3 Results

5.3.1 *Soil Carbon and Nitrogen Stocks*

Total SOC and N stocks varied considerably between land uses systems. The average C stock (0–50 cm soil depth) was significantly higher in the church forests (hereafter refers to as forest) than other land use types and croplands had the lowest C stock (Fig. 1a). The average C stocks per land use type across three sites can be ranked as forest (22.0 kg C m^{-2}) > eucalyptus plantation (15.7 kg C m^{-2}) > grazing land (9.3 kg C m^{-2}) > cropland (6.8 C kg m^{-2}), however, the difference between grazing land and cropland was not significant. Similarly, soil N stock was also showed significant differences between the land use types and followed a similar trend as that of the C stock (Fig. 1a). From forest ecosystems, the lowest C and N stocks were found in the lowland semiarid forest at Mahibere-Selassie. There was no significant difference found in soil C or N stock between Gelawdios and Tara Gedam. At Mahibere-Selassie, in contrast to all the other forest sites, the C and N stocks of cropland were significantly higher than that of the corresponding forest (Fig. 1b; $P < 0.001$). Soil organic carbon stock of the eucalyptus plantation was on average 57% and 40% higher than the respective cropland and grazing land, but was on average 29% lower than the SOC stock of the church forest (Fig. 1a). Similarly, soil N stock was 47% and 29% higher in eucalyptus soil than the cropland and grazing land soils.

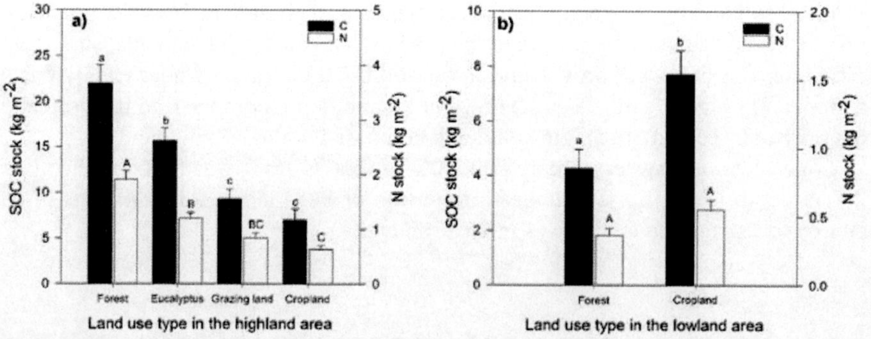

Fig. 5.1 Stocks (kg m^{-2}, mean ± SE) of soil organic carbon (filled bars) and soil nitrogen (unfilled bars) in different land use types. Shown are the pooled data of different land use systems from two highland areas (**a**) and two land use systems of the lowland area (**b**). Values are for a soil depth of 50 cm. Bars without the same letters (small case letters for filled bars and capital case letters for unfilled bars) are significantly different (mean ± SE; Scheffe, $P < 0.05$, n (highland) = 20 and n (lowland) = 10)

5.3.2 Vertical Distributions of C and N

The percentage of SOC decreased with increasing depth in all land use systems (Fig. 5.2) and a significantly lower C% was found at 30–50 cm compared to 0–10 cm.

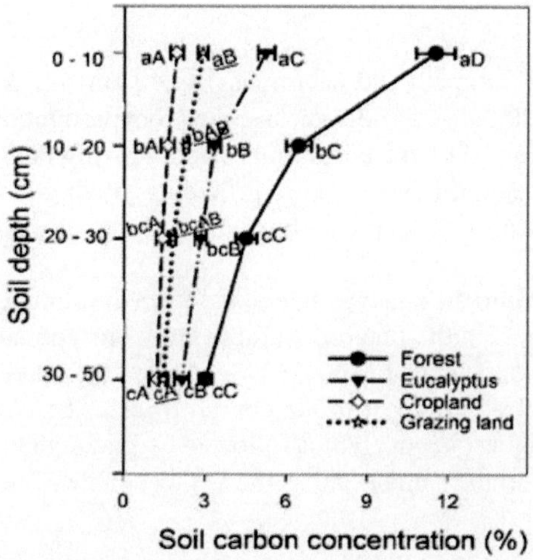

Fig. 5.2 Vertical distributions of soil C concentrations across soil depth for each land use type. Shown are the means of the pooled data of the individual sites. Different small case letters indicate significant differences of data points along soil depth within a land use type whereas capital case letters indicate differences between land use type for each depth (mean ± SE; Scheffe, $P < 0.05$, n (forest) = 30, n (eucalyptus, grazing land, cropland) = 20)

From all soil depths, the percentage C was significantly higher in forests compared to all other land use systems. Between the grazing land and cropland soils, significant differences in C% were only found in the top 0–10 cm layer. Under eucalyptus, a significantly higher soil C% was shown at all soil depth compared to the cropland, but only at 0–10 and 30–50 cm soil depth compared to the grazing land.

For all land use systems, more than 60% of total C stock was found in the upper 20 cm depth (Fig. 5.2). The largest proportion of SOC along soil layer was in the church forest and the lowest was in the cropland.

5.3.3 Strontium:Calcium and Barium:Calcium Ratios

In the Gelawdios forest site, the ratios of Sr/Ca and Ca/Ba increased with increasing soil depth (Fig. 5.3). The ratios of Sr/Ca and Ca/Ba were significantly lower in the 0–10 cm soil than in the 20–30 cm soil. Significant differences in the ratios of both Sr/Ca and Ca/Ba were found if the 0–10 cm and 30–50 cm soil depths are compared. In contrast, in the cropland soil profile, the ratios of Sr/Ca and Ca/Ba did not change with soil depth. Higher levels of Ca were determined in the 0–10 cm soil layer than the deeper soil layers. In cropland soils, the levels of Ca were similar at all soil depths.

5.3.4 Annual C and N Flux into the Soil Via Fine Roots and Leaves

Fine root production (biomass and necromass) to 40 cm soil depth for all land use systems are shown in Table 5.2. The highest annual root production was obtained from the church forest followed by the eucalyptus stand, grazing land and cropland using the in-growth core method. However, root growth in grazing land and cropland was limited to the rainy season. The annual litterfall in Gelawdios forest was 704 g m^{-2} (Table 5.2).

Total C and N flux into the soil was calculated from the annual root production and concentration of C or N in the fine roots (Table 5.2). The annual C and N input into the soil through fine roots in the top 40 cm as estimated from the in-growth core with mass balance method was highest in the church forest. The annual C flux in other land use types was highest in eucalyptus followed by grazing land and cropland. The annual N flux followed the same trend as the C flux (Table 5.2).

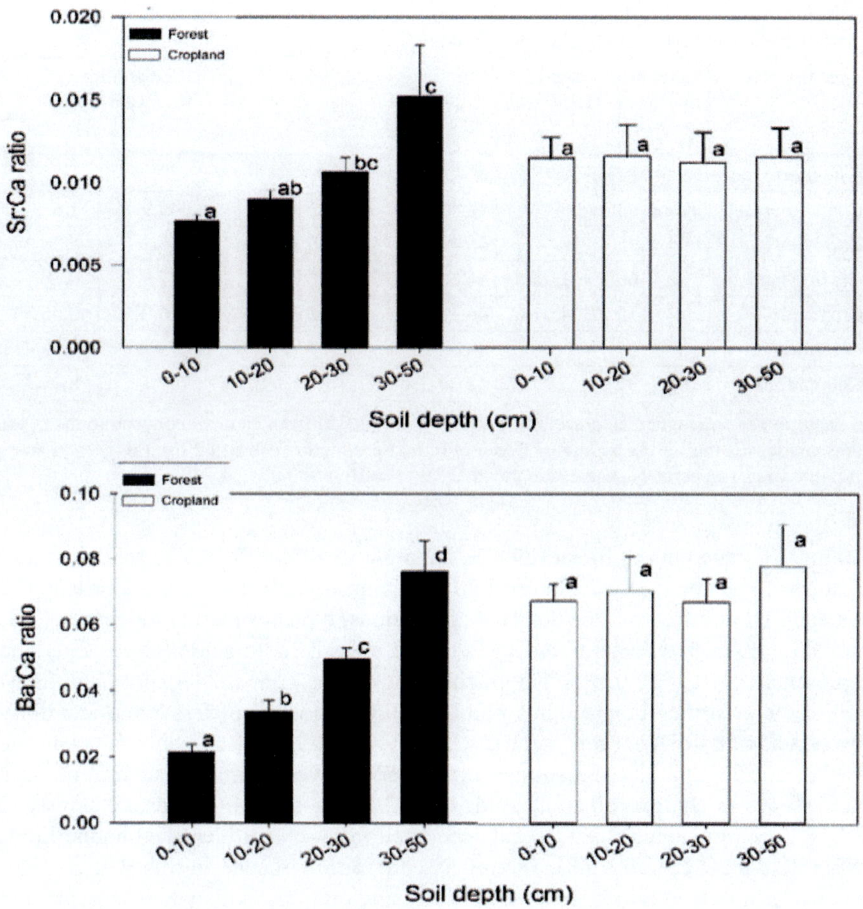

Fig. 5.3 Strontium (Sr) to calcium (Ca) and barium (Ba) to calcium ratios at different soil depths in forests (filled bars) and cropland (unfilled bars) at Gelawdios. Within a ratio and land use bars with different letters are significantly different (mean ± SE; Scheffe, $P < 0.05$, $n = 10$)

5.3.5 Composition and Distribution of Root and Leaf Biomarkers

Major products identified after base hydrolysis included a series of aliphatic lipids (*n*-alkanols, *n*-alkanoic acids, mid-chain substituted and branched acids, ω-hydroxyalkanoic acids, α-hydroxyalkanoic acids, α,ω-alkanedioic acids and glycerides), with lesser contribution from benzyls and phenols and one steroid (Table 5.3). The C-normalized concentrations of total bound lipids after base hydrolysis extraction were in general highest in the grazing land (111.3 ± 8.2 mg/g C) than the corresponding soils of cropland (97.1 ± 4.4 mg/g C), eucalyptus soil (82.0 ±

Table 5.2 Total C and N flux (g m^{-2} $year^{-1}$) to soils via fine roots in four land use systems and litterfall in the church forest at Gelawdios, Ethiopia

Land use type	Litter production (g m^{-2} $year^{-1}$)	C%	N%	Element flux (g m^{-2} $year^{-1}$)	
				C	N
Fine roots					
Church Forest	723 ± 93	48.4 ± 0.9	1.6 ± 0.3	349.9	11.6
Eucalyptus	694 ± 95	45.4 ± 0.8	0.9 ± 0.1	315.1	6.3
Grazing land	60.2 ± 11.5	46.3 ± 1.2	1.0 ± 0.0	27.9	0.60
Cropland	56.2 ± 4.6	46.3 ± 1.2	1.0 ± 0.0	26.0	0.56
Litterfall					
Church forest	704 ± 32	42.20 ± 0.48	1.27 ± 0.07	297.1	8.9

Element fluxes were calculated according to Xia et al. (2015) from element concentrations in the roots and leaves multiplied by annual fine root or leaf production estimated from in-growth cores and litter traps respectively. Values are mean ± SE; $n = 10$; $p < 0.05$

1.0 mg/g C) and church forest (29.9 ± 1.2 mg/g C) (Table 5.3). The spatial pattern of suberin-specific monomers in soil differed considerably among the four land use systems. Therefore, from the detected compounds, typical suberin biomarkers (ΣS) include ω-hydroxyalkanoic acid ($\geq C_{20}$) and α,ω-alkanoic acids ($C_{16} - C_{24}$) and represented 1.4% (322 μg/g C) in grazing land soil to 7.5% (1247 μg/g C) in forest soil of the identified bound lipids (Table 5.3). Compounds of C_{16} mono-and dihydroxy acids and diacids represented <1.4% (126–305 μg/g C; Table 5.3) and are used as cutin biomarker signatures. Biomarkers derived from both suberin or cutin (ΣSvC) are ω-Hydroxyalkanoic acids C_{16}, C_{18}, C_{18} di- and trihydroxy acids and α, ω-alkanedioic acids C_{16}, C_{18} and accounted for 3–6% of identified bound lipids (551–1035 μg/g C; Table 5.3). The relative proportion of suberin was about 2 times that of cutin in the church forest, eucalyptus and cropland soils whereas in grazing land soil, suberin and cutin concentrations were similar (Table 5.3).

5.4 Discussion

5.4.1 Effect of Land Use Change on Soil Organic Carbon Stock

There were large differences in the carbon stocks in soil between the different land use systems and differences in patterns between the highland and lowland ecosystems. In the highlands (Fig. 1a), greater SOC stocks were found in the church forest and eucalyptus plantation than in the grazing or croplands. In the lowlands, a greater SOC stock was found in the cropland than in the church forest (Fig. 1b). The overall

Table 5.3 Occurrence and quantities of compounds (μg/g C) identified from base hydrolysis of soil samples at different land use systems in Gelawdios, Ethiopia

Compound	Forest	Eucalyptus	Cropland	Grazing land
n-Alkanols (C_{16}–C_{28})[c]	104 ± 21	265 ± 12	314 ± 21	627 ± 362
n-Alkanoic acids (C_{14}–C_{30})[d]	4788 ± 1180	9998 ± 179	11,063 ± 452	12,157 ± 1173
Iso-Alkanoic acids	11[b]	38[b]	16[a]	nd
α-Alkanoic acids (C_{16}–C_{25})[e]	364 ± 60	515 ± 124	576 ± 100	571 ± 357
α,ω-alkanedioic acids (C_4–C_{20})[e]	342 ± 49	642 ± 41	589 ± 125	678 ± 69
ω-Hydroxyalkanoic acids (C_{16}–C_{30})[e]	736 ± 31	1105 ± 26	1344 ± 30	298 ± 141
Mid-chain substituted hydroxy acids	227 ± 111	592 ± 122	422 ± 160	309 ± 80
Monoacylglycerides (C_{19})[d]	238 ± 59	487 ± 128	432 ± 177	nd
Benzyles and phenols[c]	428 ± 53	1233 ± 57	1390 ± 200	2303 ± 493
Organophosphates	1503 ± 186	4511 ± 101	5555 ± 282	6759 ± 555
Steroids (ß- Sitosterol)[c]	5[a]	26[a]	27[b]	9[a]
Unknowns	21,207 ± 2232	62,542 ± 1005	75,713 ± 3719	87,606 ± 5277
Total lipids	6810 ± 1217	13,642 ± 267	14,755 ± 726	14,641 ± 2110
Total identified bound lipids	8745 ± 1404	19,412 ± 210	21,727 ± 1101	23,712 ± 2955
Total base hydrolysis	29,952 ± 3559	81,954 ± 1010	97,440 ± 4376	111,318 ± 8213
Suberin and Cutin monomers				
Suberin ΣS[f]	653 ± 56	1087 ± 63	1247 ± 15	322 ± 137
Cutin ΣC[g]	126 ± 40	268 ± 96	305 ± 90	223 ± b
Suberin or Cutin ΣSvC[h]	551 ± 153	1035 ± 187	938 ± 241	691 ± 68
Sum Suberin and Cutin ΣSC[i]	1330 ± 216	2390 ± 218	2490 ± 313	1236 ± 303
Suberin/Cutin ratio = (ΣS + ΣSvC)/(ΣC[e] + ΣSvC[f])	1.9	1.7	1.9	1.1

Values are mean ± SE, $n = 3$
nb not detected
[a] detected only from one sample only
[b] detected only from two samples only
[c] n-Alcohols and ß- Sitosterol were identified as TMS ethers
[d] Alkanoic acids were identified as methyl esters and hydroxyacids as methyl esters/TMS ethers
[e] Phenolic acids were identified as methyl esters/TMS ethers
[f] ΣS = ω-Hydroxyalkanoic acids (C20–C30) + α, ω-alkanedioic acids (C20)
[g] ΣC = C16 mono-and dihydroxy acids and diacids
[h] ΣSvC = ω-Hydroxyalkanoic acids C16, C18 + C18 di- and trihydroxy acids + α, ω-alkanedioic acids C16, C18
[i] ΣSC = ΣS + ΣC + ΣSvC

soil carbon stock in forests ranged from 4 kg SOC m^{-2} in the lowlands to 24 kg SOC m^{-2} in the highlands. These values fall within the range of estimates of global tropical means (3–41 kg C m^{-2}; Batjes, 1996) and are similar to other studies in southern Ethiopia (4–21 kg C m^{-2}; Lemenih and Itanna, 2004).

At the two highland sites, conversion of natural forests into croplands induced a strong reduction of organic carbon in the soil. At Gelawdios, the SOC stock was reduced by 50% that seems the average rate of loss of SOC is 0.23 kg m^{-2} $year^{-1}$ for the 50 years period (Fig. 1a). However, chronosequence studies have shown that the conversion of forests to cropland caused a rapid initial decrease in SOC stocks, followed by a slow decline (Wei et al. 2014; Deng et al. 2016). Similar results have also been reported for other sites in Northern Ethiopia, where cultivation land had a 58% lower SOC level compared to forest land (Gebremariam and Kebede 2010) and 63% lower SOC level in cropland compared to the forest after 30 years of cultivation period in the southern highlands of Ethiopia (Solomon et al. 2002). Similarly, studies at sites in Southern Ethiopia with a similar climate to the highland sites showed that SOC was reduced by 70% within 33 years (Lemenih et al. 2006) and 75% in a 53 year cultivation period in the top 10 cm of soil (Lemenih et al. 2005).

Above-ground and below-ground litter inputs may be a determinant factor for the high accumulation of C in the forest soil (Berg 2000; Smith 2007; Fernandez et al. 2013) compared to small carbon inputs from grazing and cropland, where potential inputs are limited and occur only during the rainy season. The reduction of carbon stock in the cropland is exacerbated by the complete removal of crop residue for cattle feeding. Malhi et al. (1999) and Balesdent et al. (1998) estimated that, after forest clearance, rates of decay of soil C were 10 times higher in cultivated soil than in forest soil. However, our data also suggest that much of the SOC has been lost in croplands by direct erosion (Tosi et al. 2016). One evidence is that the SOC level in cropland topsoil (0 – 10 cm) at Gelawdios (3.1 kg m^{-2}) was nearly comparable to the adjacent church forest carbon stock (3.7 kg m^{-2}) at 20–30 cm depth. The second evidence is elemental analysis of strontium (Sr) to calcium (Ca) and barium (Ba) to calcium ratios in soils. These elemental ratios are widely used as markers of environmental history (Bullen et al. 2005; Tabouret et al. 2010) and biogeochemical properties of the ecosystem (Kabata-Pendias 2010). The Sr/Ca and Ba/Ca ratio at 30–50 cm depth in the forest soil profile is comparable to the 0–10 cm depth of cropland soil at Gelawdios (Fig. 5.3). This shows that the top layer is clearly missing in the cropland soils. Taken together, both the soil profile distribution of C and the element ratios suggest the upper soil profile has been lost in cropland soils. Climate conditions in Ethiopia such as heavy rainfall events are known to result in a high rate of soil erosion (Bewket and Sterk 2005; Betrie et al. 2011; Shiferaw 2011).

At the highland sites, conversion of native church forest to grazing land also significantly reduced SOC stock in the soil by 53%. Bewket and Stroosnijder (2003) showed on other highland sites that grazing land had 48% lower levels of SOM than natural forest. These results are in contrast to studies of productive grasslands in tropical climates where grassland often have similar or greater SOC storage compared to forests (Conant et al. 2001) and even conversion of forest to grassland tends to increase C stock in the soil (Conant et al. 2001; Guo and Gifford 2002). However, poor management of grasslands after conversion led to a decrease in SOC even in wet tropical areas (Fearnside et al. 1998). At the highland sites in this study, the levels of fine root biomass in grasslands are only 10% of those in the church forest (Table 5.2) and thus necromass inputs must have greatly decreased. In addition, most of the grasslands are degraded due to overgrazing. Often the grazing land is completely denuded. Desta et al. (2000) estimated for the grasslands that the stocking density (23 livestock unit (LU) ha^{-1}) is ten times the carrying capacity (2–3 LU ha^{-1}).

Suberin is a characteristic biomacromolecule of roots and bark of vascular plants (Otto and Simpson 2007) and thus used to trace mainly below-ground plant origin assuming bark contributes little to soil organic carbon. Cutin is produced in the epidermal cells to form a protective barrier on leaves, flowers and fruits of vascular plants and can be used to estimate the input of organic matter originating from shoot biomass (Nierop 1998; Nierop et al. 2006). The relative proportion of suberin was about 2 times that of cutin in all land use systems except in grazing land soil (Table 5.3). This shows that in the church forest ecosystem, eucalyptus and cropland, below-ground biomass contributed about 2-times higher organic carbon than above-ground parts. In cropland, all crop residues are harvested for either cattle feeding or fuel and thus, the remaining part is crop roots that may increase the contribution of suberin compared to cutin. Since forest-derived monomers were preserved in the soil for more than 50 years, the higher suberin content in cropland seemed to be derived from former forest vegetation (Mendez-Millan et al. 2012). The lower suberin concentration in grazing land may be due to the fact that grass-derived monomers degrade rapidly compared with forest-derived monomers of the same compound classes in tropical soils (Hamer et al. 2012). The ω-hydroxy carboxylic acids and α,ω-alkanedioic acids of forest origin may have been stabilized in the soils by bonding to soil minerals.

The lowland church forest at Mahibere-Selassie has a SOC stock of ca. 20% of the highland sites. The basal area and stocking density of the tree of the forest (Table 5.1) are only a fraction of that of the highland sites and thus biomass inputs from the trees will be low. In addition, the grass cover of the church forest is regularly burnt, which potentially decreases SOC stocks (Knicker 2007). Furthermore, the soil of the church forest has shallow soil depth and is rocky compared to the adjacent cropland and all the highland forest sites. Exceptionally, cropland at Mahibere-Selassie had a 60% higher SOC stock than the church forest (Fig. 1b). The higher SOC stock may in part be due to the higher clay content of the soil (Ajami et al. 2016), but could also be due to the different farming systems in the lowlands. Ajami et al. (2016) could also show a linear relationship between SOC stock and soil clay content in loess soils in Iran after deforestation.

5.4.2 *Potentials of Soil Carbon Gain Due to Afforestation*

At two sites, Gelawdios and Tara Gedam, the soils under eucalyptus had higher SOC stocks than either the cropland or the grazing land soil. Assuming the eucalyptus plantation at Gelawdios and Tara Gedam are ca. 30 and 40 years old, the rates of SOC accumulation is 0.17 kg m^{-2} $year^{-1}$. A study in Southern Ethiopia, Tesfaye et al. (2016) determined similar levels of SOC in 28 years old *Eucalyptus saligna* plantations and represents an annual SOC accumulation of 0.18 kg m^{-2} $year^{-1}$ compared to the levels of SOC determined in cropland. These results suggest that eucalyptus can restore SOC storage in soils even under management systems using leaf litter raking and removal. However, the low bulk density of the surface layers of the natural forests was not restored. The negative ecological consequences of eucalyptus plantations also need to be considered (Martins et al. 2013).

5.4.3 Effect of Land Use Change on Carbon Input from Above- and Below-Ground Biomass

Changes in vegetation composition because of land use conversion from native church forests to grazing land or cropland alter the overall quantity and quality of fine roots production. Hence, conversion of forest to grazing land or cropland reduced fine root production by 85% and 89%, respectively (Table 5.2). On the other hand, the reverse process of these degraded lands through afforestation with exotic species (eucalyptus) increased the fine root production by 76% using the native forest as a baseline. The average production of fine roots in our study (723 g m^{-2} $year^{-1}$; Table 5.2) is comparable to estimates from mid-subtropics of China (795 g m^{-2} $year^{-1}$; Yang et al. 2004), but lower than estimates from the temperate forest in Germany (689 – 1360 g m^{-2} $year^{-1}$; Hertel & Leuschner, 2002). This comparison is based on a similar methodological approach and soil depth. Our fine root production estimates showed a tendency of the same pattern with soil C and N stocks of the soil. This suggests that fine roots are the major contributors of C and N input in the soil. Similar results have been reported by Hansson et al. (2013) that fine rootstock distribution followed the same pattern as soil carbon and nitrogen distribution in the stands. The annual C flux from fine roots in eucalyptus stand, grazing land and cropland was about 37%, 85% and 89% lower respectively compared to church forest, suggests conversion of native forest to grazing land and cropland resulted ca. 85–89% C input reduction into the soil. The result of this analysis suggests that forest protection from deforestation alone sequester about 90% more carbon compared to grazing land or croplands.

5.5 Conclusion

Removal of native church forests resulted in substantial loss of SOC stock, which appears to be due to soil erosion. Again, conversions of land use from native forest to other land use types such as grazing land or cropland strongly reduce root production (ca. 85–89%). Afforestation with eucalyptus has the potential to restore the levels of SOC storage partially but does not restore to the levels of native church forests. In the forest, eucalyptus and cropland soils, root biomass contributed 2-times that of above-ground parts as revealed by their relative proportions of suberin and cutin.

Acknowledgements This work was carried out within the project "Carbon storage and soil biodiversity in forest landscapes in Ethiopia: Knowledge base and participatory management (Carbopart)" and was funded by the Austrian Ministry of Agriculture, Forestry, Environment and Water Management under grant agreement no BMLFUW-UW.1.3.2/0122-V/4/2013.

References

Abiyu A (2012) The role of seed dispersal, exclosures, nurse shrubs and trees around churches and farms for restoration of ecosystem diversity and productivity in the Ethiopian highlands. University of Natural Resources and Life Science, Vienna

Aerts R, Van Overtveld K, November E et al (2016) Conservation of the Ethiopian church forests: threats, opportunities and implications for their management. Sci Total Environ 551:404–414. https://doi.org/10.1016/j.scitotenv.2016.02.034

Ajami M, Heidari A, Khormali F et al (2016) Environmental factors controlling soil organic carbon storage in loess soils of a subhumid region, northern Iran. Geoderma 281:1–10. https://doi.org/10.1016/j.geoderma.2016.06.017

Assaye Y, Selassie YG, Ayele B (2013) A socio-economic contribution of high land bamboo (Yushania Alpina) for household livelihood in Banja District, Northwestern Ethiopia. J Agric Biodivers Res 2:151–159

Assefa D, Rewald B, Sandén H et al (2017) Deforestation and land use strongly effect soil organic carbon and nitrogen stock in Northwest Ethiopia. CATENA 153:89–99. https://doi.org/10.1016/j.catena.2017.02.003

Balesdent J, Besnard E, Arrouays D, Chenu C (1998) The dynamics of carbon in particle-size fractions of soil in a forest-cultivation sequence. Plant Soil 201:49–57. https://doi.org/10.1023/A:1004337314970

Batjes NH (1996) Total carbon and nitrogen in the soils of the world. Eur J Soil Sci 47:151–163. https://doi.org/10.1111/j.1365-2389.1996.tb01386.x

Berg B (2000) Litter decomposition and organic matter turnover in northern forest soils. For Ecol Manage 133:13–22. https://doi.org/10.1016/S0378-1127(99)00294-7

Betrie GD, Mohamed YA, van Griensven A, Srinivasan R (2011) Sediment management modelling in the Blue Nile Basin using SWAT model. Hydrol Earth Syst Sci 15:807–818. https://doi.org/10.5194/hess-15-807-2011

Bewket W, Sterk G (2003) Assessment of soil erosion in cultivated fields using a survey methodology for rills in the Chemoga watershed, Ethiopia. Agric Ecosyst Environ 97:81–93. https://doi.org/10.1016/S0167-8809(03)00127-0

Bewket W, Sterk G (2005) Dynamics in land cover and its effect on stream flow in the Chemoga watershed, Blue Nile basin, Ethiopia. Hydrol Process 19:445–458. https://doi.org/10.1002/hyp.5542

Bewket W, Stroosnijder L (2003) Effects of agroecological land use succession on soil properties in Chemoga watershed, Blue Nile basin, Ethiopia. Geoderma 111:85–98. https://doi.org/10.1016/S0016-7061(02)00255-0

Bullen TD, Bailey SW, Thomas DB et al (2005) Identifying calcium sources at an acid deposition-impacted spruce forest: a strontium isotope, alkaline earth element multi-tracer approach. Biogeochemistry 74:63–99. https://doi.org/10.1007/s10533-004-2619-z

Capo RC, Stewart BW, Chadwick OA (1998) Strontium isotopes as tracers of ecosystem processes: theory and methods. Geoderma 82:197–225. https://doi.org/10.1016/S0016-7061(97)00102-X

Conant RT, Paustian K, Elliott ET (2001) Grassland management and conversion into grassland: effects on soil carbon. Ecol Appl 11:343–355

Deng L, Zhu G, Tang Z, Shangguan Z (2016) Global patterns of the effects of land-use changes on soil carbon stocks. Glob Ecol Conserv 5:127–138. https://doi.org/10.1016/j.gecco.2015.12.004

Derry LA, Chadwick OA (2007) Contributions from Earth's Atmosphere to Soil. Elements 3

Desta L, Kassie M, Beninc S, Penderd J (2000) Land degradation in the highlands of Amhara region and strategies for sustainable land management. In: M.A. Jabbar JP and SKE (ed) Policies for sustainable land management in the highlands of Ethiopia. International Livestock Research Institute, Addis Ababa

Don A, Schumacher J, Freibauer A (2011) Impact of tropical land-use change on soil organic carbon stocks -a meta analysis. Glob Chang Biol 17:1658–1670. https://doi.org/10.1111/j.1365-2486.2010.02336.x

Don A, Schumacher J, Scherer-Lorenzen M et al (2007) Spatial and vertical variation of soil carbon at two grassland sites—implications for measuring soil carbon stocks. Geoderma 141:272–282. https://doi.org/10.1016/j.geoderma.2007.06.003

Fearnside PM, Barbosa RI, Imbrozio Barbosa R (1998) Soil carbon changes from conversion of forest to pasture in Brazilian Amazonia. For Ecol Manage 108:147–166. https://doi.org/10.1016/S0378-1127(98)00222-9

Feng X, Simpson EJ, Wilson KP et al (2008) Increased cuticular carbon sequestration and lignin oxidation in response to soil warming. Nat Geosci 1:836–839. https://doi.org/10.1038/ngeo361

Fernandez DP, Neff JC, Huang C et al (2013) Twentieth century carbon stock changes related to Piñon-Juniper expansion into a black sagebrush community. Carbon Balance Manag 8:13. https://doi.org/10.1186/1750-0680-8-8

Gebrehana K (2015) Biodiversity and Growth Rates of Natural Forests in the Amhara Region of Ethiopia. University of Natural Resources and Life Sciences (BOKU)

Gebremariam M, Kebede F (2010) Land use change effect on soil carbon stock, above-ground biomass, aggregate stability and soil Crust: a case from Tahtay Adyabo, North Western Tigray, Northern Ethiopia. J Drylands 2:220–225

Girmay G, Singh BR, Mitiku H et al (2008) Carbon stocks in Ethiopian soils in relation to land use and soil management. L Degrad Dev 19:351–367. https://doi.org/10.1002/ldr.844CARBON

Guo LB, Gifford RM (2002) Soil carbon stocks and land use change: a meta analysis. Glob Chang Biol 8:345–360. https://doi.org/10.1046/j.1354-1013.2002.00486.x

Hamer U, Rumpel C, Dignac M-F (2012) Cutin and suberin biomarkers as tracers for the turnover of shoot and root derived organic matter along a chronosequence of Ecuadorian pasture soils. Eur J Soil Sci 63:808–819. https://doi.org/10.1111/j.1365-2389.2012.01476.x

Hansson K, Helmisaari H-S, Sah SP, Lange H (2013) Fine root production and turnover of tree and understorey vegetation in Scots pine, silver birch and Norway spruce stands in SW Sweden. For Ecol Manage 309:58–65. https://doi.org/10.1016/j.foreco.2013.01.022

Hertel D, Leuschner C (2002) A comparison of four different fine root production estimates with ecosystem carbon balance data in a Fagus-Quercus mixed forest. Plant Soil 239:237–251

Houghton RA, Goodale CL (2004) Effects of land-use change on the carbon balance of terrestrial ecosystems. Ecosyst L Use Chang 153:85–98. https://doi.org/10.1029/153GM08

Hurni H, Abate S, Bantider A et al (2010) Land Degradation and Sustainable Land Management in the Highlands of Ethiopia

Kabata-Pendias A (2010) Trace elements in soils and plants, Fourth. CRC press, USA

Knicker H (2007) How does fire affect the nature and stability of soil organic nitrogen and carbon? A review. Biogeochemistry 85:91–118. https://doi.org/10.1007/s10533-007-9104-4

Le Quéré C, Andrew RM, Canadell JG et al (2016) Global Carbon Budget 2016. Earth Syst Sci Data 8:605–649. https://doi.org/10.5194/essd-8-605-2016

Lemenih M, Itanna F (2004) Soil carbon stocks and turnovers in various vegetation types and arable lands along an elevation gradient in southern Ethiopia. Geoderma 123:177–188. https://doi.org/10.1016/j.geoderma.2004.02.004

Lemenih M, Karltun E, Olsson M (2005) Soil organic matter dynamics after deforestation along a farm field chronosequence in southern highlands of Ethiopia. Agric Ecosyst Environ 109:9–19. https://doi.org/10.1016/j.agee.2005.02.015

Lemenih M, Lemma B, Teketay D (2006) Changes in soil carbon and total nitrogen following reforestation of previously cultivated land in the highlands of Ethiopia. SINET Ethiop J Sci 28:99–108. https://doi.org/10.4314/sinet.v28i2.18245

Leppälammi-Kujansuu J, Aro L, Salemaa M et al (2014) Fine root longevity and carbon input into soil from below- and aboveground litter in climatically contrasting forests. For Ecol Manage 326:79–90. https://doi.org/10.1016/j.foreco.2014.03.039

Li J, Zhang G, Ruan L et al (2016) Sr–Nd elements and isotopes as tracers of dust input in a tropical soil chronosequence. Geoderma 262:227–234

Li X, Zhu J, Lange H, Han S (2013) A modified ingrowth core method for measuring fine root production, mortality and decomposition in forests. Tree Physiol 33:18–25

Lukac M (2012) Fine root turnover. In: Measuring roots. Springer, Berlin, pp 363–373

Malhi Y, Baldocchi DD, Jarvis PG (1999) The carbon balance of tropical, temperate and boreal forests. Plant Cell Environ 22:715–740

Martins C, Natal-da-Luz T, Sousa JP et al (2013) Effects of essential oils from Eucalyptus globulus leaves on soil organisms involved in leaf degradation. PLoS ONE 8:e61233. https://doi.org/10.1371/journal.pone.0061233

Mendez-Millan M, Dignac M-F, Rumpel C et al (2012) Contribution of maize root derived C to soil organic carbon throughout an agricultural soil profile assessed by compound specific 13 C analysis. Org Geochem 42:1502–1511

Meseret D (2016) Land degradation in Amhara Region of Ethiopia: review on extent, impacts and rehabilitation practices. J Environ Earth Sci 6:120–130

Moges Y, Kindu M (2006) Effects of fencing and ground cultivation on natural regeneration of Boswellia papyrifera in Metema Wereda, Ethiopia. J Drylands 1:45–51

Nierop KGJ (1998) Origin of aliphatic compounds in a forest soil. Org Geochem 29:1009–1016

Nierop KGJ, Jansen B, Hageman JA, Verstraten JM (2006) The complementarity of extractable and ester-bound lipids in a soil profile under pine. Plant Soil 286:269–285. https://doi.org/10.1007/s11104-006-9043-1

Nkonya E, Koo J, Kato E, Guo Z (2013) Trends and patterns of land use change and international aid in sub-Saharan Africa

Nyssen J, Temesgen H, Lemenih M et al (2008) Spatial and temporal variation of soil organic carbon stocks in a lake retreat area of the Ethiopian Rift Valley. Geoderma 146:261–268. https://doi.org/10.1016/j.geoderma.2008.06.007

Oertel C, Matschullat J, Zurba K et al (2016) Greenhouse gas emissions from soils—A review. Chem Erde 76:327–352. https://doi.org/10.1016/j.chemer.2016.04.002

Osawa A, Aizawa R (2012) A new approach to estimate fine root production, mortality, and decomposition using litter bag experiments and soil core techniques. Plant Soil 355:167–181. https://doi.org/10.1007/s11104-011-1090-6

Otto A, Shunthirasingham C, Simpson MJ (2005) A comparison of plant and microbial biomarkers in grassland soils from the Prairie Ecozone of Canada. Org Geochem 36:425–448

Otto A, Simpson MJ (2006) Sources and composition of hydrolysable aliphatic lipids and phenols in soils from western Canada. Org Geochem 37:385–407

Otto A, Simpson MJ (2007) Analysis of soil organic matter biomarkers by sequential chemical degradation and gas chromatography–mass spectrometry. J Sep Sci 30:272–282

Pan Y, Birdsey RA, Fang J et al (2011) A Large and Persistent Carbon Sink in the World's Forests. Science (80-) 333:988–993

Santantonio D, Grace JC (1987) Estimating fine-root production and turnover from biomass and decomposition data: a compartment–flow model. Can J for Res 17:900–908. https://doi.org/10.1139/x87-141

Shiferaw A (2011) Estimating soil loss rates for soil conservation planning in the Borena Woreda of South Wollo Highlands, Ethiopia. J Sustain Dev Africa 13:87–106. https://doi.org/10.1016/S0305-750X(98)00159-4

Smith P (2007) Land use change and soil organic carbon dynamics. Nutr Cycl Agroecosystems 81:169–178. https://doi.org/10.1007/s10705-007-9138-y

Soleimani A, Hosseini SM, Massah Bavani AR et al (2019) Influence of land use and land cover change on soil organic carbon and microbial activity in the forests of northern Iran. CATENA 177:227–237. https://doi.org/10.1016/j.catena.2019.02.018

Solomon D, Fritzsche F, Tekalign M et al (2002) Soil organic matter composition in the subhumid Ethiopian highlands as influenced by deforestation and agricultural management. Soil Sci Soc Am J 66:68–82. https://doi.org/10.2136/sssaj2002.6800

Tabouret H, Bareille G, Claverie F et al (2010) Simultaneous use of strontium: calcium and barium: calcium ratios in otoliths as markers of habitat: application to the European eel (*Anguilla anguilla*) in the Adour basin, South West France. Mar Environ Res 70:35–45. https://doi.org/10.1016/j.marenvres.2010.02.006

Tesfaye MA, Bravo F, Ruiz-Peinado R et al (2016) Impact of changes in land use, species and elevation on soil organic carbon and total nitrogen in Ethiopian Central Highlands. Geoderma 261:70–79. https://doi.org/10.1016/j.geoderma.2015.06.022

Tosi M, Correa OS, Soria MA et al (2016) Land-use change affects the functionality of soil microbial communities: A chronosequence approach in the Argentinian Yungas. Appl Soil Ecol 108:118–127. https://doi.org/10.1016/j.apsoil.2016.08.012

Vogt KA, Vogt DJ, Bloomfield J (1998) Analysis of some direct and indirect methods for estimating root biomass and production of forests at an ecosystem level. Plant Soil 200:71–89

Wassie A, Sterck FJ, Teketay D, Bongers F (2009) Tree regeneration in church forests of Ethiopia: effects of microsites and management. Biotropica 41:110–119. https://doi.org/10.1111/j.1744-7429.2008.00449.x

Wei X, Shao M, Gale W, Li L (2014) Global pattern of soil carbon losses due to the conversion of forests to agricultural land. Sci Rep 4:4062. https://doi.org/10.1038/srep04062

Workneh S, Glatzel G (2008) Land use changes and soil organic carbon and soil nitrogen in the Northwestern Highlands of Ethiopia. University of Natural Resources and Applied Life Sciences, Vienna

WRB (2014) World reference base for soil resources 2014. Food and Agriculture Organization of the United Nations, Rome

Xia M, Talhelm AF, Pregitzer KS (2015) Fine roots are the dominant source of recalcitrant plant litter in sugar maple-dominated northern hardwood forests. New Phytol 208:715–726

Yang Y-S, Chen G-S, Lin P et al (2004) Fine root distribution, seasonal pattern and production in four plantations compared with a natural forest in subtropical China. Ann for Sci 61:617–627

Chapter 6
Estimation and Mapping of Asabot Monastery Dry Afromontane Forest Carbon Stock Under Diverse Land-Use Scenarios

Tulu Tolla, Teshome Soromessa, Richard P. Dick, Seyoum Leta, Mekuria Argaw, Gudina Legessa, Mesfin Sahle, Merga Belina, Eyasu Elias, and Zewdu Eshetu

Abstract Even though the Ethiopian Orthodox Tewahido Church is acknowledged in the preservation patches of forest in the highlands, the church's role in conserving the Dry Afromontane ecosystem is not well documented. This study estimate and map carbon stock under three future land-use scenarios for sustainable management in the Asebot Dry Afromontane forest of Ethiopia. The study quantifies land use land cover (LULC), developed future land-use scenarios, and estimated carbon stock at the plot level. Three simulated LULC scenarios, such as Business As Usual (BAU), forest disturbance, and optimistic conservation, were developed for 2047. The LULC trends indicate that the forest and grassland cover decreased; meanwhile, the woodland and shrubland increases. In an optimistic conservation scenario, the forest, woodland, and shrubland cover would increases by 50% from the current coverage. Whereas, in a forest disturbance scenario, the woody vegetation covers would decrease. The average total carbon stock estimated from the field data ranges from 129 to 355.6 tons/ha. Because of the changes in LULC, the carbon stock amount decreased from 1986 to 2017 and will continue in the BAU scenario. In a forest disturbance scenario, the carbon stock would be decreased by 12.7% from 2017, while in an optimistic

T. Tolla (✉)
Kotebe Metropolitan University, Addis Ababa, Ethiopia

T. Soromessa · S. Leta · M. Argaw · G. Legessa · M. Belina · E. Elias · Z. Eshetu
Addis Ababa University, Addis Ababa, Ethiopia
e-mail: teshome.soromessa@aau.edu.et

S. Leta
e-mail: seyoum.leta@aau.edu.et

M. Argaw
e-mail: ensc@sc.aau.edu.et

R. P. Dick
Ohio State University, Columbus, OH, USA
e-mail: dick.78@osu.edu

M. Sahle
Wolkite Univeristy, Wolkite, Ethiopia

M. Kindu et al. (eds.), *State of the Art in Ethiopian Church Forests and Restoration Options*, https://doi.org/10.1007/978-3-030-86626-6_6

conservation scenario increases by 12.4%. The study shows that the Dry Afromontane forest, such as Asebot Monastery, has a positive role in climate change mitigation under optimistic conservation. Therefore, proper forest management is essential for enhancing ecosystem services in Dry Afromontane forests and contributing to the restoration of the downstream's degraded landscape.

Keywords Church forest · Land use change · Future scenario · Total carbon · Climate change regulation

6.1 Introduction

According to FAO Forest Resource Assessment (FRA), the global forests and other woodland accounts for 31% of the total land areas (FAO 2010). The dry (42%), moist (33%), and wet (25%) tropical and subtropical forests were accounted for 40% of the African dry forest ecosystems (Hasnat and Hossain 2020). However, the African Dry Afromontane forest ecosystem supplies goods and services to the surrounding community and lower basin country becoming under severe exploitation and degradation threat (Blackie et al. 2014). In Africa, the forest and agricultural sectors, including conversion to croplands and shifting cultivation, account for 75% of the region's total emissions (Canadell et al. 2009). Similarly, land clearing and degradation turn the valuable carbon sink into a significant greenhouse gas source. Consequently, about five million hectares of the forest may have been lost annually in Africa from 2005 to 2015, releasing nearly 2 billion tons of CO_2eq equivalent each year, or 13% of annual global emissions from forestry and agriculture (IPCC 2007; Sohngen and Brown 2008).

Land use land cover (LULC) change is the alteration of Earth's terrestrial surface mainly by anthropogenic activities (Reusing 2000). The main agents of global LULC changes were the interactions of climate, biogeochemical cycles, biodiversity, ecosystem processes, and human activities, and might be occurring at various spatial and temporal scales (IGBP 1999). The anticipated LULC change deforestation and forest degradation influence the amount of biomass and carbon stored in vegetation (Pearson et al. 2005; IPCC 2007). This contributes 0.3–3.0 Pg C year^{-1} of the total fossil fuel emissions of the globe (Sohngen and Brown 2008).

The Ethiopian Dry Afromontane forest is described by an altitude range from 1500 to 3400 m above sea level (Teketay 1996). This agro-ecological zone is suitable for human inhabitation accompanied by sedentary agriculture, extensive cattle herding activities, and socio-political instability. In this zone, the forest and woodland are the most threatened and least protected ecosystems (Mertz et al. 2007; Brink and Eva 2009). This is mainly due to population increase, climate change, and weak environmental governance and policy frameworks in the agro-ecological zone (Teketay 1996; FAO 2010).

Land conversion is the primary driver of ecosystem service depletion or enhancement (Furst et al. 2013). A more sustainable land management strategy could ensure the continued provision of goods and services. The detection of changes becomes an essential process in managing and monitoring natural resources (Mejebi 2008). By spatially modeling the future scenarios, we can better understand the impact and uncertainties of various LULC planning and policies in a wide range of potential futures before they occur (Gibson and Quinn 2017; Kindu et al. 2018). Scenarios and models can effectively address relationships between nature, nature's benefits to people, and good quality of life (IPBES 2016). They support several researchers to engage in scenario analysis to address the effects of future land use change on ecosystem services and human well-being for decision-making. In-depth information about the various impacts on ecosystem services, such as climate regulation, is required to evaluate land management strategies' sustainability.

In Ethiopia's highlands, sacred groves are associated with the Ethiopian Orthodox Tewahedo Church (EOTC) are known as "church forests" (Wassie et al. 2007; Aerts et al. 2016). In the North-central parts of Ethiopia, these church forests are virtually the only patches of forest on the degraded agricultural landscape (Aerts et al. 2006; Wassie et al. 2010; Reynolds et al. 2017). Church communities manage their forests largely autonomously, and management varies from strict protection to weak protection with poorly controlled harvesting of trees (Aerts et al. 2016; Orlowska and Klepeis 2018). Despite the relevance of church forests widely acknowledged in biodiversity conservation and other ecological functionality (Cardelus et al. 2019), many studies did not apply the detailed assessment of ecosystem services such as climate regulation. However, this information could enhance planners' and decision-makers' understanding to integrate these forest patches into more comprehensive conservation strategies and policy frameworks. Moreover, although several past studies have evaluated the conservation value of church forests in central and northern Ethiopia (e.g., Aerts et al. 2006; Wassie et al. 2010, Cardelús et al. 2017, 2019), to date, there is no information on the contribution of church forests to the conservation of biodiversity and ecosystem services in the other parts of Dry Afromontane forest.

Although numerous spatial scenario simulation models have provided valuable insights into the LULC, future LULC patterns and changes are still poorly understood in the Ethiopian landscapes (Kindu et al. 2018; Sahle et al. 2019). Similarly, information on the regular monitoring and dynamics of LULC changes is not much available on the Dry Afromontane forests in Ethiopia. This making challenging to manage the natural resources, monitor environmental changes, and predict appropriate future conditions. Therefore, analyzing the forest conditions' status and developing appropriate scenarios are very important since the Dry Afromontane forests are found in the agricultural landscape with several socio-economic changes.

Asebot forest is among the few remnant Dry Afromontane forest relicts remaining over the vast land of the Harerge highlands of Oromia Regional State (Worku and Zewede 2008). The forest is known for its high diversity in vegetation composition and trees aged over eight hundred years. This forest is also unique as it is situated at a strategic place buffering the arid and semi-arid rangelands of the Somali and Afar regions from that of Western Harerge Zone farmlands of Oromia. This study aims to quantify and map carbon stock under diverse future land-use scenarios for sustainable management in the Asebot Dry Afromontane forest of Ethiopia. This study is the first in the semi-arid regions and could contribute to understanding the future effects of LULC dynamics and climate regulation in Dry Afromontane forest in Ethiopia in general and church forests in particular.

6.2 Material and Methods

6.2.1 Descriptions of the Study Area

Asebot Dry Afromontane Monastery forest is located in Central Eastern parts of Ethiopia at Miso district of West Hararghe Zone, Oromia Regional State (Fig. 6.1). Geographically, the study site is located at 40° 31′ to 40° 42′ E and 90° 12′ to 90° 21′ N. Until recently; the forest is conserved by Asebot Debre Wegege Aba Samuel and Kidest Selassie Monastery as a church forest for a long time. During the Derg regime (1974–1991), the government has owned the forest and leaving the forest patches around the two churches to be possessed by the Monastery. Currently, except for the forest surrounding the church compounds, the administration is undertaken by Hallideghe Asebot National Park. The topography of the study site is characterized by rigid mountains. The elevation ranges between 1080 and 2447 m above sea level, subject to immediate changes in facing. The Asebot Dry Afromontane

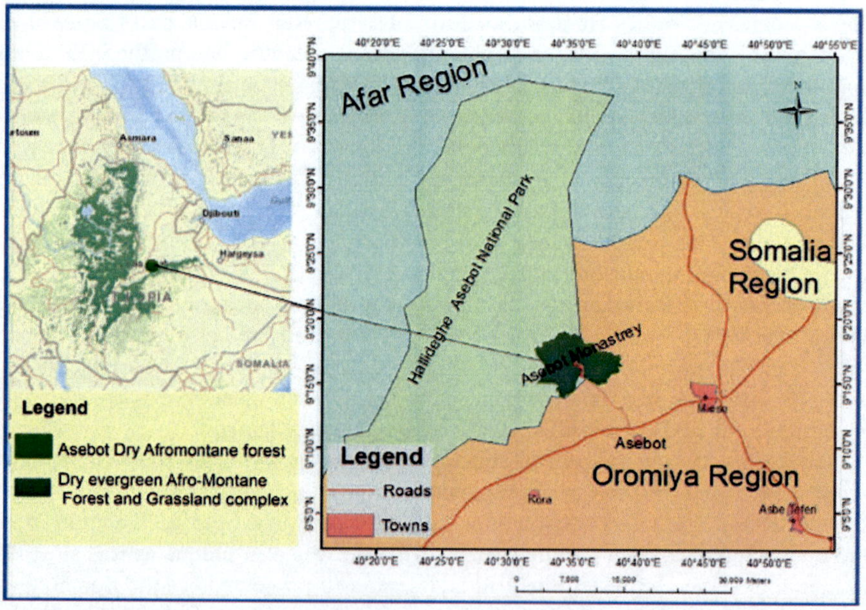

Fig. 6.1 Location map of Asabot Dry Afromontane forest

Monastery forest's climatic condition is part of evergreen dry Afromontane ecology and characterized by 451–1055 mm annual rainfall and 11.9–31.9 °C daily average temperature (Worku and Zewede 2008). Asebot forest is mostly covered with the East African Dry Afromontane vegetation formation and the high lying areas with relatively higher rainfall. The upland forest includes *Juniperus procera, Podocarpus falcatus, Olea europaea, Acacia abyssinica, and Croton marcrostachyus species.* Many seasonal streams emerge from the top mountain, but no water source is available in the forest cover during the dry season. In the mountains, only the monastery community is living. However, a large number of people are living around the Asebot forest and practicing subsistent agriculture. The Oromia region communities are more agrarians with limited livestock production, and their mode of life was sedentary. However, those from Somali (*Issa*) and Afar regions part of the forest depends on animal husbandry and hence are pastoralists. There is also boundary conflict between *Issa* and Oromo peoples living around the forest. Lack of clear ownership right on the forest, encouraging severe encroachment and its consequent deforestation from all sides (Worku and Zewede 2008).

6.2.2 Land Use Land Cover Assessment

Land use land cover analysis was conducted based on Landsat 5 images of 1986 and 2001 and Landsat 8 (OLI) of 2017, which had a similar 30 m spatial resolution. All the imageries were acquired in the same month between January 19 and 30 to reduce seasonal discrepancies. The images we used were taken during the dry season to reduce spectral value similarities between LULC's. The images fall on the same scene (167/54) and do not have image mosaic effects.

The raw digital number (DN) values converted to top-of-atmosphere reflectance to reduce an atmospheric effect following Chander et al. (2009) for Landsat 5 and the method described in the USGS for Landsat 8 (OLI). The Minnaert correction, one of the non-Lambertian topographic correction methods, was considered to correct the images' terrain effects by using QGIS 3.16 topographic correction tools. The USGS SRTM DEM with a spatial resolution of 30 m is used as an input to correct the images. Georeferencing by using the topographic images of the study area was done to precise the images' spatial location. The final steps in image restoration and enhancement techniques used in this study were contrast-stretching to improve image visualization.

The image classification in this study started with the iso cluster unsupervised classification method. The unsupervised classification was done before the fieldwork and supports collecting reference points representing the various LULC, and places can lead to misclassifications by using GPS on the field. These reference locations were overlayed on the raw images and support supervised classification. Supervised LULC classification analysis was done by using the maximum likelihood method of ERDAS Imagine software. Five major LULC were identified in the study site (in the field) considering the vegetation cover's vertical structure (Di Gregorio and Jansen

Table 6.1 Land use land cover classification of Asebot Dry Afromontane forest

Forest land	The natural or planted trees in areas that covered a minimum size of 0.5 ha
Woodlands	The open stands of trees in areas mainly dominated by Acacia Spp
Shrubland	Small trees mixed with grasses and less dense than forests
Grazing land	All areas covered by grasses
Built up	The area is used for church buildings, residential sites for the monastery communities, and other facilities of the monastery

1996) (Table 6.1). Classified images were compared from two periods, 1986–2000, 2000–2017, and 1986–2017.

The LULC accuracy was assessed by using 70, 80, and 80 reference points of the classified images of the years 1986, 2000, and 2017, respectively. The number of reference points was adjusted approximately to the area covers of each LULC. The 1986 reference points were generated from the topographic map with known places. For the year 2000, the reference points were extracted from the topographic map in areas where no change has occurred and from the raw Landsat images. The reference points for the year 2017 were generated from the recorded location using hand-held GPS and the recent high-resolution images of Google Earth Engine. The overall accuracy of the classification is 85%, 85%, and 91.2% in 1986, 2000, and 2017, respectively (Table 6.2). The kappa coefficient showed a high level of agreement—0.81 in 1986, 0.81 in 2000, and 0.91 in 2017.

Table 6.2 Accuracy assessment matrices for the classified images of 1986 (a), 2000 (b), and 2017(c)

(a)

LULC	Forest	Woodland	Shrubland	Grassland	Built up	User accuracy (%)
Forest	17	1	2	0	0	85
Woodland	0	18	1	1	0	90
Shrub land	0	0	18	1	0	95
Grass land	1	1	1	7	0	70
Built up	0	0	0	0	0	0
Producer accuracy (%)	94	90	82	78	0	60

Over all accuracy = 85% Kappa coefficient = 0.8

(b)

LULC	Forest	Woodland	Shrub land	Grassland	Built up	User accuracy (%)
Forest	18	1	1	0	0	90

(continued)

Table 6.2 (continued)

(a)

Woodland	0	18	1	1	0	90
Shrub land	1	1	16	1	1	80
Grass land	1	1	1	7	0	70
Road	0	0	0	0	0	100
Built up	0	0	1	0	5	80
Producer accuracy (%)	90	86	80	78	90	68

Over all accuracy = 85% Kappa coefficient = 0.8

(c)

LULC	Forest	Woodland	Shrub land	Grassland	Built up	User accuracy (%)
Forest	18	2	0	0	0	90
Woodland	0	18	2	0	0	90
Shrub land	0	0	19	1	0	95
Grass land	1	0	0	9	0	90
Road	0	0	0	0	0	100
Built up	0	0	0	0	4	80
Producer accuracy (%)	95	90	90	90	100	

Over all accuracy = 91.2% Kappa coefficient = 0.91

6.2.3 Future LULC Scenarios

Based on the forest landscape's existing situation, three future scenarios have been defined to predict LULC demand for 2047, namely Business as Usual (BAU), forest disturbance, and optimistic conservation scenarios based on own assumptions (Table 6.3). These scenarios were developed for the year 2047 to match with the REDD strategy for resource management on climate change mitigation scenarios.

The BAU scenario was designed mainly based on an assumption of a continuation of LULC conversion rates of the past 30 years in the forest landscape (Table 6.3). The scenario was developed by employing Land Change Modeler (LCM) 2.0 for ArcGIS (Clarklabs.org). The transition suitability map, which was used to predict the LULC in 2017 and to simulate the distribution in 2047, was generated based on the main transitions that occurred among the LULC categories from 1986 to 2017. The LCM algorithms integrate the functions of the cellular automata (CA) filter and Markov process, using conversion tables and conditional probabilities from the conversion map applied to simulate and forecast the states of LULC change. Kappa statistics were used to assess the accuracy of the forecasted 2047. The location probability of each land-use type was addressed based on the logistic regression analysis. Driving factors such as elevation, slope, distance to roads, and settlement areas were considered to

Table 6.3 Summary of assumptions considered for the development of three scenarios in 2047

BAU scenario	An optimistic conservation scenario	Forest disturbance scenario
Continuation of historical LULC changes	• Forest area increases due to natural vegetation regenerations and native tree plantation on large areas • There would be good coordination in forest management • Awareness creation and incentive mechanisms would be established • Too much livestock grazing and timber harvesting would be minimized and frequent fire breaks can be controlled • As a result, forest, woodland, and shrubland would be increased by 50% from the existing cover. Grassland would be replaced by other land covers	• The existing challenges such as open grazing, timber harvesting, and frequent fire outbreaks could be aggravated due to a lack of proper forest management • Regional conflicts would lead to further forest disturbance • As a result, a 50% diminishing assumed in the forest, woodland, and shrubland covers and grassland would be increased

evaluate the sub-model. LULC map to evaluate its agreement with the actual 2017 LULC map. During the assessment, the 2017 LULC map acted as the reference map (reality), while the simulated map was the comparison map.

In an optimistic conservation scenario, the forest area increases due to natural vegetation regenerations, and indigenous tree plantations would occur in the degraded areas and grasslands (Table 6.3). In this strict conservation and reforestation scenario, a 50% increase in forest, woodland, and shrubland coverage would be expected. The assumption is based on the National Forest Sector Development Program (NFSDP) in Ethiopia that planned to expand forest by 15.7%, in 2015 to 20% by 2020 and 30% by 2025 (Ministry of Environment, Forest and Climate Change [MEFCC]. This continuous progress plan of the country would lead to a 50% forest expansion in 2047. There is an assumption of good coordination in forest management between the monastery and government and other conservation initiatives in an optimistic conservation scenario. Awareness creation and incentive mechanisms would be established for the monastery and the surrounding communities for decreasing illegal livestock grazing, timber harvesting, and frequent fire breaks. InVEST proximity-based scenario generator model was used for creating the scenario. The proximity-based scenario generator creates a set of contrasting land use change maps that convert habitat in different spatial patterns. The forest cover increases by 50% at the expense of woodland and grassland. The assumption of woodland to forest conversion was large since the trend showed that woodland proxy to the forest had a very high conversion probability than shrubland. The forest expansion in the grassland is due to reforestation. The woodland expansion by 50% is at the expense of the existing shrubland coverage area based on the assumption that there would be a high conversion probability due to strict conservation.

The forest disturbance scenario assumes that the existing challenges such as overgrazing, timber harvesting, and frequent fire outbreaks would be aggravated due to a lack of proper forest management (Table 6.3). As Asebot Dry Afromontane forest is found in the junction of three regions and there are historical conflict trends, the problem would be worsened in the future, assuming the country's current instability has its own contributions. As a result, the existing forest covers converted to woodland, the woodland to shrubland, and the shrubland to grassland. The large area would be converted to grassland. InVEST proximity-based scenario generator model was considered for LULC transitions, and a 50% diminishing would be occurred in forest, woodland, and shrubland covers according to the nearest conversion probabilities.

6.2.4 Carbon Stock Estimation

6.2.4.1 Vegetation Biomass Data Collection

A mixed spatial stratified-systematic sampling method was used for vegetation sampling to ensure full coverage of the altitude gradients' environmental variation and habitat heterogeneity (Kent and Coker 1992). The strict procedure was not followed in selecting the direction of transect and sampling quadrats due to the mountain's complexity (Supplementary material). Sampled quadrats lied in the study sites' clustered altitude (lower (1087–1600 m), middle (1601–2000 m), and upper (2001–2447 m)). Data on woody vegetation were collected from 20 m × 20 m (400 m^2) size sampling quadrats of 150 m and 100 m distance between transect and quadrat. Five (5 m × 5 m) subplots were established for shrubs sampling inside the 20 m × 20 m quadrat, one at each corner and the other at the center. Data on seedling and sapling were collected from five (2 m × 2 m) subplots, one at each corner and the other at the center of the main quadrates.

The diameter breast height (1.3 m height) of all woody plants was recorded using diameter tape, while the height was measured using a Hagan hypsometer and Clinometer. The diameter of individual trees was measured at breast height using diameter tape (Aynekulu et al. 2011). Individual tree diameter was measured and recorded if the tree was buttressed at breast height. Visual estimation was used, where topographic features were difficult to measure trees and shrubs and their heights. Environmental variables such as altitude, slope, and geographical coordinates of each plot were measured using Garmin global position system and satellite image (Kent and Coker 1992). Specimens of all the woody plants were pressed, dried, and taken to the National Herbarium of Addis Ababa University for identification and storage. Wood plant nomenclature was made using Flora of Ethiopia and Eritrea references (Friis and Demissew 2001).

6.2.4.2 Above-Ground and Below-Ground Biomass Estimation

The total biomass and the biomass per hectare were estimated using the biomass of different vegetation layers (trees, shrubs, and herbaceous layers). Before calculating the total biomass, different layers were converted to the same unit (Mg ha^{-1}). The total biomass was calculated as the sum of all biomass layers for each plot and averaged over all plots. Total above-ground biomass was the sum of tree biomass, shrub biomass, and herb biomass. The allometric equation developed by Brown was used to estimate above-ground biomass (Brown 2002; Navar 2009).

$$Y = 34.4703 - 8.0671(\text{DBH}) + 0.6589\left(\text{DBH}^2\right);$$

where, Y was above-ground biomass (ABG), DBH was diameter at breast height.

Below-ground biomass (BGB) was estimated from AGB using the relationship derived for the tropical forest as root-shoot ratios (R/S), which is 25% of above-ground biomass (Cairns et al. 1997).

The shrubs, herbaceous layers, and litter biomass were estimated according to Pearson et al. (2005).

$$\text{SHLB} = \frac{\text{wt. field}}{\text{A}} * \frac{\text{wt.(Subsample, dry)}}{\text{wt.(Subsample, fresh)}} * \frac{1}{10,000}$$

where, SHLB was shrubs, herbaceous layers, and litter biomass (Mg ha^{-1}); wt. the weight of the wet field sample; A is the size of the area in which sample was collected (ha); wt. (subsample, fresh) was the weight of the fresh subsample taken to the laboratory for moisture content determination (100 g); wt. (subsample, dry) was the weight of the oven-dry subsample of litter taken to the laboratory (g).

Tree biomass conversions to carbon stocks were considered using a common proxy based on the assumption that 50% of the biomass was corrected to carbon (Malhi et al. 2004). In this study estimation of carbon was made by multiplying biomass by 0.47, and taking into consideration the multiplication factor 3.67 of the CO_2 equivalent (Pearson et al. 2005). The unity of carbon stock was estimated to be metric ton per hector (Mg ha^{-1}).

6.2.4.3 Soil Carbon Stock Estimation

Composite soil samples in the depth of 20 cm were collected from the vegetation sampling quadrat. Soil carbon concentration was determined using the Potassium Dichromate method (Walkley and Black 1934). Bulk density was determined by weighing a 105 °C dried soil sample of known volume (125 cm^3). The bulk density of soil (BD) was calculated using the formula:

$$BD = S/V,$$

where S is the oven-dry weight of the soil sample, and V is the sample volume.

The soil organic carbon density for a sampling plot was estimated using the formula developed by (Yu et al. 2009):

$$\text{SOC(ton/ha)} = \frac{\%\text{OC} * BD * D}{100} * 10,000\ \text{m}^2\ \text{ha}^{-1};$$

where OC (%) is the percent of organic carbon, *BD* (in Mg m^{-3}) is the bulk density of soil sample, *D* (in m) is the given soil depth.

The unit metric tons per hectare was used for measuring soil organic carbon because mega-gram (Mg) was equal to metric tons. The carbon stock density was converted into tons of CO_2 equivalent by multiplying it by 44/12 or 3.67, which was the molecular ratio of CO2 to carbon to understand the climate change mitigation potential of the study area (Pearson et al. 2005).

6.2.4.4 Carbon Stock Mapping

Carbon stock was quantified and mapped using InVEST carbon storage and sequestration model (He et al. 2016). InVEST model used maps of land use land cover and ground survey-based carbon stock estimated for the four-carbon pools (above-ground biomass, below-ground biomass, soil, and dead organic matter). The model estimated the amount of carbon stored in a landscape or the amount of carbon sequestered over time. The model also summarizes results into raster outputs of storage, value, as well as aggregate totals. The net change in carbon storage over time (sequestration and loss) was produced by providing a current and future LULC map. In this study, the developed future scenarios were considered to estimate the differences between the then and forthcoming alternate future landscape carbon stocks.

6.3 Results

6.3.1 Land Use Land Cover Change in Asebot Afromontane Forest

The LULC analysis from satellite images indicated that forest coverage decreased from 1483 ha (22.7%) in 1986 to 1332 ha (20.4%) in 2001 and then to 1147 ha (17.5%) in 2017 (Fig. 6.2). While the woodland coverage area increased from 1854 ha (28.4%) to 2214 ha (33.9%) in the past 30 years. In similar to woodland, shrubland increased from 2739 ha (41.9%) in 1986 to 2809 ha (43%) in 2001 and 2843 ha (43.5%) in 2017. The trend shows that grassland coverage area decreased from 456 ha (7%) in 1986 to 308 ha (4.7%) in 2017. Even though the built-up area coverage is very small,

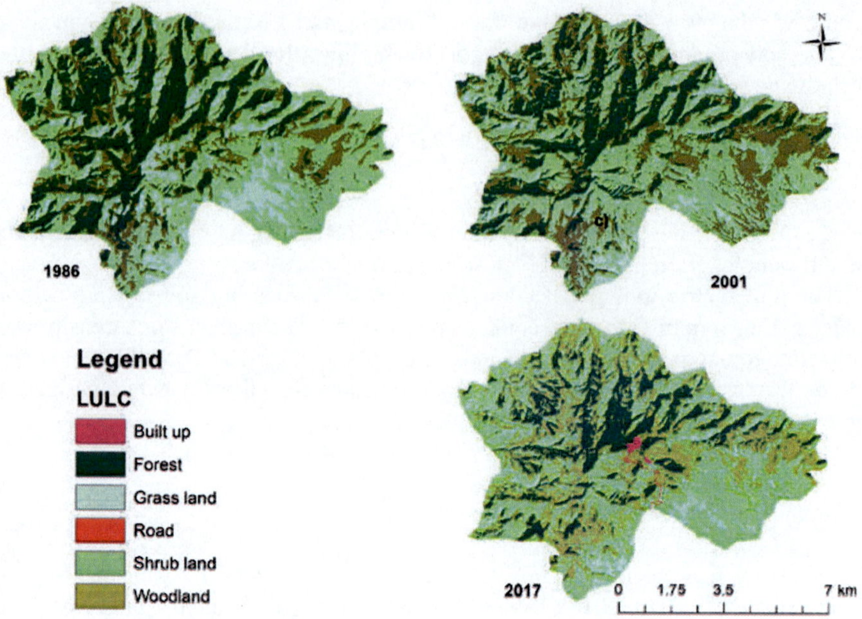

Fig. 6.2 The LULC map of 1986, 2000, and 2017 in Asebot Dry Afromontane Monastery forest

its size has been continuously increasing from 0.06% in 1986 to 0.15% in 2000 and 0.37% in 2017.

Coverage of forest and grassland decreased significantly from 1986 to 2017 by 22.7% and 32.5%, respectively, whereas that of woodland, shrubland, and built-up area increased by 19.42%, 3.8%, and 500%, respectively. The grassland was converted to shrubland due to high shrubs expansion in the vicinity. The open grasslands could allow for growing new propagation in natural mechanisms. The forest conversion rate from 1986 to 2000 was lower than the period between 2000 and 2017.

6.3.2 Simulation of Land Use Land Cover Patterns Under Different Scenarios

The LULC probability conversion was varied except for grassland. The conversions of woodland and shrubland were very dynamic, and the probabilities of their resistances were found to be 69% and 74%, respectively. Out of the total forest cover, 23% converted to woodland while the remaining 77% was stagnant. The future LULC change prediction was considered the changing probabilities of the land uses and

Table 6.4 The area coverage of the simulated LULC at different scenarios in 2047

LULC	1986 (ha)	2001 (ha)	2017 (ha)	BAU scenario (ha)	Forest disturbance scenario (ha)	An optimistic conservation scenario (ha)
Forest	1483	1332	1147	723	648	1648
Grassland	456	318	308	224	1709	0
Shrubland	2739	2809	2843	2904	2520	1779
Woodland	1854	2067	2214	2658	1635	3085
Built up	4	10	24	24	24	24

significant factors that could ably determine the suitability of each land-use type's location.

In the BAU scenario, forest and grassland cover would decrease from 17.5% (1147 ha) to 11.1% (723 ha) and 4.7% (308 ha) to 3.4% (224 ha) in the period between 2017 and 2047, respectively (Table 6.4). On the contrary, woodland, and shrubland cover would increase from 33.9% to 40.6% and 43.5% to 44.4%, respectively. The BAU scenario trend indicated that the Asebot Afromontane forest could lose its stagnant under the existing management system unless sustainable management is not employed.

In the forest disturbance scenario, the forest coverage area would decrease from 1147 to 648 ha (Table 6.4). Similarly, the shrubland and woodland would also decrease in this scenario (Table 6.4). Meanwhile, the grassland cover would increase significantly from 308 ha in 2017 to 1709 ha in the forest disturbance scenario.

In the optimistic conservation scenario, the LULC transition is the opposite of the disturbance scenario, and the forest coverage would increase from 1147 to 1648 ha (Table 6.4). In comparison, the woodland cover increases from 2214 ha in 2017 to 3085 ha in 2047. The shrubland would decreases from 2843 to 1779 ha. The total grassland coverage area would be replaced by forest, woodland and shrubland (Fig. 6.3).

6.3.3 *Carbon Stock*

The carbon stock amount varies according to the LULC. The highest average above-ground carbon stock is estimated in forest cover, which is 192.6 tons/ha. The woodland has 91 tons/ha carbon stock. The low carbon stock is estimated in shrubland and grassland with an average of 45 and 35.2 ha, respectively.

The soil carbon is highest in forest cover, with an average of 123.3 tons/ha. In the woodland cover, 116.2 tons/ha is estimated. About 99.5 and 86.7 tons/ha soil carbon stock is estimated in shrubland and grassland, respectively.

The amount of carbon in deadwood is low in the Asebot Afromontane Monastery forest, with 1.12 tons/ha in the forest and 0.45 tons/ha in woodland. The highest

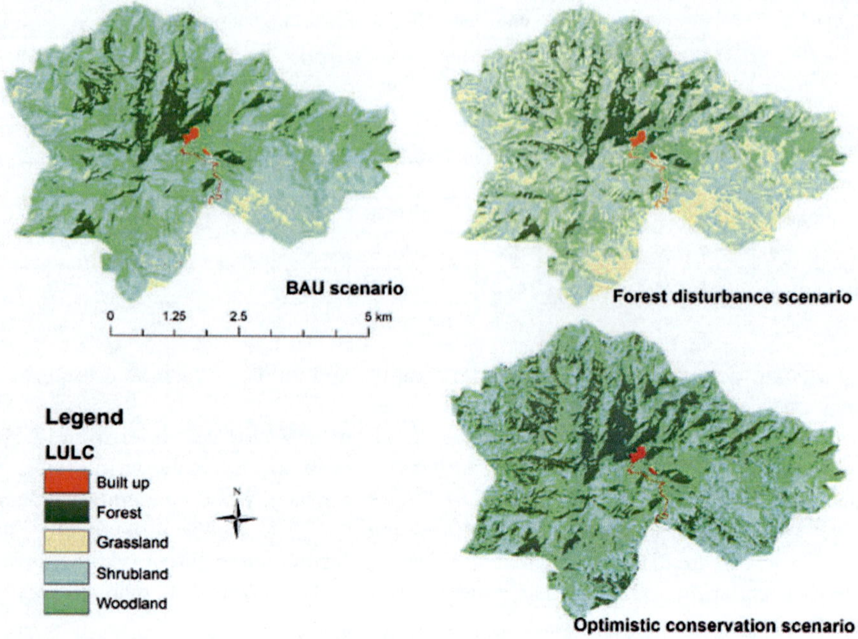

Fig. 6.3 The LULC simulated at different scenarios in Asebot Dry Afromontane Monastery forest

carbon stock in biomass and soil leads to the highest total carbon stock and is estimated in the forest (355.6 ha) (Fig. 6.4).

The woodland and shrubland covers have a total carbon stock of 225.9 and 153.5 tons/ha, respectively. The grassland (129 tons/ha) has the lowest total carbon stock among the LULC covers in the forest.

The Asebot Dry Afromontane Monastery forest's total carbon stock in 1986, 2000, and 2017 were 1,425,292 Mg, 1,412,654 Mg, and 1,384,003 Mg, respectively. The trend indicates that carbon stock decreased between 1986 and 2017. The BAU shows that the carbon stock in the year 2047 would be 1,332,092 Mg. This decreasing carbon stock trend was attributed to the conversion of forest and woodland cover to other land cover types. Carbon stock change estimation of the year 2047, based on 1986 and 2017 baseline data of various land cover conversion scenarios, showed different trends. In the forest disturbance scenario, about 1,206,913 Mg carbon stock was estimated in the forest. Meanwhile, the largest carbon stock of 1,555,918 Mg estimated in an optimistic conservation scenario.

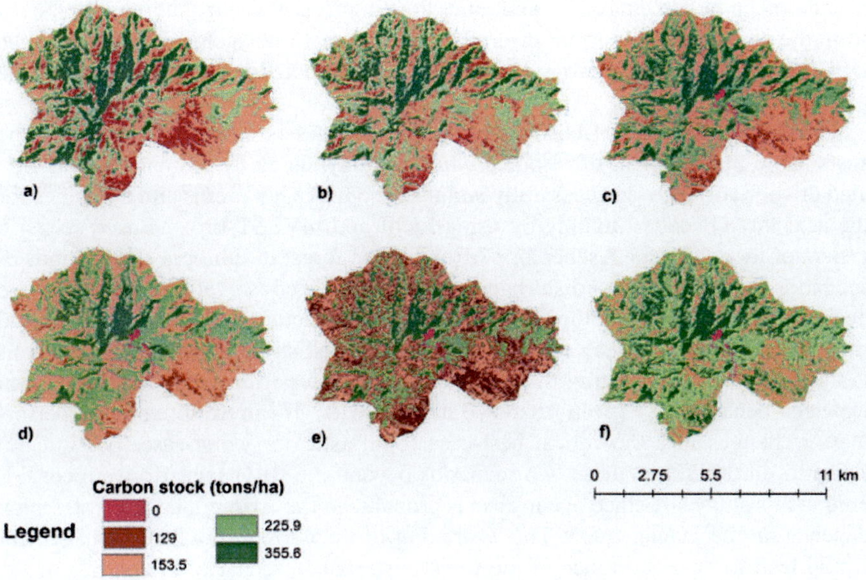

Fig. 6.4 Map of carbon stock in historical LULU and simulated scenarios in Asebot Dry Afromontane Monastery forest **a** 1986 **b** 2000 **c** 2017 **d** BAU scenario **e** Disturbance scenario **f** An optimistic conservation scenario)

6.4 Discussion

The Ethiopian mountains have particularly experienced intense land cover conversions, mainly due to demographic pressures and the consequent expansion of croplands and household energy demand (Gessesse and Bewket 2014; Muke 2019; Getachew and Melesse 2012). Similarly, the foremost underlying causes of Asebot Monastery forest LULC change are the increase in the human, leading to high demand for timber and firewood consumption. An overgrazing and charcoal production are also considered as the primary driver of LULC change in the vicinity due to the area found in the moisture stress agroecology. Frequently, fire outbreaks are occurring in the forest landscape, which in most cases anthropogenic factors.

The other potential LULC change driver on the forest is the lack of proper ownership. Large parts of the forest conservation shifted from Asebot Monastery to state, then to the regional office, and now became a part of Hallideghe Asebot National Park. All of the listed organizations did not develop well conservation strategic plan and managed it in a fragmented manner. The decrease in the forest density and conversion between land covers can be associated with a lack of proper management. The monastery communities perceived that the forest is a part of the monastery and managed the forest with all their efforts. However, the management is beyond their capacity, and the communities intervened in the forest for timber, forage, and charcoal. While fire outbreaks are occurring within the forest, the quick response is undertaken by the church communities arriving from the surrounding cities, even from the capital. If the current discrepancies in forest management continue in the Dry Afromontane forest, large degradations are expected in the future than in the past.

Understanding future LULC patterns and changes is an essential tool for the sustainable management of resources. In this study, using LULC and their associated driver datasets, we successfully simulate the LULCs patterns and dynamics for the next three decades, mainly by using LCM and InVEST Proxy-Based scenario generator models in the Asebot Dry Afromontane forest in Ethiopia. Three kinds of scenarios, i.e., BAU, forest disturbance, and optimistic conservation scenarios, were designed. The BAU scenario was validated using the actual LULC map of 2017 and showed an overall accuracy of 85% with a Kappa statistic of 0.8. Scenario analysis is the process of forecasting the expected value of a performance indicator or an expected behavior in a given period (Balaman 2019). It can explore system performance changes in a theoretical best-case (optimistic) or worst-case (pessimistic) scenario. In this study, these two scenarios beyond the BAU scenario were considered in this study site since occurrence is probable in the Afromontane forest, which depends on the management. The worsening of the current loosened management would lead to the occurrence of the forest disturbance scenario. In contrast, if the current management situations changed, the probability of the best-case (optimistic) scenario is high.

The average total carbon stock estimated in this study is 355.6 Mg ha^{-1} (192.6 Mg ha^{-1} AGB and 123.3 Mg ha^{-1} in SOC) for the forest, 225.9 Mg ha^{-1} for woodland, 153 Mg ha^{-1} for shrubland, and 129 Mg ha^{-1} for grassland. The total carbon stocks estimated in the woody plants of the dry evergreen Afromontane forest of Mount Zequalla monastery forest was 348.8 Mg ha^{-1} (Girma et al. 2014), and that of the estimated carbon stocks of some selected church forest in Addis Ababa was 291.77 Mg ha^{-1} (Tolla et al. 2013). According to Brown and Gaston (1995), the potential and actual biomass density estimates ranged from 33 to 412 Mg ha^{-1} (10^6 g ha^{-1}) and 20 to 299 Mg ha^{-1}, respectively, for very dry lowland to moist lowland forests and from 78 to 197 Mg ha^{-1} and 37 to 105 Mg ha^{-1}, respectively, for montane-seasonal to montane-moist forests. While IPCC (2008) default value of 72 Mg ha^{-1} for Sub Saharan African tropical dry forests. The Asebot forest's carbon stock is existing in the upper limits of AGB carbon stock for Dry Montane tropical forests.

The land cover scenario results indicated that the highest carbon stock was attributed to the optimistic conservation scenario, whereas the lowest carbon stock was to the forest disturbance scenario. The decreasing trend of carbon stock was attributed to the conversion of forest and woodland cover to shrubland and grassland land cover types that evidenced how deforestation and forest degradation influenced the amount of biomass and carbon stored in vegetation (Araújo et al. 2005; Meehl et al. 2007; Smithson 2002). The carbon stock under future land-use management scenarios depends on the implementations of the best scenarios. As forest and woodland areas increased, carbon stock also increased. It is mainly governed by the protection of laws for nature reserves and other legally binding ecological zones. If the laws and other legally binding are not respected and seriously threatened, the current land-use scenario might be going in worst directions. This is because the LULC affects the GHG emissions and carbon stocks associated with soil and vegetation (Feddema et al. 2005; Schulp et al. 2008).

6.5 Conclusions

In BAU, the forest and woodland coverage of Asebot Monastery Dry Afromontane forest would decrease and affects mountain ecosystem regulation services. The trend affects Asebot monastery's dry Afromontane forest climate mitigation potential because the decrease in forest and woodland cover reduces carbon stock and increases carbon emission. If the forest cover conversion continued, it would be a driving factor rather than contributing to climate change mitigation. However, if the optimistic conservation scenario would be in position, the ecosystem regulation service's contribution might be realized. This study illustrated that forest management has its own effects on the future sustainability of ecosystem services. In addition to the effects of the mountains' topographic features, the monastery conservation approach has a generous contribution to conservation. The effects can be observed in high forest areas nearest to the monastery churches. Therefore, enhancing forest conservation in consultation with the monastery would enable sustainable management of the Asabot Dry Afromontane Monastery forest and ecosystem service enhancement. Furthermore, awareness creations, provision of alternative energy sources, and enhancing livelihood income alternatives to the surrounding community could contribute to the forest's sustainability.

References

Aerts R, Van Overtveld K, Haile M, Hermy M, Deckers J, Muys B (2006) Species composition and diversity of small Afromontane forest fragments in northern Ethiopia. Plant Ecol 187:127–142

Aerts R, Van Overtveld K, November E, Wassie A, Abiyu A, Demissew S, Daye DD, Giday K, Haile M, TewoldeBerhan S, Teketay D, Teklehaimanot Z, Binggeli P, Deckers J, Friis I, Gratzer G, Hermy M, Heyn M, Honnay O, Paris M, Sterck FJ, Muys B, Bongers F, Healey JR (2016) Conservation of the Ethiopian church forests: threats, opportunities, and implications for their management. Sci Total Environ 551–552:404–414

Araújo MB, Pearson RG, Thuiller W, Erhard M (2005) Validation of species–climate impact models under climate change. Glob Change Biol 11:1504–1513

Aynekulu E, Denich M, Tsegaye D, Aerts R, Neuwirth B, Boehmer HJ (2011) Dieback affects forest structure in a dry Afromontane forest in northern Ethiopia. Jo Arid Environ 75:499–503

Balaman ŞY (2019) Chapter 5 - Uncertainty issues in biomass-based production chains. In: Balaman ŞY (ed) Decision-making for biomass-based production chains. Academic Press, pp 113–142

Blackie R, Baldauf C, Gautier D, Gumbo D, Kassa H, Parthasarathy N, Paumgarten F, Sola P, Pulla S, Waeber P, Sunderland T (2014) Tropical dry forests: the state of global knowledge and recommendations for future research. Discussion Paper. Bogor, Indonesia: CIFOR

Brink AB, Eva HD (2009) Monitoring 25 years of land cover change dynamics in Africa: A sample based remote sensing approach. Appl Geogr 29:501–512

Brown S (2002) Measuring carbon in forests: current status and future challenges. Environ Pollut 116:363–372

Brown S, Gaston G (1995) Use of forest inventories and geographic information systems to estimate biomass density of tropical forests: application to tropical Africa. Environ Monit Assess 38:157–168

Cairns MA, Brown S, Helmer EH, Baumgardner GA (1997) Root biomass allocation in the world's upland forests. Oecologia 111:1–11

Canadell J, Raupach M, Houghton R (2009) Anthropogenic CO_2 emissions in Africa. Biogeosciences 6:463–468

Cardelús CL, Woods CL, Mekonnen AB, Dexter S, Scull P, Tsegay BA (2019) Human disturbance impacts the integrity of sacred church forests, Ethiopia. PLoS ONE 14(3):e0212430

Chander G, Markham BL, Helder DL (2009) Summary of current radiometric calibration coefficients for Landsat MSS, TM, ETM+, and EO-1 ALI sensors. Remote Sens Environ 113(5):893–903

Di Gregorio A, Jansen LJM (1996) Part I - Technical document on the Africover land cover classification scheme. In: FAO. Africover Land Cover Classification, pp 4–33; 63–76.

FAO.(2010) Global forest resources assessment main report. FAO forestry paper 163, FAO, Rome

Feddema JJ, Oleson KW, Bonan GB, Mearns LO, Buja LE, Meehl GA, Washington WM (2005) The importance of land-cover change in simulating future climates. Science 310(5754):1674–1678

Friis I, Demissew S (2001) A review of existing maps and the need for a new map for the Flora of Ethiopia and Eritrea. Biodiversity Research in the Horn of Africa Region: Proceedings of the Third International Symposium on the Flora of Ethiopia and Eritrea at the Carlsberg Academy, Copenhagen, August 25–27. Kgl. Danske Videnskabernes Selskab, 399

Furst C, Helming K, Lorz C, Muller F, Verburg PH (2013) Integrated land use and regional resource management—a cross-disciplinary dialogue on future perspectives for a sustainable development of regional resources. J Environ Manage 127(Supplement):S1–S5

Gessesse B, Bewket W (2014) Drivers and implications of land use and land cover change in the central highlands of Ethiopia: evidence from remote sensing and socio-demographic data integration. Ethiopian J Soc Sci Human 2:1–23

Getachew HE, Melesse AM (2012) The impact of land use change on the hydrology of the Angereb watershed, Ethiopia. Int J Water 1:4. https://doi.org/10.5772/56266

Gibson D, Quinn J (2017) Application of anthromes to frame scenario planning for landscape- scale conservation decision making. Land 6(2):33

Girma A, Soromessa T, Bekele T (2014) Forest carbon stocks in woody plants of Mount Zequalla Monastery and it's variation along altitudinal gradient: implication of managing forests for climate change mitigation. Sci Technol Arts Res J 3:132–140

Glenday J (2006) Carbon storage and emissions offset potential in an East African tropical rainforest. For Ecol Manage 235:72–83

Hasnat GNT, Hossain MK (2020) Global-overview-of-tropical-dry-forests. In: Handbook of research on the conservation and restoration of tropical dry forests. IGI Global

Hungate BA, Johnson DW, Dijkstra P, Hymus G, Stiling P, Megonigal JP, Pagel AL, Moan JL, Day F, Li J (2006) Nitrogen cycling during seven years of atmospheric CO_2 enrichment in a scrub oak woodland. Ecology 87:26–40

IGBP (The International Geosphere-Biosphere Programme) (1999) Land use and land cover change (LUCC) Implementation Strategy. IGBP Report No. 48/IHDP Report No Stockholm, Sweden

IPBES (2016) Summary for policymakers of the methodological assessment of scenarios and models of biodiversity and ecosystem services of the intergovernmental science-policy platform on biodiversity and ecosystem services. In: Ferrier S, Ninan KN, Leadley P, Alkemade R et al (eds) Secretariat of the intergovernmental science-policy platform on biodiversity and ecosystem services, Bonn, Germany

IPCC (2007) Mitigation. Annex I-Glossary) Contribution of working group III to the fourth assessment report of the intergovernmental panel on climate change. In: Metz B, Davidson OR, Bosch PR, Dave R, Meyer LA (eds) Cambridge University Press, Cambridge

Kent M, Coker P (1992) The description of vegetation in the field. Vegetation description and analysis: a practical approach. 2nd ed, Wiley-Blackwell, 428 p

Kindu M, Thomas S, Martin D, Demel T, Thomas K (2018) Scenario modeling of land use/land cover changes in Munessa-Shashemene landscape of the Ethiopian highlands. Sci Total Environ 622–623:534–546

Malhi Y, Baker TR, Phillips OL, Almeida S, Alvarez E, Arroyo L, Chave J, Czimczik CI, Fiore AD, Higuchi N (2004) The above-ground coarse wood productivity of 104 Neotropical forest plots. Glob Change Biol 10:563–591

Meehl GA, Stocker TF, Collins WD, Friedlingstein P, Gaye T, Gregory JM, Kitoh A, Knutti R, Murphy JM, Noda A (2007) Global climate projections, the physical science basis. Contribution of Working Group I to the Fourth Assessment Report of the Intergovernmental Panel on Climate Change, pp 747–846

Mejebi DJJ (2008) An assessment of the various types of land degradation within the River Kubanni Basin of Zaria, Nigeria. In: Bisong FE (ed) Geography and the millennium development goals: translating vision into reality, ultimate index books. Publishers, Calabar, pp 64–70

Mertz O, Ravnborg HM, Lövei GL, Nielsen I, Konijnendijk CC (2007) Ecosystem services and biodiversity in developing countries. Biodivers Conserv 16:2729–2737

Ministry of Environment, Forest and Climate Change (MEFCC) (2018) Volume III: Synthesis report 1. National Forest Sector Development Program, Ethiopia

Muke M (2019) Reported driving factors of land-use/cover changes and its mounting consequences in Ethiopia: a review. Afr J Environ Sci Technol 13:273–280

Murphy PG, Lugo AE (1986) Ecology of tropical dry forest. Annu Rev Ecol Syst 17:67–88

Navar J (2009) Allometric equations for tree species and carbon stocks for forests of northwestern Mexico. For Ecol Manage 257:427–434

Orlowska I, Klepeis P (2018) Ethiopian church forests: a socio-religious conservation model under change. J Eastern Afr Stud 12(4):674–695

Pearson T, Walker S, Brown S (2005) Sourcebook for land use, land-use change and forestry projects. Winrock International and the BioCarbon Fund of the World Bank, 57

Reusing M (2000) Change detection of natural high forests in Ethiopia using remote sensing and GIS techniques. Int Arch Photogram Rem Sens 33:1253–1258

Reynolds TW, Stave KA, Sisay TS, Wassie AE (2017) Changes in community perspectives on the roles and rules of church forests in northern Ethiopia: evidence from a panel survey of four Ethiopian Orthodox communities. Int J Commons 11(1):355–387

Sahle M, Saito O, Fürst C, Demissew S, Yeshitela et al (2019) Future land use management effects on ecosystem services under different scenarios in the Wabe River catchment of Gurage Mountain chain landscape, Ethiopia. Sustain Sci 14:175–190

Schulp CJE, Nabuurs GJ, Verburg PH (2008) Future carbon sequestration in Europe: effects of land use change. Agric Ecosyst Environ 127(3–4):251–264

Sohngen B, Brown S (2008) Extending timber rotations: carbon and cost implications. Climate Policy 8:435–451

Swetnam RD, Fisher B, Mbilinyi BP, Munishi PK, Willcock S, Ricketts T, Mwakalila S, Balmford A, Burgess ND, Marshall AR (2011) Mapping socio-economic scenarios of land cover change: A GIS method to enable ecosystem service modelling. J Environ Manage 92:563–574

Teketay D (1996) Seed ecology and regeneration in dry Afromontane forests of Ethiopia, Thesis, SLU, Umeaa, Sweden

Tulu T, Makuria A, Zewdu E (2013) Estimation of carbon stock in church forests: implications for managing church forest to help with carbon emission reduction. In: Leal Filho W, Mannke F, Mohee R, Schulte V, Surroop D (eds) Climate-smart technologies. Climate change management. Springer, Berlin

Walkley A, Black IA (1934) An examination of the Degtjareff method for determining soil organic matter, and a proposed modification of the chromic acid titration method. Soil Sci 37:29–38

Wassie A (2007) Ethiopian Church Forests: opportunities and challenges for restoration. PhD thesis, Wageningen Univeristy, Wageningen, The Netherlands. ISBN: 978-90-8504-768-1

Wassie A, Sterck FJ, Bongers F (2010) Species and structural diversity of church forests in a fragmented Ethiopian Highland landscape. J Veg Sci 21:938–948

Williams CA, Hanan NP, Neff JC, Scholes RJ, Berry JA, Denning AS, Baker DF (2007) Africa and global carbon cycle. Carbon Balance Manag 2:3

Worku A, Zewede W (2008) The 2008 Fire incidence in Asebot forest: cause, impact and its implication for sustainable management of the ever-fragmented dry Afromontane forests in Ethiopia. Field report. Ethiopian Institute of Agricultural Research, Forestry Research Center, Addis Ababa, Ethiopia

Yu Y, Guo Z, Wu H, Kahmann JA, Oldfield F (2009) Spatial changes in soil organic carbon density and storage of cultivated soils in China from 1980 to 2000. Global Biogeochem Cycles 23:1–11

Chapter 7
Aboveground and Belowground Carbon Pools for Some Selected Native and Introduced Tree Species of Abune Teklehayimanot Church Forest, Welayita Sodo, Southern Ethiopia

Mehari Alebachew, Abeje Eshete, and Mindaye Teshome

Abstract Quantifying the amount of carbon pools in church forest ecosystems enables us to understand about the role of church forest for climate change mitigation and adaptation. Therefore, this study was conducted with the goal of estimate the amount of carbon stored for some commonly grown planted trees in the Abune Teklehyimanot Church, Welayita Sodo, Southern Nations, Nationalities and Peoples Regional State, Welayita town, Ethiopia. In 2018, a total of 23 temporarily sample plots with an area of 110 to 578 m^2 each were established, laid out on transects along altitudinal gradients with a distance of 100 m between plots. All trees with diameter at breast height (DBH) $\geq$ 2 cm were identified and height (m) was measured using diamater tape, calliper and vertex (digital height measurement instrument). Aboveground and belowground biomass was calculated using Chave et al. (Global Change Biology 3177–3190, 2014) and IPCC (National Greenhouse Gas Inventories Program, IGES, Japan, 2006), respectively, and converted into carbon density using the default factor. Data analysis made using descriptive statistics. The aboveground biomass of the natural forest ranged from 241.41 ± 5.50 t ha^{-1} for *Eucalyptus camaldulensis* to 1.60 ± 0.00 for *Jacaranda mimosifolia* at Teknik na Muya mender forest 24 patch. The belowground biomass ranged from 27.23 ± 0.62 t ha^{-1} for *Eucalyptus camaldulensis* at Teknik na Muya mender forest patch to 0.01 ± 0.00 t ha^{-1} for *Juniperus procera.* The mean ecosystem carbon stock density of the sampled plots in the planted forest ranged from 140.70 ± 3.21 t C to 0.06 ± 0.005 ha^{-1} for *Eucalyptus camaldulensis* and *Juniperus procera*, respectively. In conclusion, fast growing introduced tree species accumulated more carbon stock than native tree species in the studied period of time. There was variation in carbon pools among a species and forest patches. We recommend forest carbon-related awareness creation for local people and promotion of the local knowledge as a possible option for sustainable

M. Alebachew (✉) · M. Teshome
Central Ethiopia Environment and Forest Research Centre, Box 30708, Addis Ababa, Ethiopia

A. Eshete
Ethiopian Environment and Forest Research Institute, Box 24536 code 1000, Addis Ababa, Ethiopia

M. Kindu et al. (eds.), *State of the Art in Ethiopian Church Forests and Restoration Options*, https://doi.org/10.1007/978-3-030-86626-6_7

forest management. This forest needs management intervention including enrichment planting of economically important native tree species and fencing to avoid free grazing by livestock. Better attention of development practitioners, policy makers and church communities may help to improve the woody species composition of this forest and conservation role of Ethiopian Orthodox church.

Keywords Carbon stock · Church forest · Forest patch · Total ecosystem carbon

7.1 Introduction

The natural forests of Ethiopia are classified into 12 different major vegetation types based on altitudinal gradient (Friis et al. 2010). The structure and species composition of the natural vegetation types are diverse due to the presence of wider physiographi and climatic landscapes in the country. The natural forest of Ethiopia supply most of the wood products consumed within the country and diverse non-wood forest products. They also provide various ecosystem services such as watershed protection, biodiversity conservation and carbon sequestration.

The national carbon stock of Ethiopia was estimated to be 153 Tera gram (Tg) C by Houghton (1998), 867 Mg ha^{-1} C by Gibbs et al. (2007) and 2.5 Mg ha^{-1} by Sisay (2010). The natural high forest carbon stock ranges between 101 and 200 Mg ha^{-1} C ha^{-1} (Brown 1997; Moges et al. 2010; Temam 2010; Tsegaye 2010). Site-specific carbon stock studies should have been done (IBC 2005; Moges et al. 2010). However, relatively little study about forest carbon assessment has been made except few studies by (Chave et al. 2001; Abel et al. 2014; Hassen 2015). Periodic forest inventories and continuous monitoring and assessment works in the country are lacking, eventhough, they are most useful in order to evaluate the magnitude of carbon fluxes between aboveground biomass (AGB) and the atmosphere (Girma et al. 2014). Accordingly, also noted that, carbon pool varies between vegetation types and soil types, there are limited number of studies except few works by (Melese et al. 2014; Hassen 2015; Gebre 2015; Tesfaye 2015; Yahya 2015; Wedajo 2018).

The church forests are administered by the Ethiopian Orthodox Tewahido Church (EOTC), one of the oldest churches in the world and founding member of the World Council of Churches Wassie (2002). These forests comprise characteristic trees of dry afromontane forests and are considered as sanctuaries of different plant and animal species (Wassie et al. 2005). In addition, church compounds are considered as an in situ conservation and hot spot sites for biodiversity resources, mainly indigenous trees and shrubs of Ethiopia, which, in turn, give prestige for the religious sites (Wassie 2002). The church forests can serve us the sources of carbon pools. Assessment of above and belowground carbon pools in these forest types is important to show its role for climate change mitigation and adaptation. It is therefore essential to conduct periodic assessment of the various carbon pools of the targeted monastery.

A total of 16 woody species (6 introduced and 10 native tree species) belonging to 14 families were recorded in the Welayita Sodo monastery forest (Mindaye Teshome and Nesibu Yahya 2018). Accordingly, the most dominant introduced tree species were *Eucalyptus camaldulensis* contributing while native tree species were *Juniperus procera, Podocarpus falcatus, Dracena fragrans* and *Olea africana* ssp. *cuspidata.* A serious of studies have been conducted focusing on woody species composition, regeneration status and socioeconomic importance and soil seed bank status (Wassie et al. 2005, 2006), effects of microsites and management on tree regeneration (Wassie et al. 2009), and effects of livestock exclusion on tree regeneration (Wassie et al. 2009) of church forests have been well documented in the past few decades for various church forests in Northern Ethiopia. However, detailed information is scanty on biomass production and carbon storage potential of Wolayita Debere Menekerat Abune Tekele Hayimanot monastery forests in Southern Nations Nationalities and Peoples Regional State (SNNPR) in Southern Ethiopia. Therefore, there is a need to generate relevant information in order to ensure the conservation, management and sustainable utilization of these forest resources. This study was aimed to determine, biomass production and carbon assimilation in Wolayta Debere Menekerat Abune Tekele Haymanot monastery forest for native and introduced tree species. This information may be help to complement the ongoing efforts to develop decision support tools that guide the sustainable management and utilization for planted, native and church forests in Ethiopia.

7.2 Material and Methods

7.2.1 Description of the Study Area

Wolayita Debere Menekerat Abune Tekele Haymanot monastery is located in Wolayita Zone of the Southern Nations, Nationalities and People's Regional State of Ethiopia between 037° 35′ 30″—037° 58′ 36″ E Longitude and 06° 57′ 20″—07° 04′ 31″ N latitude (Fig. 7.1). It is located about ca. 390 km towards the southwest direction from the capital city, Addis Ababa. The altitude ranges between 2086 and 2093 m. The area has a bimodal rainfall pattern with mean annual precipitation of 1355 mm. The temperature ranges between 17.7 and 21.7 °C with a mean temperature of 19.7 °C. The dominant soil type of the landscape is Nitosol and Vertisols vary in pH from 5 to 6. Small-scale Mixed crop-livestock production farming system predominates in the study are, but there are also some pastoralists in the lowlands (Teshome 2017).

7.2.2 Reconnaissance Survey

Discussion was held with Welayita Sodo church communities and surrounding administratives for awareness creation and issuance of authorizations to work in

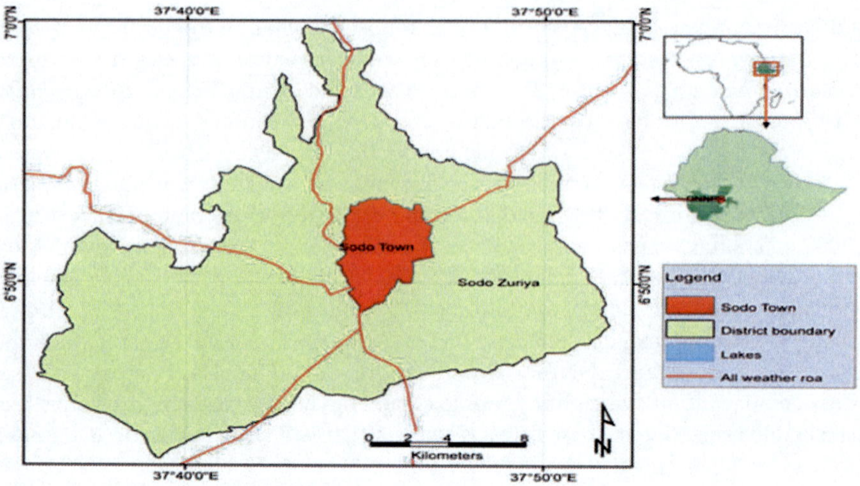

Fig. 7.1 Location map of the study site. The data was obtained from national meteorology agency of Ethiopia

this forest. Then physical observation to view the vegetation structure, composition and physiognomy. Three forest was further stratified into three forest patches (Gedamsefer, Kedanemehiret holy water and Tenik na Muya mender) for further in-depth study based on species type. Then an inventory was conducted in these three forest patches.

7.2.3 *Sampling Design*

A systematic sampling approach was used to carry out the inventory. A total of 23 temporarily sample plots from 110 to 578 m^2 each were established (6 plots in Gedam sefer, 11 plots in Kedanemehiret holy water and 6 plots in Tenikna Muya Mender Church forest patches) based on Neyman optimal proportional allocation formula in the natural forest (Kangas and Maltamo 2006; Köhl et al. 2006). The centre of the first plot was laid out systematically using Silva compass, 100 m away from the outer edge following North direction to avoid edging effect. To attain 90°, blunt stick was used in the plots. The plots were laid out along 100 m horizontal distance, starting from the highest to the lowest ridges of the mountains using a measuring tape, GPS and compass. The distance between two consecutive transect lines was 500 m.

7.2.3.1 Field Data Collection

Before the actual measurements started, all trees found in the border of each plots were marked and numbered. Tree measurements were conducted in the main plot 110 to 578 m^2. Within each strips, the local names were recorded; diameter at breast height (DBH $\geq$ 2 cm) and height (m) of all trees were measured using a diameter tape and Haglöf Vertex IV ultrasonic hypsometer.

7.2.4 *Data preparation*

Stand characteristics such as mean tree dimensions (diameter and height), stand density (N ha^{-1}), stand basal area (m^2 ha^{-1}) and volume (m^3 ha^{-1}) were calculated using inventoried data. Total volume was calculated using the conventional volume equation because local volume equations not available for these species (Atta-Boateng and Moser 1998):

$$V = \pi * \left(\frac{\text{DBH}}{2}\right)^2 * h * f \tag{7.1}$$

where V: tree volume, DBH: diameter at breast height, h: total height, f: form factor (0.42) (Atta-Boateng and Moser 1998): BA (Basal area) calculated using the formula:

$$\text{BA} = \pi * \left(\frac{\text{DBH}}{2}\right)^2 \tag{7.2}$$

where BA: Basal area, DBH: Diameter at breast height, π: 3.142.

7.2.4.1 Aboveground Biomass

Aboveground biomass was calculated using Chave et al. (2014) (Eq. 7.3).

$$\begin{aligned} \text{AGB} &= 0.0673 * \left(\rho * H * \text{DBH}^2\right)\text{\^{}}0.976 \\ \text{BGB} &= 0.24(\text{AGB}) \end{aligned} \tag{7.3}$$

where AGB = aboveground biomass (in kg), D = DBH (in cm), H = height (in m), and ρ = basic wood density (in g cm^{-3}) and BGB = belowground biomass. The wood density data information was obtained from global wood density data base (Zanne et al. 2009), World Agroforestry Centre (ICRAF) wood density database (www.worldagroforestry.org) and Wood Technology Research Centre, Addis Ababa (Desalegn et al. 2012).

Accumulated aboveground and total carbon density was calculated following Eqs. (7.4 and 7.5) (IPCC 2006), (Gibbs et al. 2007; Ponce-Hernandez 2004):

$$ACD = AGB{*}0.47 \tag{7.4}$$

$$BCD = ACD * 0.24 \tag{7.5}$$

where ACD: Aboveground carbon density (t C ha^{-1}), BCD: Belowground carbon density (t C ha^{-1}). The accumulated aboveground and total carbon density for each tree were calculated separately in each plot; then, figures for the carbon density of each tree were summed up to give plot accumulated carbon density and converted to per hectare.

7.2.4.2 Total Forest Ecosystem Carbon Estimation

The total carbon stock (carbon density) was calculated by summing up above and belowground carbon stocks of each carbon pools of the forest ecosystem following Pearson et al. (2005) and then converted into tonnes of CO_2 equivalent by multiplying it by 3.67 as developed (where one tone of CO_2 = 3.67 of CO_2) by Pearson et al. (2007). Carbon stock density of the study area was calculated using (Eq. 7.6):

$$TECD = ACD + BCD \tag{7.6}$$

where: TECD is total ecosystem carbon density (t C ha^{-1}), ACD: aboveground carbon density (t C ha^{-1}), BCD: belowground carbon density (t C ha^{-1}) (Table 7.1).

7.3 Results

7.3.1 *Aboveground Carbon, Basal Area and Number of Stems*

The aboveground biomass (t), basal area (m^2) and the volume (m^3) for the sampled species in 2018 ranged from 241.41 ± 5.50 to 1.60 t ha^{-1}; 17.57 ± 0.01 to 0.05 m^2 ha^{-1} and 118.06 ± 0.10 to 0.18 ± 0.00 for *Eucalyptus camaladulensis* and *Juniperus procera*, respectively (Table 7.2). The number of stems ranged from 3715 to 25 N ha^{-1}. The highest number of stems recorded for *Eucalyptus camaldulensis* followed by *Cupressus lusitanica.* While the lowest number of stems 25 N ha^{-1} was recorded for *Grevillea robusta.*

The diameter at breast height, diameter at stump height (up to cm height of the stem) and total height of the measured trees ranged from 10.43 ± 9.295 to 39 ± 5.41, 11.8 ± 9.53 to 6.58 ± 6.37 and 8.08 ± 3.0 to 6.09 ± 5.34, respectivery, for

Grevillea robusta and *Eucalyptus camaldulensis*, respectively (Table 7.2). The basal area ranged from 17.57 ± 0.01 to 0.05 ± 0.00 m^2 ha^{-1} for *Eucalyptus camaldulensis* and *Jacaranda mimosifolia*, respectively (Table 7.2). The largest trees were:- *Eucalyptus camaldulensis* and *Cupressus lusitanica*. However, the thinnest trees were:- *Jacaranda mimosifolia*, *Juniperus procera* and *Grevillea robusta* (Table 7.2).

7.3.2 Aboveground Biomass and Belowground Carbon Biomass

The aboveground and belowground biomass varied among species $P \leq 0.05$ (Table 7.2). The mean aboveground and below biomass carbon for the sampled church forest ranged from 113.46 ± 2.59 to 0.05 and 27.23 ± 0.62 to 0.01 t C ha^{-1} for *Eucalyptus camaldulensis* and *Juniperus procera*, respectively. The ACD and BCD were also highest for Gedam Sefere forest patch and lowest in Teknik na Muya mender. Overall, ACD and BCD affirmed temporal variation among species and forest patch (Table 7.2).

7.3.3 Do Total Ecosystem Carbon Affected by Species and Forest Patch of the Church Forest?

The results affirmed that the total ecosystem carbon is influenced by species and forest patch
. The TECD ranged from 140.70 ± 3.21 to 0.06 ± 0.005 t C ha^{-1}. The highest total ecosystem carbon stock found under Teknik na Muya mender forest patch (140.70 ± 3.21 t C ha^{-1}) and Kedanemehiret holy water (136.79 ± 2.23 t C ha^{-1}) forest patch for *Eucalyptus camaldulensis* species in both cases (Table 7.2). However, the lowest total ecosystem carbon density recorded under Teknik na Muya mender (0.06 ± 0.005 t C ha^{-1}). At species level, the lowest TECD 0.06 ± 0.005 t C ha^{-1} recorded for *Juniperus procera* followed by *Jacaranda mimosifolia* (0.93 ± 0.00 t C ha^{-1}). The total ecosystem carbon density for Gedam sefer forest patch ranged from 10.25 ± 0.31 to 2.26 ± 0.03 t C ha^{-1} for Kidane mihret holy water forest patch ranged from 6.62 ± 0.15 to 0.30 ± 0.06 t C ha^{-1}. Moreover, the total Tenkir na Muya mender ranged from 10.65 ± 0.14 to 0.008 ± 0.00 t C ha^{-1}. *Eucalyptus camaldulensis* has the highest ecoystsem carbon density in all the sampled plots. However, *Jacaranda mimosifolia* has the lowest total ecosystem carbon density (Table 7.1). The sum of total ecosystem carbon stock density for the natural and planted forest was 84.167 t (308.72 t CO_2 eq). The carbon sequestration potential of *Eucalyptus camaldulensis* was 9.42 t C ha^{-1} yr^{-1} and or 34.57 CO_2 eq.yr^{-1}.

Table 7.1 Aboveground biomass, belowground biomas, aboveground and belowground carbon for Abune Teklehyimanot Church, Welayita Sode

Name of forest patch	Strip no	Size (m^2)	Spp	AGB.kg	BGB.kg	AGB.tonne	ACD.toneC	BCD.tonne	TEC D
Gedam sefer	1	370	*EC*	17,583.90 ± 533.72	4220.14 ± 128.09	17.58 ± 0.53	8.26 ± 0.25	1.99 ± 0.06	10.25 ± 0.31
	2	380		6585.67 ± 164.57	1580.56 ± 39.50	6.59 ± 0.17	3.10 ± 0.08	0.74 ± 0.02	3.84 ± 0.10
	3	380		5241.76 ± 78.15	1258.02 ± 18.76	5.24 ± 0.08	2.46 ± 0.04	0.59 ± 0.01	3.06 ± 0.05
	4	387		6771.37 ± 213.27	1625.13 ± 51.19	6.77 ± 0.21	3.18 ± 0.10	0.76 ± 0.02	3.95 ± 0.12
	5	380		3876.60 ± 55.62	930.38 ± 13.40	3.88 ± 0.06	1.82 ± 0.03	0.44 ± 0.01	2.26 ± 0.03
Kedanemehiret holy water	1	360	*EC*	11,352.24 ± 192.71	2724.54 ± 46.25	11.35 ± 0.19	5.34 ± 0.09	1.28 ± 0.02	6.62 ± 0.11
	2	250		8654.87 ± 277.74	2077.17 ± 66.49	8.66 ± 0.28	4.07 ± 0.13	0.98 ± 0.03	5.04 ± 0.16
	3	125		7242.57 ± 504.24	1738.22 ± 121.02	7.24 ± 0.50	3.40 ± 0.24	0.82 ± 0.06	4.22 ± 0.29
	4	125		713.52 ± 158.17	171.24 ± 37.96	0.71 ± 0.16	0.34 ± 0.07	0.08 ± 0.02	0.42 ± 0.09
	5	140		2117.42 ± 117.87	508.18 ± 28.29	2.12 ± 0.12	1.00 ± 0.06	0.24 ± 0.01	1.23 ± 0.07
	6	140		3366.90 ± 267.50	808.06 ± 64.00	3.37 ± 0.27	1.58 ± 0.13	0.38 ± 0.03	1.96 ± 0.16
	7	180		514.32 ± 104.62	123.44 ± 25.11	0.51 ± 0.11	0.24 ± 0.05	0.06 ± 0.01	0.30 ± 0.06
	8	110		958.05 ± 59.04	229.93 ± 14.17	0.96 ± 0.06	0.45 ± 0.03	0.11 ± 0.01	0.56 ± 0.03
	9	110		1415.49 ± 92.78	339.73 ± 22.27	1.42 ± 0.09	0.67 ± 0.04	0.16 ± 0.01	0.83 ± 0.05
	10	110		2392.65 ± 150.33	574.24 ± 36.08	2.39 ± 0.15	1.13 ± 0.07	0.27 ± 0.02	1.39 ± 0.09
Teknik na Muya mender	1	430	*EC*	5728.15 ± 169.10	1374.76 ± 40.59	5.73 ± 0.17	2.69 ± 0.08	0.65 ± 0.02	3.34 ± 0.10
	2	315		4055.76 ± 80.08	973.38 ± 19.22	4.06 ± 0.08	1.91 ± 0.04	0.46 ± 0.01	2.36 ± 0.05
	3	578		15,341.63 ± 148.91	3681.99 ± 35.74	15.34 ± 0.15	7.21 ± 0.07	1.73 ± 0.02	8.94 ± 0.16
	4	501		18,280.54 ± 202.41	4387.33 ± 48.58	18.28 ± 0.20	8.59 ± 0.10	2.06 ± 0.02	10.65 ± 0.14
	5	448		10,346.68 ± 112.49	2483.20 ± 27.00	10.35 ± 0.11	4.86 ± 0.05	1.17 ± 0.01	6.03 ± 0.09
	6	315		8700.86 ± 161.45	2088.21 ± 38.75	8.70 ± 0.16	4.09 ± 0.08	0.98 ± 0.02	5.07 ± 0.09

(continued)

Table 7.1 (continued)

Name of forest patch	Strip no	Size (m^2)	Spp	AGB.kg	BGB.kg	AGB.tonne	ACD.toneC	BCD.tonne	TEC D
	1	430	*CL*	1901.19 ± 82.03	456.29 ± 19.69	1.90 ± 0.08	0.89 ± 0.04	0.22 ± 0.01	1.11 ± 0.05
	3	578		618.77 ± 53.82	148.50 ± 12.92	0.62 ± 0.06	0.29 ± 0.33	0.07 ± 0.01	0.36 ± 0.03
	3	578	*GR*	322.32 ± 175.08	77.36 ± 42.02	0.32 ± 0.18	0.15 ± 0.08	0.04 ± 0.02	0.19 ± 0.10
	3	578	*JP*	106.29 ± 8.53	25.51 ± 2.05	0.11 ± 0.01	0.05 ± 0.004	0.012 ± 0.00	0.06 ± 0.00
	4	501		13.15 ± 0.00	3.16 ± 0.00	0.01 ± 0.00	0.006 ± 0.00	0.001 ± 0.00	0.008 ± 0.00
	3	578	*JM*	92.52 ± 7.40	22.25 ± 1.75	0.09 ± 0.007	0.04 ± 0.004	0.01 ± 0.001	0.05 ± 0.004
	4	501		29.37 ± 2.11	7.05 ± 0.00	0.03 ± 0.002	0.01 ± 0.001	0.003 ± 0.00	0.02 ± 0.001

Where: *AGB* Aboveground biomass, *ACD* Aboveground carbon density, *BGB* Belowground biomass, *BCD* Belowground carbon density, *EC Eucalyptus camaldulensis*, *CL Cupressus lusitanica*, *DBH* diameter at breast height, *DSH* diameter at stump height, *Ht* total height, *GR Grevillea robusta*, *JP Juniperus procera* and *JM Jacaranda mimosifola*, *Kg* Kologram ,m^2 metre squared, *SPP* Species, and TCED: Total ecosystem carbon density

Table 7.2 Growth and biomass for Abune Teklehyimanot Church, Welayita Sodo

Spp	Forest patch	Mean DBH cm	Mean DSH.cm	Mean Ht.m	Vol (m^3 ha^{-1})	BA (m^2 ha^{-1})	AGB (t ha^{-1})	ACD (t C ha^{-1})	BGB (t C ha^{-1})	BCD (t C ha^{-1})	TECD)(t C ha^{-1})
EC	Gedam Sefrer	5.51 ± 4.58	6.92 ± 5.56	6.26 ± 4.21	118.06 ± 0.10	16.43 ± 0.12	211.17 ± 224.20	99.25 ± 0.11	50.68 ± 0.05	23.82 ± 0.03	123.07 ± 0.13
	Kedanemehiret holy water	5.39 ± 5.41	6.58 ± 6.37	6.09 ± 5.34	144.58 ± 0.13	16.28 ± 0.011	234.72 ± 3.82	110.32 ± 1.80	56.33 ± 0.92	26.48 ± 0.43	136.79 ± 2.23
	Teknik na Muya mender	6.27 ± 5.58	7.65 ± 6.71	7.24 ± 5.62	148.64 ± 0.11	17.57 ± 0.01	241.41 ± 5.50	113.46 ± 2.59	57.94 ± 1.32	27.23 ± 0.62	140.70 ± 3.21
CL	Teknik na Muya mender	8.85 ± 5.49	10.36 ± 6.02	7.91 ± 4.13	19.88 ± 0.07	3.37 ± 0.01	19.63 ± 0.08	9.23 ± 0.04	4.71 ± 0.02	2.21 ± 0.01	11.44 ± 0.05
GR		10.43 ± 9.29	11.8 ± 9.53	7.93 ± 4.26	4.06 ± 0.13	0.68 ± 0.02	5.57 ± 0.17	2.62 ± 0.08	1.34 ± 0.04	0.63 ± 0.02	3.25 ± 0.10
JP		7.6 ± 0.89	8.58 ± 0.94	6.82 ± 1.79	1.23 ± 0.006	0.42 ± 0.001	1.84 ± 0.01	0.05 ± 0.00	0.03 ± 0.002	0.01 ± 0.00	0.06 ± 0.005
JM		5.82 ± 1.86	7.00 ± 2.11	8.08 ± 3.0	0.18 ± 0.00	0.05 ± 0.00	1.60 ± 0.00	0.75 ± 0.00	0.38 ± 0.00	0.18 ± 0.00	0.93 ± 0.00

Where: *AGB* Aboveground biomas, *ACD* Aboveground carbon density, *BA* basal area, *BGB* Belowground biomass, *BCD* Belowground carbon density, *BA* basal area, *C* carbon, *cm* centimetre, *DBH* Dimater at breast height, *EC Eucalyptus camaldulensis, CL Cupressus lusitanica, DBH* diameter at breast height, DSH; diameter at stump height, ha^{-1} per hectare *Ht* total height, *GR Grevillea robusta, JP Juniperus procera, JM Jacaranda mimosifola, m* metre, TECD: total ecosystem carbon density, and *t* tonne

7.4 Discussion

The present study investigated the carbon pools of the Teklehayimanot Church Forest, Wolayita Sodo, Southern Ethiopia forest along species and forest patch. In this study, we found lower aboveground biomass in the Teknik na Muya mender forest patch; this might be due to intense anthropogenic effect, which resulted in few but big trees with higher above and belowground biomass. In line with, similar reports are also reported by (Shumi 2009; Hassen 2015; Tesfaye 2015; Tesfaye et al. 2016, 2019).

The aboveground biomass carbon density reported (113.46 ± 2.59 t C ha^{-1}) in Teknik na Muya mender forest is also lower than Gara Mukitar dry afromontane forest Arba Minch Riverine Forest, Egdu forest and Gera Moist Afromontane forest, 142.47 and 174.02 t C ha^{-1} (Hassen 2015; Meles et al. 2014), Chato forest (301.86 t C ha^{-1}) (Iticha 2017), Ades forest too (259.17 t C ha^{-1}) (Kassahun et al. 2015) and Chilimo dry afromontane forest (200 t C ha $^{-1}$) (Tesfaye et al. 2019). The highest ACD (113.46 ± 2.59 t C ha^{-1}) under na Muya mender forest patch might be due to lower disturbance. Similar findings are also reported for other Ethiopian forests too (Girma et al. 2004; Hassen 2015; Iticha 2017).

Moreover, the mean accumulated biomass capacity of the Teklehayimanot Church Forest is also lower than the reports of Eticha et al. (2017), Zequala monastery (475.51 t C ha^{-1}) and Semien mountain national park forest (994.16 t C ha^{-1}) in Central and Northern Ethiopia, respectively (Girma et al. 2014; Yelemfrhal et al. 2014) and Chilimo dry afromontane forest (Tesfaye et al. 2019). Species is a major factor in carbon stock in the studied church forest, among the five studied species studied in Welayita Sodo Abune Teklehayimanot Church Forest such as *Eucalyptus camaldulensis* is found to be the best in above and belowbiomass carbon accumulation potential. This might be due to its fast growing nature. The higher carbon stock in the the aforementioned species in all the sampled period might be due to higher litter fall and fast growing nature of the species. The carbon stock density was lower in *Juniperus procera* and *Jacaranda mimosifolia*; this might be due to slow growing nature of the species and lack of appropriate forest management practices to enhance their growth and productivity (Tesfaye et al. 2020). In our study, introduced tree species stored more carbon than native tree species. This showed to have more carbon accumulation potential of introduced tree species in shorter period of time. On the contrary, native tree species accumulated less biomass and carbon due to slow growing rate with the same time period. The C stock density under native the aforementioned monastery was higher than those reported in other regions (Beets et al. 2002; Harms et al. 2005; Twongyirwe et al. 2013) and suggests two management strategies for improving carbon stock accumulation potential. The first is to maintain and preserve the exitsed church forest as other African tropical forests do (Lewis et al. 2009), and the second is to recover abandoned cropland and degraded—lands by establishing tree plantations in the nearby areas if any and the third is to improve the silvicultural management to improve productivity and yield.

7.5 Conclusions

The aboveground and belowground biomass of the Abune Teklehayimanot Church Forest, Wolayita Sodo, was higher than Teknik na Muya mender for highest altitudinal gradients and Teknik na Muya mender in all the measurement times. The aboveground and belowground biomass and carbon was highest for *Eucalyptus camaldulensis* in all the sampled plots followed by *Cupressus lusitanica*. However, *Juniperus procera* and J*acaranda mimosifolia* have the lowest aboveground and belowground biomas and carbon in the sampled plots and forest patches. In general, fast growing introduced tree species accumulated more carbon biomass than native tree species. The highest mean ecosystem carbon stock density for introduced and native tree species was found under Welayita Sodo Abune Teklehayimanot monastery. For maintaining higher carbon stock density in the study area, other land use types should be converted into plantations. Forest management options should be applied to improve productivity. We recommend forest carbon-related awareness creation for local people and promotion of the local knowledge can be regarded as a possible option for sustainable forest management.

References

Adugna F, Teshome S, Mekuria A (2013) Forest Carbon Stocks and Variations along Altitudinal Gradients in Egdu Forest: Implications of Managing Forests for Climate Change Mitigation. Sci Technol Art Res J 2(4):40–46

Atta-Boateng J, Moser JR (1998) A method for classifying commercial tree species of an un evenaged mixed species tropical forest growth and yield model construction. For Ecol Manage 104:89–99. http://www.fao.org/forestry/25518/en/

Beets PN, Oliver GR, Clinton PW (2002) Soil carbon protection in Podocarp/hardwood forest and effects of conversion to pasture and exotic pine forest. Environ Pollut 116:S63–73. PMID: 11833919 (PubMed- Indexed for MEDLINE)

Brown S (1997) Estimating biomass and biomass changes of tropical forests: a primer : FAO Forestry Paper 134, FAO. United Nations, Rome

Chave J, Rieâra B, Dubois M (2001) Estimation of biomass in a neotropical forest of French Guiana: spatial and temporal variability. J Trop Ecol 17:79–96

Chave J, Réjou-Méchain M, Búrquez A, Chidumayo E, Colgan MS, Delitti WBC, Vieilleden G (2014) Improved allometric models to estimate the aboveground biomass of tropical trees. Global Change Biol 3177–3190. https://doi.org/10.1111/gcb.12629

Desalegn G, Abega M, Teketay D, Gezahgne A (2012) Commercial timer species in Ethiopia: characteristics and uses—a handbook for forest industries, construction and energy sectors, foresters and other stakeholders. Addis Ababa University Press, Addis Ababa

Friss I, Demissew S, van Bruegel L (2010) Atlas of potential vegetation of Ethiopia. Biologiske skrifter (Biol Skr Dan Vid Selsk) 58:307

Gebre TM (2015) Biomass and soil carbon stocks along elevation gradeints of woodland ecosystems. The case of Liben district, South Ethiopia. MSc Thesis, WGCF, Shashemene, Ethiopia

Gibbs HK, Brown S, Niles JO, Foley JA (2007) Monitoring and estimating tropical forest carbon stocks: making REDD a reality. Environ Res Lett 2:045023

Girma A, Soromessa T, Bekele T (2004) Forest carbon stocks in woody plants of mount zequalla monastery and its variation along altitudinal gradient: implication of managing forests for climate change mitigation. Sci Technol Arts Res J 3(2):133–141

Girma A, Soromessa T, Bekele T (2014) Forest carbon stocks in woody plants of Mount Zequalla Monastery and its variation along altitudinal gradient: Implication of managing forests for climate change mitigation. Sci Technol Arts Res J 3(2):132–140

Harms BP, Dalal RC, Cramp AP (2005) Changes in soil carbon and soil nitrogen after tree clearing in the semi—arid range lands of Queensland. Aust J Bot 53:639–650. https://doi.org/10.1071/BT04154

Hassen N (2015) Carbon stock along altitudinal gradient in Gera Moist Evergreen Afromontane forest, South Western Ethiopia, MSc Thesis, AAU, Addis Ababa, Ethiopia

Houghton RA (1998) The annual net flux of carbon to the atmosphere from changes in land use 1850–1990 Tellus B 51:298–313

IBC (2005) National biodiversity strategy and action plan. Institute of Bodiversity Conservation, Addis Ababa, Ethiopia

IPCC (2006) Guidelines for national greenhouse gas inventories (Vol 4, AFOLU). In: Egsleston HS, Buendia L, Miwa K, NgaranT, Tanabe K (eds) National greenhouse gas inventories program. IGES, Japan

Iticha B (2017) Ecosystem carbon storage and partitioning in Chato Afromontane forest. Its climate change mitigation and economic potential. Int J oEnviron Agric Biotechnol (IJEA B) 2(4):2450–1878

Kangas A, Maltamo M (2006) Forest inventory methodology and applications: managing forest ecosystems 10. Springer, Dordrecht, the Netherlands

Kassahun K, Soromessa T, Belliethathan S (2015) Forest carbon stock in woody plants of Ades Forest, Western Haraghae Zone of Ethiopia and its variation along environmental factors: Implications for climate change mitigation. J Nat Sci Res 5(21). www.iiste.org

Köhl M, Magnussen SS, Marchett M (2006) Sampling methods, remote sensing and GIS multi resource forest inventory. Springer, Berlin

Meles BW, Kelbessa E, Soromessa T (2014) Forest carbon stocks in woody plants of arba minch ground water forest and its variations along environmental gradients. Sci Technol Arts Res J 3(2):141–147

Moges Y, Eshetu Z, Nune S (2010) Ethiopian forest resources: Status and future Management options in view of access to carbon finances. Ethiopian Climate Research and Networking Report, UNDP, Addis Ababa, Ethiopia

Pearson T, Walker S, Brown S (2005) Source book for land use, land—use change and forestry projects

Pearson T, Brown S,Birdsey R (2007) Measurement guidelines for the sequestration of forest carbon. Northern Research Station, Department of Agriculture, Washington, DC

Ponce-Hernandez R (2004) Assessing carbon stocks and modeling win–win scenarios of carbon sequestration through land-use changes. FAO, UN, Rome

Shumi G (2009) The structure and regeneration status of tree and shrub species of Chilimo forest—Ecological suatainability indicators for participatory forest management (PFM) in Oromia, Ethiopia. MSc Thesis, University of Dresden, Germany

Sisay N (2010) Land use, land use change and forestry (LULUCF): Afforestation and Reforestation. In: Clean development mechanisms investor's Guide, DNA office the Environment Protection Authority, EPA, Addis Ababa, Ethiopia

Temam A (2010) Impact of disturbance on carbon stocks in HaranaBulluk natural forests, Bale Zone, SW Ethiopia. MSc Thesis, Wondogenet College of Forestry, Shashemene, Ethiopia

Tesfaye AM (2015) Forest management options for carbon stock and soil rehabilitation in Chilimo dry afromontane forest, Ethiopia. Ph.D. Thesis, INIA- Palencia, University of Valladolid, Palencia, Spain

Tesfaye MA, Bravo F, Ruiz-Peinado R, Pando V, Bravo-Oviedo A (2016) Impact of Changes in land use, species and elevation on soil organic carbon and total nitrogen in Ethiopian Central Highlands. Geoderma 261: 70–79. 0016–7061/© 2015 Elsevier B.V. All rights reserved

Tesfaye AM, Gardi O, Tesfaye B, Jűrgen B (2019) Temporal variation of ecosystem carbon pools along altitudinal gradient and slope: the case of Chilimo dry afromontane natural forest,Central Highlands of Ethiopia. Int J Ecol Environ. https://doi.org/10.1186/s41610-019-0112-9

Teshome and Nesibu (2019) Progress report of welayita sodo Monastery forest, Addis Ababa, Ethiopia

Tsegaye T (2010) Bale Eco-region sustainable management programme. http://www.pfmpfarmsos.org/publication.html

Twongyirwe R, Sheil D, Majaliwa JGM, Ebanyat P, Tenywa MM, Van Heist M, Kumar L (2013) Variability of soil organic carbon stocks under different land uses: A study in an afromontane landscape in South-Western Uganda. Geoderma 193–194, 282–289. https://doi.org/10.1016/j.Geoderma.2012.09.005

Wassie A (2002) Opportunities, constraints and prospects of ethiopian orthodox tewahedo church I conserving forest resources. The case of churches in South Gonder, Northern Ethiopia, MSc Thesis, Swedish University of Agricultural Sciences. Skinnskatterberg, Sweden

Wassie A, Teketay D, Powell N (2005) Church forests in North—gonder administrative zone, Northern Ethiopia. Forest Trees Livelihood 15:349–373

Wassie A, Streck FJ, Teketay D andBongers T (2009) Effects of livestock exclusion on tree regeneration on church forests of Ethiopia. For Ecol Manag 257(3):765–772

Wodajo A (2018) Variation in carbon stock along environmental gradients in Gara—Muktar Forest, West Herargae Zone, Southern Ethiopia, MSC Thesis, Haromaya University, Dire Dawa, Ethiopia

Yahya (2015) Carbon stock along altitudinal gradient of Araba Gugu Dry afromontane forest, MSC Thesis, HAFl, Switzerlands

Yelemfrhat T, Soromessa T, Bayable E (2014) Forest carbon stocks in lowland area of Semien Mountains National Park: Implication forclimate change mitigation. Sci Technol Arts Res J 3(3):29–36

Zanne A, Lopez-Gonzalez G, Coomes D, Ilic J, Jansen S, Lewis L, Miller B, Swenson G, Wiemann C, Chave J (2009) Global wood density database [http://hdl.handle.net/10255/dryad.235]

Part III
Structure and Diversity of Ethiopian Church Forests

Chapter 8
Floristic Composition, Diversity, Population Structure and Regeneration Status of Woody Species in Four Church Forests in Ethiopia

Shiferaw Alem, Nesru Hassen, Mindaye Teshome, Kibruyesfa Sisay, Zelalem Teshager, Nesibu Yahya, and Abeje Eshete

Abstract The Ethiopian orthodox churches play a pivotal role in conserving and protecting forest resources and they harbour diverse flora and fauna. Despite this fact, there is scant knowledge on the floristic composition, diversity and regeneration status of woody plants in most of the church forests in Ethiopia. Therefore, the objectives of this study were to: (1) evaluate the species composition, structure and diversity of woody plants in the Bole Bulbula Tekle Haymanot Church Forest (BBTH) in Addis Ababa, the Mantogera Estifanos Church Forest (ME) in Amhara National Regional State (NRS), the Emba Kidist Arsema Church Forest (EKA) in Tigray NRS and the Wolayta Debere Menekerat Abune Tekele Haymanot Church Forest (WDM) in Southern Nations, Nationalities and Peoples (SNNP) NRS and (2) assess the regeneration status of woody plants in the aforementioned four church forests. To achieve those objectives, the vegetation data were collected using a nested quadrat plot design, measuring 20×20 m. Depending on the proportion of the area coverage of each church forest, a total of 30, 15, 17 and 15 sample plots were taken to collect the vegetation data in BBTHM, ME, EKA and WDM church forests, respectively. Shannon–Wiener diversity index was used for the diversity analysis and densities of trees and seedlings were determined. The results revealed that the number of woody species recorded in ME (38) > BBTH (32) > EKA (18) > WDM (16). The diversity of woody species in the church forests of BBTH, ME, EKA and WDM were 1.25, 2.15, 1.8 and 0.52, respectively. The densities (stems ha^{-1}) of woody species in BBTH, ME, EKA and WDM were 4300, 716, 235 and 533, respectively. The number of seedlings in EKA was higher than in the other studied church forests. Depending on their geographical locations, the results generally indicated that the studied church

S. Alem (✉) · N. Hassen · M. Teshome · K. Sisay · N. Yahya
Central Ethiopia Environment and Forest Research Center, P. O. Box 33042, Addis Ababa, Ethiopia

Z. Teshager
Bahir Dar Environment and Forest Research Center, Bahir Dar, Ethiopia

A. Eshete
Ethiopian Environments and Forest Research Institute, P. O. Box, 24536, Code 1000, Addis Ababa, Ethiopia

M. Kindu et al. (eds.), *State of the Art in Ethiopian Church Forests and Restoration Options*, https://doi.org/10.1007/978-3-030-86626-6_8

forests harboured a variety of species, which are important for genetic conservation and development purposes in Ethiopia. Hence, appropriate attention must be given to conserve and develop these church forests in the country. All of the studied church forests exhibited poor natural regeneration. Therefore, it is recommended to apply appropriate silvicultural treatments and management practices, such as weeding, hoeing and loosening the ground under the forest canopy, removing the thick litters that covers the ground under the forests, thinning, pruning, enrichment plantations and controlling livestock movements to enhance the regeneration of native woody species in the studied church forests.

Keywords Church forests · Conservation · Diversity · Floristic composition · Regeneration

8.1 Introduction

Sacred groves, symbolized globally with different names and purposes like church forests, fetish forests and sacred forests, are homes to different fauna and flora (Zegeye et al. 2011). Ethiopia has a total of around 35,000 churches and monasteries, some of which are 1660 years old (Taye 1998). Wassie (2002) stated that the Ethiopian Orthodox Tewahido Church is one of the oldest Christian churches in the world, with some of the earliest church buildings dating to 300 A.D. The church has over 45 million followers, more than 400,000 clergies and several Archbishops and dioceses (Wondmagegnehu and Motovu 1970). For centuries, followers of the Ethiopian Orthodox Church have conserved patches of native trees around church buildings as sacred sanctuaries for church communities (Wassie 2007). The church has a long history of planting, protecting and preserving forests (Wassie 2002).

Aerts et al. (2016) indicated that about 1940 churches in the Ethiopian highlands are covered with forests, occupying a total area of 39,000–57,000 ha. A recent inventory using high-resolution satellite imagery revealed more than 8,000 church forests in the Amhara NRS alone, ranging from <1 to over 100 ha (Reynolds et al. 2015). Reynolds et al. (2015) further stated that church forests can be found at virtually every latitude, longitude and elevation and in every agroecology in the Amhara NRS. Findings indicated that the churches, located in the Central and Northern Highlands of Ethiopia, contain natural forest vegetation rich in biodiversity in their yards (Taye 1998; Wassie 2002). Wassie et al. (2010) in his study on 28 church forests identified a total of 168 woody species, including 160 indigenous to Ethiopia. The church forests provide also essential ecosystem services that include both the material provisions (timber and non-timber forest products), non-material (spiritual and cultural values) and support services (nutrient cycling, water storage, carbon storage and pollination) (Bhagwat 2009; Powledge 2006).

In Ethiopia, most of the remnant dry Afromontane forests are biodiversity hotspot areas that support the protection and conservation of threatened native flora and fauna (Cardelus et al. 2013). They are also critical conservation areas that provide ecosystem services. The forests located around churches comprise local as well as global hotspots in biodiversity and some of the species are also endemic (Zegeye et al. 2011). Some studies that describe the forest vegetation of different church forests in Ethiopia are available (Zegeye et al. 2011; Wassie 2002; Wassie et al. 2009; Taye 1998). Bole Bulbula Tekle Haymanot Church Forest (BBTH) in Addis Ababa, Mantogera Estifanos Church Forest (ME) in Amhara NRS, Emba Kidist Arsema Church Forest (EKA) in Tigray NRS and Wolayta Debere Menekerat Abune Tekele Haymanot Church Forest (WDM) in Southern Nations, Nationalities and Peoples (SNNP) NRS are some of the churches that harbour the dry Afromontane forests in Ethiopia. However, the flora of these church forests is not, yet, studied and documented in the country. Different studies also indicated that each of the church forests in Ethiopia is unique and do have less similarity, suggesting that for the conservation and management of the species, data on each church forest have to be generated independently. Thus, the objectives of this study were to: (1) determine the species composition, structure and diversity of woody plants and (2) assess the regeneration status of woody plants in the aforementioned four church forests in Ethiopia.

8.2 Materials and Methods

8.2.1 Description of the Study Area

For this study, four church forests, which are located in Addis Ababa, Amhara, Tigray and Southern Nations, Nationalities and Peoples (SNNP) NRSs were selected (Fig. 8.1). The geographical locations, altitudes and size of the four church forests are presented in Table 8.1. All of these church forests are situated at more than 2000 m (Table 8.1). The size of the church forests ranged from 5 to 41 ha (Table 8.1). The vegetation of all the church forests is described as dry evergreen montane forests of Ethiopia. The mean annual rainfalls and temperatures of each of the study sites are presented in Fig. 8.2. The forests of the churches of WDM and BBTH maped from google earth is presented in Fig. 8.3 and Fig. 8.4 respectively.

8.2.2 Method of Vegetation Data Collection

A nested quadrat plot design that had an area of 400 m^2 (20 * 20 m) was used to collect vegetation data in each of the selected church forests. Depending on the proportion of the area coverage of each church forest, a total of 30, 15, 17 and 15 sample plots were sampled to collect the vegetation data in BBTHM, ME, EKA and WDM church forests, respectively. The distances between plots within a transect line and transects were 100 m each. In each plot, the identity of species, as well as number, diameter

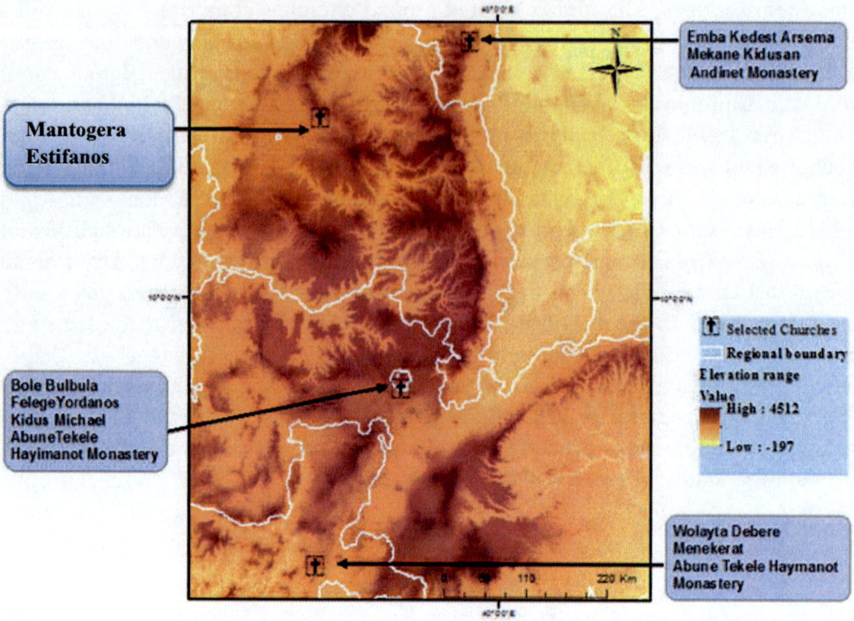

Fig. 8.1 The geographical location of the studied church forests

at breast height (DBH) and height of stems, were recorded. The height and DBH of individuals of each tree species were measured with a Suunto-hypsometer and calliper, respectively. Seedling (height ≤50 cm) and sapling data were collected in the five plots, covering 4 m^2 (2 * 2 m) and established at the four corners and centre of the quadrat. Species identification in the field was undertaken based on expert knowledge and by using tree identification field guide manuals (Bekele-Tessema et al. 2007; Fichtle and Adi 1994). Specimens, which were difficult to identify in the field were recorded using the local names of the study area. The nomenclature of the species follows Flora of Ethiopia and Eritrea (Edwards et al. 1995, 1997, 2000; Hedberg and Edwards 1989).

8.2.3 *Data Analyses*

Shannon–Wiener Diversity Index (*H´*) was used to determine the diversity of woody plants of each of the studied church forests and it was computed using the formula described in Kent and Coker (1994) as follows:

Table 8.1 Geographical location, altitude and size of the different church forests

No	Name of the church	Longitude	Latitude	Altitude (m)	Area (ha)	National regional state
1	Emba Kedest Arsema Mekane Kidusan Andinet Monastery (EKA)	39° 38′ 59.7″	13° 01′ 26.7″	2419	41	Tigray
2	Mantogera Estifanos Church (ME)	37° 47′ 06″	12° 9′ 45″	2177	5	Amhara
3	Bole Bulbula Tekele Hayimanot Monastery (BBTH)	38° 48′ 19.9″	08° 56′ 54.8″	2212	21	Addis Ababa
4	Wolayta Debere Menekerat Abune Tekele Haymanot Monastery (WDM)	37° 45′ 35.5″	06° 52′ 37.7″	2122	25	SNNPR

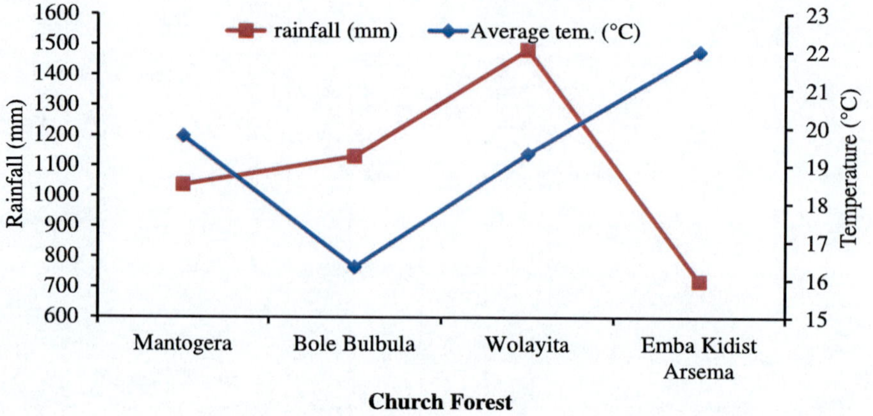

Fig. 8.2 Mean annual rainfall and temperature of the studied four church forests in Ethiopia

Fig. 8.3 Map of the church forests downloaded from Google Earth (left) and photos (top and bottom right side) that showed the forests in Bole Bulbula Felege Yordanos Kidus Michael AbuneTekele Hayimanot Monastery. (Photos: Shiferaw Alem, February 10, 2019)

$$H' = \sum_{i=1}^{s} pi \ln pi$$

where S = number of species, pi = proportion of the individual species to the total and ln = natural logarithm.

Equitability (evenness) of species was also calculated using the formula

$$H'/H'\max$$

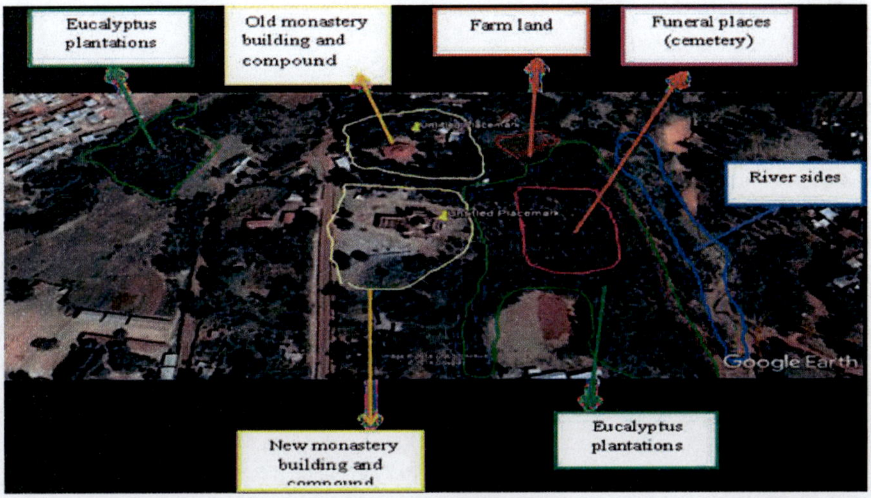

Fig. 8.4 Wolayta Debere Menekerat Abune Tekele Haymanot Monastery shows the different land uses in the church forest (*Source* google earth, accessed on July 20, 2019)

where H' max = ln (natural logarithm) of S (number of species).

The data collected in each major quadrat and church forest were analyzed to determine the population structure of the studied church forests.

The Importance Value Index (IVI), was calculated for each species (Kent and Coker 1994) using the following formulae:

$$\text{Relative density}(\%) = \frac{\text{Density of a species}}{\text{Total density}} * 100$$

$$\text{Relative frequency}(\%) = \frac{\text{Frequency of a species}}{\text{Total frequency}} * 100$$

$$\text{Relative dominance}(\%) = \frac{\text{Basal area of a species}}{\text{Total basal area}} * 100$$

where, Basal Area (B) = (BA = $\frac{\pi D^2}{4}$) $\pi = 3.14$ and D = diameter at breast height.

The vegetation structures of the church forests (i.e. the distribution of height and diameter classes) were grouped into different categories and used to assess the regeneration status of the woody species.

8.3 Results

8.3.1 *Floristic Composition, Diversity and Density of Woody Species*

A total of 82 woody species were recorded in the studied four church forests (Table 8.2). The results showed that a total of 38, 18, 16 and 32 species of trees, shrubs and climbers, representing different families were recorded in ME, EKA, WDM and BBTH church forests, respectively (Table 8.2). Out of the total recorded woody species, two species (*Cupressus lusitanica* and 'Agustizima') in ME and one species (*Eucalyptus camaldulensis*) in EKA were exotic. Similarly, out of the total woody species recorded, seven in WDM and two in BBTH were exotics.

Table 8.2 Lists of woody species identified in the different church forests

No	Species[a]	Family	Mantoger estifanos	Bolie Bulbula church	Emba kidist arsema	Wolayita debremenkert
1	*Acokanthera schimperi* (A.DC.) Schweinf	Apocynaceaea	X		X	
2	*Albizia gummifera*	Fabaceae	X			X
3	*Allophylus abyssinicus* (Hochst.) Radlk	Sapindaceae	X	X		
4	*Asparagus racemosus* Willd	Asparagaceae			X	
5	*Brucea antidysenterica* Swiss Chard	Simaroubaceae	X			
6	*Calpurnia aurea* subsp. *indica* Brummitt	Fabaceae	X			
7	*Capparis tomentosa* Lam	Capparidaceae	X			
8	*Carissa spinarum* L	Apocynaceae	X	X		
9	*Casuarina equisitifolia* L	Casuarinaceae				X
10	*Celtis africana* Burm.f	Cannabaceae	X			
11	*Clausena anisata* (Willd.) Benth	Rutaceae	X			
12	*Clematis hirsuta* Guill. & Perr	Ranunculaceae		X		
13	*Clutia lanceolata* Forssk	Euphorbiaceae	X			
14	*Coffea arabica* L	Rubiaceae	X			
15	*Cordia africana* L	Boraginaceae		X	X	X
16	*Croton macrostachyus* Hochst. ex Delile	Euphorbiaceae	X	X		X
17	*Cupressus lusitanica* L	Cupressaceae	X	X		X
18	*Dodonaea angustifolia* L.f	Sapindaceae	X		X	

(continued)

Table 8.2 (continued)

No	Species[a]	Family	Mantoger estifanos	Bolie Bulbula church	Emba kidist arsema	Wolayita debremenkert
19	*Dovyalis abyssinica* (A.Rich.) Warb	Salicaceae		X		
20	*Dracena* sp.	Asparagaceae				X
21	*Entadopsis abyssinica* (Steud. ex A. Rich.) Gilbert & Boutique	Fabaceae	X			
22	*Eucalyptus camaldulensis* Dehnh	Myrtaceae			X	X
23	*Eucalyptus globulus* Labill	Myrtaceae		X		
24	*Euclea racemosa* subsp. *schimperi* (A.DC.) F.White	Ebenaceae	X	X		
25	*Euclea schimperi* (A.DC.) Dandy	Ebenaceae				
26	*Euphorbia abyssinica* J.F.Gmel	Euphorbiaceae	X		X	
27	*Ficus carica* L	Moraceae	X			
28	*Ficus sur* Forssk	Moraceae		X		
29	*Ficus vasta* Forssk	Moraceae				X
30	*Foeniculum vulgare* Mill	Apiaceae			X	
31	*Galiniera saxifraga* (Hochst) Bridson	Rubiaceae	X			
32	*Grevillea robusta* A.Cunn. ex R.Br	Proteaceae		X		X
33	*Grewia ferruginea* Hochst. ex A.Rich	Malvaceae	X			
34	*Helinus mystacinus* (Ait.) E. Mey. Ex Steud	Rhamnaceae	X			
35	*Hibiscus macranthus* Hochst. ex A. Rich	Malvaceae	X			

(continued)

The results showed that the diversity values of the four church forests were in the order of ME > EKA > BBTH > WDM while their evenness values were in the order of EKA > BBTH > ME > WD (Fig. 8.5). The findings also showed that the densities (trees ha^{-1}) in the studied were in the order of BBTH (4300) > WD (533) > ME (716) > EKA (235) while the densities of seedlings ha^{-1} exhibited the order of ME > EKA > BBTH > WDM (Fig. 8.6).

In the ME Church Forest, *Acokanthera. schimperi*, *Olea europaea*, *Calpurnia aurea*, *Juniperus procera,* and *Premna schimperi* had tree densities (stems ha^{-1}) of 296.4, 135.7, 128.6, 67.9 and 39.3, respectively. Woody species, such as *Mimusops kummel*, *Grewia ferruginea* and *Croton macrostachyus*, had the lowest densities of

Table 8.2 (continued)

No	Species[a]	Family	Mantoger estifanos	Bolie Bulbula church	Emba kidist arsema	Wolayita debremenkert
36	*Jacaranda mimosifolia* D.Don	Bignoniaceae		X		X
37	*Jasminum abyssinicum* Hochst. Ex Dc	Oleaceae	X			
38	*Jasminum grandiflorum* (R.Br.ex.Fresen.) P.S.Green	Oleaceae			X	
39	*Juniperus procera* Hochst ex. Engl	Cupressaceae	X		X	X
40	*Justicia schimperiana* T.Anders	Acanthaceae	X			
41	*Maesa lanceolata* Forssk	Primulaceae			X	
42	*Maytenus addat* (Loes.) Sebsebe	Celastraceae		X		
43	*Maytenus arbutifolia* (Hochst. ex A.Rich.) R.Wilczek	Celastraceae		X		
44	*Maytenus obovata* Hook. f	Celastraceae	X			
45	*Millettia ferruginea* Hochst	Fabaceae		X		X
46	*Mimusops kummel* Bruce ex A.DC	Sapotaceae	X			
47	*Myrsine africana* L	Myrsinaceae		X		
48	*Olea europaea ssp. cuspidata* (Mill.) P. Green	Oleaceae	X		X	X
49	*Osyris quadripartita* Decn	Santalaceae			X	
50	*Paveta abyssinica* Fres	Rubiaceae	X			

(continued)

trees in the ME (Fig. 8.7). About 68% of the tree/shrub species recorded in ME had tree densities of less than 10 stems ha^{-1}. In BBTH, *Vachellia abyssinica* was the abundant tree species. In the WDM, the density of trees ranged between 1 and 1361 stems ha^{-1} with a mean value of 95 stems ha^{-1}. The top five abundant tree species in WDM were *Eucalyptus camaldulensis* (1361 stems ha^{-1}), *Cupressus lusitanica* (69 stems ha^{-1}), *Juniperus procera* (25 stems ha^{-1}), *Olea europaea* subsp. *cuspidata* (13 stems ha^{-1}) and *Grevillea robusta* (13 stems ha^{-1}), representing 98% of the total densities of trees. Exotic trees contributed 96% of the total densities of trees in WDM. The top five dominant tree species in the WDM were *Eucalyptus camaldulensis,* contributing 42% of the total basal area, followed by *Juniperus procera* (16%), *Podocarpus falcatus* (14%), *Dracena fragrans* (8%) and *Olea europaea* subsp. *cuspidata* (7%).

Table 8.2 (continued)

No	Species[a]	Family	Mantoger estifanos	Bolie Bulbula church	Emba kidist arsema	Wolayita debremenkert
51	*Persea americana* Mill	Lauraceae				X
52	*Physalis peruviana* L	Solanaceae			X	
53	*Phytolacca dodecandra* L'Hér	Phytolaccaceae	X	X		
54	*Pittosporum viridiflorum* Sims	Pittosporaceae	X	X		
55	*Podocarpus falcatus* (Thunb.) Endl	Podocarpaceae				X
56	*Premna schimperi* Engl	Lamiaceae	X			
57	*Premna viburnoides var. schimperi* (Engl.) Pic.Serm	Lamiaceae	X			
58	*Prunus africana* (Hook.f.) Kalkman	Rosaceae				X
59	*Pterolobium stellatum* (Forssk.) Brenan	Fabaceae				
60	*Rhamnus staddo* A. Rich	Rhamnaceae		X		
61	*Rhus natalensis* Bernh. ex C.Krauss	Anacardiaceae		X	X	
62	*Rhus vulgaris* Meikle	Anacardiaceae			X	
63	*Rosa abyssinica* R.Br	Rosaceae		X		
64	*Rubus steundneri* Schweinf	Rosaceae				
65	*Schefflera abyssinica* (Hochst. ex A.Rich.) Harms	Araliaceae	X			
66	*Sida schimperiana* Hochst. ex A. Rich	Malvaceae		X		

(continued)

8.3.2 Regeneration Status of Woody Plants

In ME, the total densities of seedlings were 10,130 stems ha^{-1}. In ME, the densities of seedlings (ha^{-1}) of *Justicia schimperiana, Maytenus ovatus, Euclea racemosa, Acokanthera schimperi, Calpurnia aurea, Carissa spinarum* and *Olea europaea* were 2660, 2180, 1430, 1030, 620, 400 and 350, respectively. Seedlings of *Mimusops kummel, Juniperus procera, Vachellia abyssinica* and *Albizia gummifera*, about 39% of the reported tree/shrub species, were not recorded in the studied plots.

In BBTH, most of the woody species had poor natural regeneration. Even the abundant tree species recorded, i.e. *Vachellia abyssinica* had a very poor natural regeneration. For most of the species in this church forest, the densities of seedlings were $< 10\ ha^{-1}$.

Only 19 of the woody species recorded in EKA accounted for a total of 364 seedlings ha^{-1}. Except for the three tree species, which had more than 50 seedlings ha^{-1}, the remaining tree species had less than five seedlings ha^{-1}.

Table 8.2 (continued)

No	Species[a]	Family	Mantoger estifanos	Bolie Bulbula church	Emba kidist arsema	Wolayita debremenkert
67	*Sida tenuicarpa* Vollesen	Malvaceae				
68	*Spathodea campanulata* P.Beauv	Bignoniaceae		X		
69	*Vachellia seyal* Delile	Fabaceae		X		
70	*Vachellia abyssinica* Hochst. Ex Benth	Fabaceae	X	X	X	
71	*Vernonia amygdalina* Delile	Asteraceae		X		
72	Agustizima*		X			
73	Alenquat*			X		
74	Anfar mesaye*			X		
75	Demza*			X		
76	Harets**				X	
77	Naca*			X		
78	Shanshura*		X			
79	Yegeb areg*		X			
80	Zelemko *			X		
81	Qeleminto**				X	
82	Unidentified species			X		

[a] Name of a species in: * = Amharic language and ** = Tigrigna language

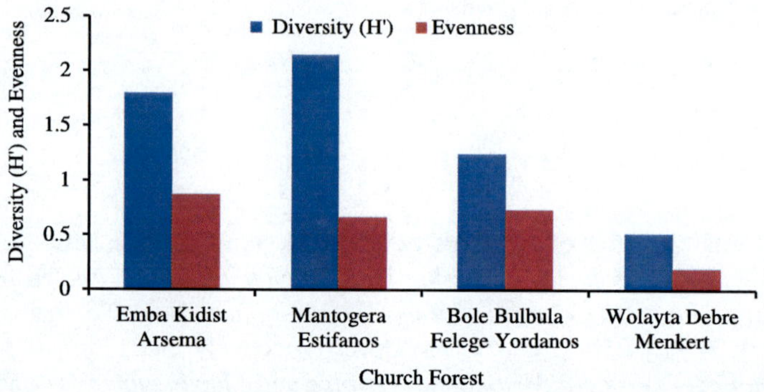

Fig. 8.5 Diversity (*H'*) and evenness of the studied church forests

In WDM, most of the recorded tree species, including *Milletia ferruginea, Albizia gummifera, Ficus vasta, Croton macrostachyus* and *Cordia africana,* lacked seedlings and were also represented by very few adult tree species.

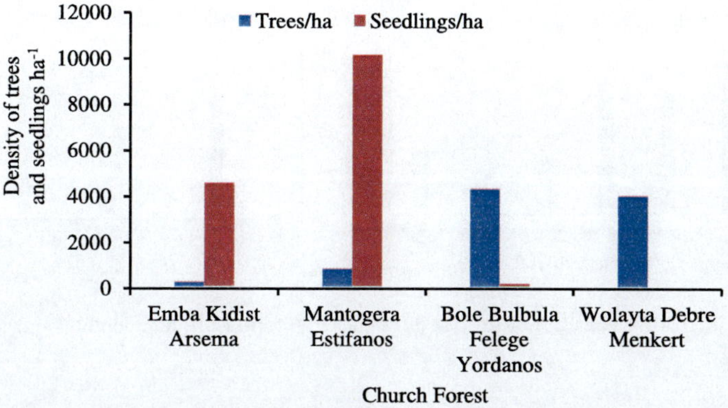

Fig. 8.6 Densities (ha^{-1}) of trees and seedlings in the studied church forests

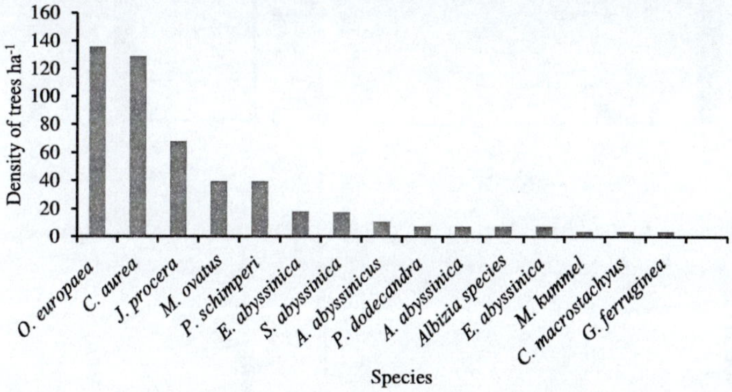

Fig. 8.7 Densities of trees (ha^{-1}) for all woody species in the Mantogera Estifanos Church Forest

8.3.3 Population Structure of Woody Species

The DBH-based population structure of the woody species in BBTH showed that the forest had reversed J-shaped structure (Fig. 8.8), which is ecologically considered to represent healthy populations. However, the height-based population structure indicated the existence of hampered natural regeneration (Fig. 8.8).

Although EKA seems to have reversed J-shaped distribution of the woody species (Fig. 8.9), the growth of seedlings and saplings to bigger-sized trees is much hampered, suggesting that the population structure has been affected by disturbance.

In WDM, the diameters of the recorded trees ranged from 2 to 149 cm whereas their corresponding heights ranged from 2.2 to 37.5 m.

The results from ME indicated that the relative density of the tree species in the diameter classes of 1–5 cm, 5–10 cm, 10–15 cm and 15–20 cm were 6%, 45.4%, 18.8% and 10.6%, respectively (Fig. 8.10). Similarly, the number of species that had

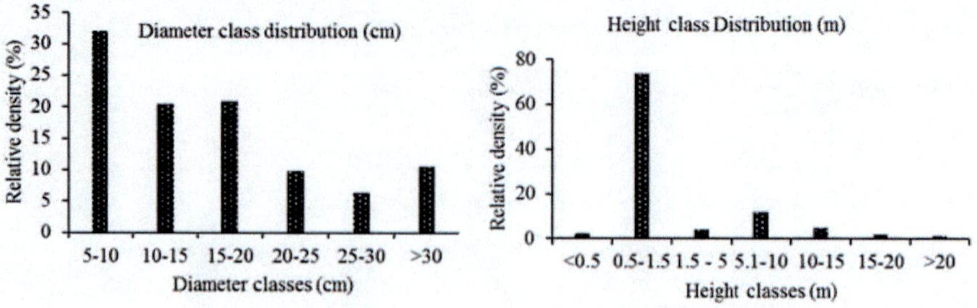

Fig. 8.8 The relative densities of trees at different DBH (cm) (left) and height (m) (right) classes in BBTHM

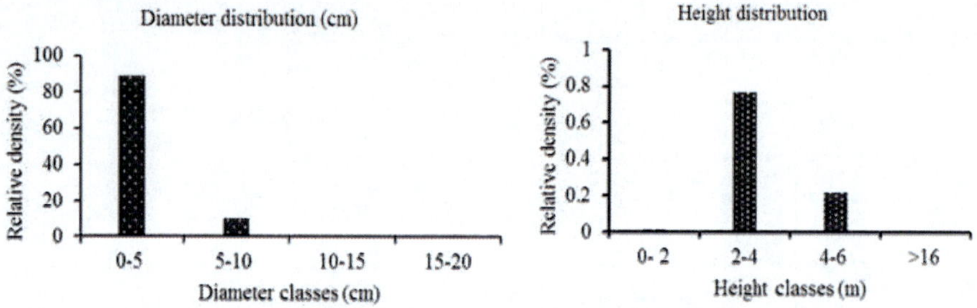

Fig. 8.9 Diameter (cm) (left) and height (m) (right) class distribution of woody species in EKA

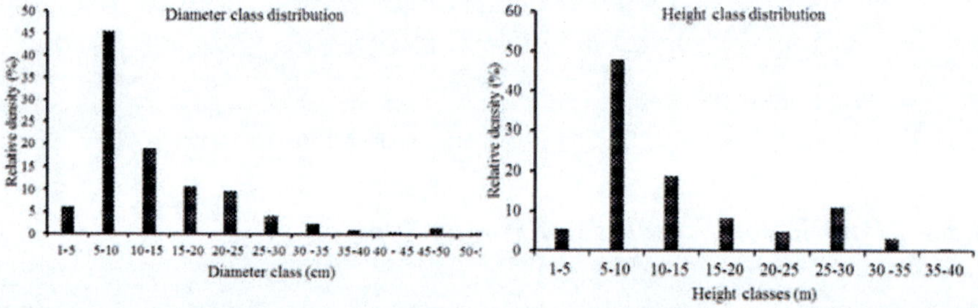

Fig. 8.10 Diameter (cm) (left) and height (m) (right) class distributions of woody species in Mantogera Estifanos Church Forest

diameter classes of 1–5 cm, 5–10 cm, 10–15 cm and 15–20 cm were 3, 13, 7 and 11, respectively (Fig. 8.10). The results of the height distribution in the ME showed that the relative densities of the species in the height classes of 1–5 m, 5–10 m, 10–15 m and 15–20 m were 5.5%, 47.7%, 18.8% and 8.3%, respectively (Fig. 8.10). Similarly, the number of species that were recorded in the height classes of 1–5 m, 5–10 m, 10–15 m and 15–20 m, were 2, 13, 12 and 8, respectively.

Table 8.3 The relative density (RD), frequency (RF), Basal area (BA) and Dominance (RD) values and IVI of the different woody species in BBTH

No	Species	RD	RF	BA	RDO	IVI
1	*Vachellia abyssinica*	72.35	72.35	19.52	0.09	144.80
2	*Euclea schimperi*	8.01	8.01	7.07	0.03	16.05
3	*Croton macrostachyus*	6.98	6.98	12.83	0.06	14.02
4	*Eucalyptus globulus*	2.84	2.84	22.95	0.11	5.80
5	*Grevillea robusta*	2.07	2.07	16.65	0.08	4.21
6	*Rhus natalensis*	1.81	1.81	9.14	0.04	3.66
7	*Maytenus arbutifolia*	1.55	1.55	6.82	0.03	3.13
8	*Milletia ferruginea*	0.78	0.78	11.20	0.05	1.60
9	Zelemko *	0.78	0.78	6.30	0.03	1.58
10	*Rhamnus staddo*	0.52	0.52	7.40	0.04	1.07
11	*Maytenus addat*	0.52	0.52	7.25	0.04	1.07
12	*Vachellia seyal*	0.26	0.26	21.20	0.10	0.62
13	*Vernonia amygdalina*	0.26	0.26	17.40	0.08	0.60
14	*Phytolacca dodecandra*	0.26	0.26	10.70	0.05	0.57
15	*Jakaranda mimosifolia*	0.26	0.26	8.00	0.04	0.56
16	*Ficus sur*	0.26	0.26	7.70	0.04	0.55
17	Demza*	0.26	0.26	7.60	0.04	0.55
18	Unidentified species	0.26	0.26	7.30	0.04	0.55

* Local name of the species in Amharic Language

8.3.4 Importance Value Index (IVI)

The IVI ranged from 0.55 to 144.8 in BBTH (Table 8.3) and *Vachellia abyssinica* and *Ficus sur* had the highest and lowest IVI values, respectively. The relative frequency, density and dominance values of woody species recorded in EKA ranged from 1.96 to 29.41, 0.66–52.32 and 0–56.14%, respectively (Table 8.4). The IVI in EKA ranged from 2.6 to 137.9 and relative to other species, *J. procera* exhibited the highest relative frequency, density, dominance values, hence, the highest IVI (Table 8.4).

The result showed that the IVI of woody species in ME ranged from 3.51 to 51.42 (Table 8.5). *Acokanthera schimperi* exhibited the highest IVI (51.42) than all the species recorded in the forest (Table 8.5). *Olea europaea, C. aurea, J. procera* and *P. schimperi* had IVI values of 40.8, 25.95, 17.6 and 17.4, respectively (Table 8.5). About, 47.6% of the recorded species in ME had IVI values of less than 10, which may also show that they are not evenly distributed in the forest. The IVI of *O. europaea*, which is the second-highest value, in the forest suggests its relatively high ecological importance.

Table 8.4 The relative density (RD), frequency (RF) and dominance (RDO) and IVI of woody species in EKA

No	Species	RF	RD	RDO	IVI
1	*Rhus natalensis*	5.88	5.3	4.56	15.74
2	*Rhus vulgaris*	11.76	3.97	4	19.74
3	*Cordia africana*	1.96	1.99	1.69	5.63
4	*Olea africana*	15.69	14.57	14.4	44.66
4	*Euphorbia abyssinica*	1.96	0.66	0	2.62
5	*Vachellia abyssinica*	1.96	0.66	0.62	3.24
6	*Eucalyptus camaldulensis*	3.92	1.99	1.64	7.55
7	*Osyris quadripartita*	5.88	3.31	2.76	11.95
8	*Carissa spinarum*	1.96	0.66	0.54	3.17
9	*Maesa lanceolata*	1.96	0.66	0.53	3.16
10	*Acokanthera schimperi*	1.96	1.32	1.26	4.55
12	Kelem-tsos*	1.96	0.66	0.69	3.31
13	*Foeniculum vulgare*	1.96	1.99	1.61	5.56
14	*Cupressus lusitanica*	9.8	9.27	9	28.07
15	*Juniperus procera*	29.41	52.32	56.14	137.87
16	*Jasminum grandiflorum*	1.96	0.66	0.54	3.17

* Local name of the species in Tigrigna Language

8.4 Discussion

The average number of woody species recorded in the studied church forests was 27. This shows that the church forests are poor in the richness of woody species. This result concurs with the findings of Aerts et al. (2016) that found an average of 25 woody plants in a study involving 78 church forests. Similarly, the number of woody species recorded in the studied church forests was also comparable with the findings of other similar studies in the church forests of Dengolt, Gibtsawit, Debresena, Hiruy, Zahra, Ascha, Yimrehane Kiristos, Shewa, Mahbere Sellassie and Tara Gedam church forests located in different parts of the country that had 16–39 woody species/church forest (Agdew and Mezgebe 2019; Abunie and Dalle 2018; Habtamu 2017; Gedefaw and Soromessa 2014; Wassie et al. 2005). However, the present study result on the number of woody species recorded in the studied church forests was relatively lower than those reported from Gelawdiwos church forests (Wassie et al. 2005), Debrelibanos Monastry Forest (Shiferaw et al. 2019; Demie 2015), Sesa Mariam Monastery Forest (Meshesha et al. 2015), Sellassie Monastery Forest (Abunie and Dalle 2018) and Aba Asrat Monsatry Forest (Bayeh 2013).

The differences of findings from the current study with results from other similar studies on the number of identified woody species could be associated with differences in disturbances. In their study on the ecological status of church forests in

Table 8.5 Relative density, frequency and dominance and IVI of trees and shrubs in ME

No	Species	RD	RF	RDo	IVI
1	*Brucea antidysenterica*	0.87	2.33	13.68	16.88
2	*Agustizima**	0.44	2.33	1.84	4.61
3	*Premna schimperi*	4.8	11.63	0.92	17.35
4	*Dengayseber**	0.87	2.33	0.74	3.94
5	*Allophylus abyssinicus*	1.31	2.33	1.62	5.26
6	*Phytolacca dodecandra*	0.87	2.33	5.49	8.69
7	*Mimusops kummel*	0.44	2.33	2.57	5.34
8	*Schefflera abyssinica*	0.87	2.33	13.68	16.88
9	*Celtis africana*	1.75	4.66	7.17	13.58
10	*Grewia ferruginea*	0.44	2.33	0.74	3.51
11	*Acokanthera schimperi*	36.24	13.96	1.22	51.42
12	*Croton macrostachyus*	0.44	2.33	4.93	7.7
13	*Vachellia abyssinica*	0.87	2.33	0.64	3.84
14	*Albizia gummifera*	0.87	4.65	5.34	10.86
15	*Entadopsis abyssinica*	0.87	2.33	2.47	5.67
16	*Euphorbia abyssinica*	2.18	4.65	2.15	8.98
17	*Olea africana*	16.59	13.96	10.28	40.83
18	*Juniperus procera*	8.3	4.65	4.6	17.55
19	*Maytenus ovatus*	4.8	4.65	2.19	11.64
20	*Galiniera saxifraga*	0.44	2.33	12.79	15.56
21	*Calpurnia aurea*	15.72	9.31	0.92	25.95

* Local name of the species in Amharic language

the northern highlands of South Gondar, Ethiopia, Cardelu´s et al. (2019) found that disturbance was high across all church forests (56%) and was negatively associated with trees species richness and seedling richness and density.

Only a few species (less than five trees/shrubs) were common to all the studied four church forests. This may show that each of the studied church forests is unique when considering woody species conservation and when management strategies for in situ/ex situ conservation have to be developed.

The woody species diversity values of the four church forests were relatively lower than other similar church forests in Ethiopia. Zegeye et al. (2011) in the Tara Gedam Church Forest found a diversity value of 2.98. Meshesha et al. (2015) reported a diversity value of 3.81 in Sesa Mariam Monastery Church Forest and Abunie and Dalle (2018) found a diversity value of 2.88 in Yemrehane Kirstos Church Forest of Lasta Woreda. Demie (2015) and Meshesha et al. (2015) reported diversity values of 3.35–3.81 in Debrelibanos Monsatry Forest. Habtamu (2017) recorded diversity values of 3.01–3.45 in Sellassie Monastry Church Forest. This differences on the diversity of woody species of the current study result with others could have resulted from disturbances, size of the church forests, lack of regeneration of woody species and selective logging problems.

Lower diversity of woody species in WDM than the other three studied church forests could be associated with the expansion of the *Eucalyptus* plantations in the compounds of the church. In WDM, the density of *Eucalyptus camaldulensis* was 1361 stems ha $^{-1}$, which may show that most of the area is occupied by this species. Other similar studies have reported the expansion of exotic plantations in other church forests in Ethiopia. For example, Mekonnen (2019) reported that 38 (82.6%) out of the 46 studied church forests had exotic woody species (924 exotic woody stems from 20 different species). Accordingly, the genus *Eucalyptus* represented the most abundant exotic tree plantations with 791 stems, representing 85.6% of the total stems of all woody exotic species. It is also indicated that native tree and seedling species richness, abundance, diversity and evenness in the different studied church forests decreased significantly with increasing proportions of exotic woody species in the forest (Mekonnen, 2019).

The results on the densities of trees (ha^{-1}) in EKA and BBTH were lower than the other studied church forests in Ethiopia. Wassie et al. (2005) recorded 1109–2250 woody stems ha^{-1} in seven church forests, Shiferaw et al. (2019) recorded 1960 woody stems ha^{-1} and Abunie and Dalle (2018) found 898 woody stems ha^{-1} in Yimirihane Kirstos Church Forest. This result may show that some management strategies are necessary to increase the density and growth of woody species in the studied church forests to enhance their sustainability.

The higher number of woody stems in BBTH than the other studied church forests could be associated with the higher density of shrub species in the church forest than the other church forests. It was observed that shrubs dominantly covered BBTH while there were a few numbers of matured trees. The study further showed that in all the studied church forests, only a few species were dominant. For example, the dominant woody species were *V. abyssinica* in BBTH, *J. procera* and *O. europaea* in EKA and *O. europaea* and *A. schimperi* in ME. This may show that the different church forests are favouring the recruitment of only very few and certain species. For some very dominant species (e.g. *V. abyssinica* in the BBTH), no seedlings were recorded. For example, about 39% of the recorded tree/shrub species in ME did not have seedlings. In WDM, *J. procera* and *P. falcatus* showed relatively healthy regeneration status. However, most of the recorded tree species, including *Millettia ferruginea, Albizia gummifera, Ficus vasta, Croton macrostachyus* and *Cordia africana* lacked natural regeneration and were represented by very few numbers of adult tree species.

About 47.6% of the recorded species in ME had IVI values of less than 10, which may show that most of the species are not evenly distributed in the forest. In most of the studied church forests, there was a problem of regeneration for most of the recorded species. This poor regeneration of the woody species could be associated with the impacts of exotic species, such as *Eucalyptus*, poor soil seed bank and overgrazing. Mekonnen (2019) indicated that native tree and seedling species richness, abundance, diversity and evenness in the different studied church forests decreased significantly with increasing proportions of exotic woody species.

In their soil seed bank studies involving seven church forests, Wassie and Teketay (2006) recorded large quantities of persistent seeds of herbaceous species in the soil while only five (6%) of the 91 woody species recorded in the standing vegetation were represented in the soil seed banks, which they attributed to the removal of mature trees that could have provided seeds. Shiferaw et al. (2019) indicated that grazing intensity had significant effects on vegetation patterns and regeneration in the Debrelibanos Monastery Forest.

Lack of protection of the church forests could also be another cause of poor regeneration since they are susceptible to disturbances. Wassie et al. (2009) found that in the fenced plots, more seeds germinated, seedling survival was higher and seedlings grew faster than in the open area. Hence, they concluded that for effective natural regeneration, controlling livestock grazing is necessary. Woods et al. (2017) indicated that the density and species richness of seedlings was significantly higher in church forests fenced with a wall and they recommended that construction of walls around churches would be an effective conservation tool.

Invasive species were also observed in the different church forests although their intensities varied. The expansion of invasive species could also be hindering the regeneration of woody species. For example, the expansion of *Lantana* species in WDM (Fig. 8.11) that covers the ground widely may prohibit the availability of light, which, in turn, could affect the recruitment of woody species.

Tilahun et al. (2015) indicated the destructive impacts of invasive alien species in church forests found in Kewet and Yifratana Gidim Districts where *Lantana camara* was replacing most of the shrubs at a very rapid rate. Invasive species are capable of takeover indigenous herbs and shrub species quickly (Tilahun et al. 2015; Liang et al. 2016).

Fig. 8.11 *Lantana* species expanding in the Wolayta Debre Menkert Church Forest. (Photo: Shiferaw Alem, 25/06/2019)

8.5 Conclusions and Recommendations

The present study showed that the studied church forests harboured a variety of woody species, ranging from 16 to 40 species and their diversity ranged from 0.52 to 2.15. The average density of woody species was 1446 stems ha^{-1}. Even though the different church forests are unique and important for biodiversity conservation, poor natural regeneration was observed in the church forests. Expansion of plantations of exotic species, such as *Eucalyptus,* and the invasion of church forests by invasive species is becoming common. Therefore, the following management interventions are recommended.

To enhance recruitment of woody species in the church forests, different silvicultural management interventions, such as weeding, hoeing to loosen the ground, removing the thick litter layer, removing the dense shrub layers through thinning, enrichment planting and burying mature seeds in the soil and enhancing their recruitment through assisted natural regeneration (e.g. weeding, spot hoeing, avoiding grazing and trampling and watering) are recommended.

It was also observed that the studied church forests were not fenced and open grazing is very common. Excluding livestock by fencing the church forests will help to enhance the growth of naturally regenerated seedlings as browsing and trampling will be reduced.

Direct sowing of selected woody species, through burying of the seeds at a shallow depth in the soil, is also recommended as it is becoming one of the options in the restoration of degraded tropical forests.

There is an expansion of exotic plantations, mainly of *Eucalyptus* species, in most of the studied church forests. The expansion of *Eucalyptus* species in the church forests could hinder the regeneration of woody species due to the high nutrient and moisture demanding nature of the species. Therefore, proper management plans in church forests, such as planting exotic species at the periphery of the churches (as a buffer zone plantation), instead of planting them in areas where native woody species are growing, is recommended. The churches plant *Eucalyptus* species to gain income and, therefore, finding alternative income generation activities is necessary to sustain the conservation of woody species in the churches. For example, expansion of honeybee management practices as well as enhancement and management of non-timber forest products within the church forests, such as an expansion of coffee plantations with different agroforestry systems, is recommended since it could enhance their income and, indirectly, halt the expansion of *Eucalyptus*. These measures could help in the conservation of native woody species growing in the church.

The other observed problem that could hinder the recruitment of woody species in the studied church forests was the expansion of invasive species, such as *Lantana* species. Management of the invasive species through uprooting, cutting or griddling could help reduce their expansion. Therefore, a series of monitoring and management undertakings are recommended to control the expansion of invasive species.

Finally, it is observed that most of the studied church forests were not fenced. Various studies that were carried out in different church forests reported that the churches, which had fences exhibited good natural regeneration, diversity of species and densities of trees. Therefore, to enhance conservation of the indigenous woody species, fencing of the studied church forests is recommended.

Acknowledgements Funding for this study was provided by the Ethiopian Orthodox Tewahido Church. We also thank the Ethiopian Environment and Forest Research Institute for the provision of field vehicles for the data collection. The leaders of each of the respective study churches were helpful in the provision of information and supporting data collection and, hence, we extend our thanks to all of them. Lastly, our acknowledgement goes to the different individuals who supported us during vegetation data collection in the different church forests.

Authors' Contributions In this paper, NH, MT, ZT, NY and AE designed and conducted the field research. SA, NH, MT and KS analyzed the data and drafted the manuscript. SA interpreted the results and helped in manuscript writing. All authors read, revised and approved the final manuscript.

Competing Interests The authors declare that they have no competing interests.

References

Abunie AA, Dalle G (2018) Woody species diversity, structure, and regeneration status of Yemrehane Kirstos Church Forest of Lasta Woreda, North Wollo Zone, Amhara Region, Ethiopia. Int J Res 2018:1–8. https://doi.org/10.1155/2018/5302523

Aerts R, Overtveld KV, November E, Wassie A, Abiyu A, Demissew S, Daye DD et al (2016) Conservation of the Ethiopian Church Forests: threats, opportunities and implications for their management. Sci Total Environ 551(1):404–414. https://doi.org/10.1016/j.scitotenv.2016.02.034

Agidew ST, Mezgebe AH (2019) Woody species composition analysis of Shawo Sacred forest, Ethiopia. Int J Appl Sci Biotechnol 7(1):12–21. https://doi.org/10.3126/ijasbt.v7i1.21978

Bayeh G (2013) Floristic composition and diversity analysis of vegetation at Aba Asrat Monastery forest, East Gojjam, Ethiopia. M.Sc thesis, Addis Ababa University

Bekele-Tessema A, Beánie A, Tengnas B (2007) Useful trees and shrubs for Ethiopia: identification, propagation and management for 17 agro-climatic zones. Regional Land Management Unit, Nairobi

Bhagwat SA (2009) Ecosystem services and sacred natural sites: reconciling material and non-material values in nature conservation. Environ Values 18:417–427. https://doi.org/10.3197/096327109X12532653285731

Cardelu´s CL, Woods CL, Bitew MA, Dexter S, Scull P, Tsegay BA (2019) Human disturbance impacts the integrity of sacred church forests, Ethiopia. PLoSONE 14(3):e0212430

Cardelús CL, Scull P, Hair J, Baimas-George M, Lowman MD, Eshete AW (2013) A preliminary assessment of Ethiopian sacred grove status at the landscape and ecosystem scales. Diversity 5(2):320–334. https://doi.org/10.3390/d5020320

Demie G (2015) Floristic composition and diversity of sacred site and challenges towards sustainable forest management: the case of remnant forest patch of Debrelibanos Monastery, Ethiopia. J Nat Sci Res 5(15):171–181

Edwards S, Tadesse M, Hedberg I (eds) (1995) Flora of Ethiopia and Eritrea, vol 2. Addis Ababa University, Addis Ababa

Edwards S, Demissew S, Hedberg I (eds) (1997) Flora of Ethiopia and Eritrea, vol 6. Addis Ababa University, Addis Ababa

Edwards S, Tadesse M, Hedberg I (2000) Flora of Ethiopia and Eritrea, vol 2. Addis Ababa Addis Ababa University, Addis Ababa

Fichtl R, Adi A (1994) Honeybee flora of Ethiopia. Margraf Verlag, Weikersheim

Gedefaw M, Soromessa T (2014) Status and woody plant species diversity in Tara Gedam Forest, Northern Ethopia. J Sci Technol Arts 3(2):113–118

Habtamu B (2017) Woody species, diversity in Mahbere Sellassie Monastery Forest, Northwestern Ethiopia. M.Sc. thesis, Gonder University

Hedberg I, Edwards S (eds) (1989) Flora of Ethiopia, vol 3. Addis Ababa University, Addis Ababa

Kent M, Coker P (1994) Vegetation description and analysis: a practical approach. John Wiley and Sons, London

Liang J, Reynolds T, Wassie A, Collins C, Wubalem A (2016) Effects of exotic Eucalyptus spp. Plantations on soil properties in and around sacred natural sites in the northern Ethiopian Highlands. AIMS Agric Food 1(2):175–193. https://doi.org/10.3934/agrfood.2016.2.175

Mekonnen AB (2019) Distribution and ecological impact of exotic woody plant species inside sacred groves of North-western Ethiopia. Biodivers Conserv 28:2845–2859. https://doi.org/10.1007/s10531-019-01799-4

Meshesha WB, Tsegay BH, Telake BB (2015) Survey on composition of perennial vegetation in Sesa Mariam Monastery, North-Western Ethiopia. BMC Res Notes 8:22. https://doi.org/10.1186/s13104-015-1562-5

Powledge F (2006) The millennium assessment. Bioscience 56:880–886. https://doi.org/10.1641/0006-3568(2006)56[880:TMA]2.0.CO;2

Reynolds T, Shimekach T, Wassie A, Lowman M (2015) Sacred natural sites provide ecological libraries for landscape restoration and institutional models for biodiversity conservation. GSDR Brief. Global Sustainable Development Report (GSDR), The United Nation

Shiferaw W, Bekele T, Demissew S (2019) Anthropogenic effects on floristic composition, diversity and regeneration potential of the Debrelibanos Monastery forest patch, central Ethiopia. J Res 30(6):2151–2161. https://doi.org/10.1007/s11676-018-0782-7

Taye B (1998) Biodiversity conservation: experience of the Ethiopian Orthodox Tewahido Church. M.Sc. Forestry Graduates Magazine, 1994–1998

Tilahun A, Terefe H, Soromessa T (2015) The contribution of Ethiopian Orthodox Tewahido Church in Forest management and its best practices to be scaled up in North Shewa Zone of Amhara Region, Ethiopia. Agric Forest Fisheries 4(3):123–137

Wassie A, Teketay D (2006) Soil seed banks in church forests of northern Ethiopia: implications for the conservation of woody plants. Flora 201:32–43. https://doi.org/10.1016/j.flora.2005.04.002

Wassie A, Teketay D, Powell N (2005) Church forests in north gonder administrative zone, northern Ethiopia. For Trees Livelihoods 15:349–373. https://doi.org/10.1080/14728028.2005.9752536

Wassie A, Sterck FJ, Teketay D, Bongers F (2009) Effects of livestock exclusion on tree regeneration in church forests of Ethiopia. For Ecol Manag 257:765–772. https://doi.org/10.1016/j.foreco.2008.07.032

Wassie A, Sterck F, Bongers F (2010) Species and structural diversity of church forests in a fragmented Ethiopian Highland landscape. J Veg Sci 21(5):938–948. https://doi.org/10.1111/j.1654-1103.2010.01202.x

Wassie A (2002) Opportunities, constraints and prospects of Ethiopian Orthodox Tewahido churches in conserving forest resources: the case of churches in south Gonder, Northern Ethiopia. M.Sc. thesis, Swedish University of Agricultural Sciences

Wassie A (2007) Ethiopian Church Forests: opportunities and challenges for restoration. Dissertation, Wageningen University

Wondmagegnehu A, Motovu J (1970) The Ethiopian Orthodox Church. The Ethiopian Orthodox Mission, Addis Ababa, Ethiopia

Woods CL, Cardelu´s CL, Scull P, Wassie A, Baez A, Klepeis P (2017) Stone walls and sacred forest conservation in Ethiopia. Biodivers Conserv 26:209–221. https://doi.org/10.1007/s10531-016-1239-y

Zegeye H, Teketay D, Kelbessa E (2011) Diversity and regeneration status of woody species in Tara Gedam and Abebaye forests, northwestern Ethiopia. J Res 22(3):315–328. https://doi.org/10.1007/s11676-011-0176-6

Chapter 9
Diversity, Regeneration Status, and Socio-Economic Importance of Tara Gedam, Abebaye and Fach Forests, South Gondar, Northwestern Ethiopia

Haileab Zegeye

Abstract The study was conducted in Tara Gedam, Abebaye, and Fach forests to investigate the floristic composition, diversity, population structure, regeneration status and socio-economic importance of the forests, and the anthropogenic factors affecting them. A total of 64 plots (21 from Tara Gedam, 9 from Abebaye, 34 from Fach), measuring 20 m × 20 m (400 m^2) each and established along line transects approximately at 100 m intervals, were used to collect vegetation data. A general survey consisting of field observations, key-informant interviews, and Focus Group Discussions (FGDs) was carried out to collect socio-economic data. A total of 263 vascular plant species belonging to 198 genera and 79 families were identified from the study areas. The diversity and evenness of woody species in Tara Gedam (TG), Abebaye (AB), and Fach (FC) forests were 2.98 and 0.65 (TG), 1.31 and 0.31 (AB), and 3.53 and 0.72 (FC), respectively. Woody species having low Importance Value Index (IVI) values, such as *Rhus retinorrhea* and *Ficus sycomorus*, and those exhibiting poor regeneration status (as indicated by their Diameter at Breast Height [DBH] class distributions), such as *Acacia abyssinica* and *Premna schimperi*, need high priority for conservation. The local communities were dependent on the forests for fuelwood, construction material, charcoal, timber, farm implements, food, medicines, fodder, and bee forage. At present, however, the forests are being destroyed due to livestock grazing/browsing, tree cutting for various purposes, farmland expansion, human settlements, urbanization, and fire incidences. Therefore, effective conservation and management interventions are urgently needed to ensure the long-term maintenance of the forest ecosystems, and benefit the local communities through sustainable utilization of the forests.

Keywords Conservation · Diversity · Ethiopia · Importance Value Index (IVI) · Regeneration status · Species richness

---

H. Zegeye (✉)
Department of Biology, Faculty of Natural and Computational Sciences, Debre Tabor University, P.O. Box 272, Debre Tabor, Ethiopia

M. Kindu et al. (eds.), *State of the Art in Ethiopian Church Forests and Restoration Options*, https://doi.org/10.1007/978-3-030-86626-6_9

9.1 Introduction

Ethiopia is endowed with rich biodiversity and endemism, which is attributed to its topographic, edaphic, climatic and cultural diversity. The country is rich in biological resources, and many plant and animal species are endemic. Ethiopia and Eritrea possess about 6027 vascular plant species (including subspecies), of which about 10.74% are endemic (Kelbessa and Demissew 2014). Indeed, Ethiopia is recognized internationally as one of the major centers of biodiversity and endemism (Edwards and Kelbessa 1999). It is also encompassed by two of the 34 global biodiversity hotspots (Eastern Afromontane Biodiversity Hotspot, Horn of Africa Biodiversity Hotspot).

It is believed that about 35% of the total land area of Ethiopia was covered with natural forests in the beginning of the 20th century (EFAP 1994). However, the forest cover has been declining from time to time. The forest cover of the country was 16% in the early 1950s, 3.6% in the early 1980s, and 2.7% in 1989 (EFAP 1994). The remaining natural forests of Ethiopia are only small patches mostly confined to inaccessible areas and sacred places. Furthermore, these remnant forests are dwindling at an alarming rate mainly due to human activities. The major drivers of deforestation and forest degradation are agricultural expansion, overexploitation, urbanization, fire incidences, invasive alien species, and villagization/resettlement (EFAP 1994; Bekele 1994; Zegeye 2017).

There are three state forests in Libokemkem district of South Gondar Administrative Zone (SGAZ), namely Tara Gedam, Abebaye, and Fach forests, which were established as protected areas during the Derg regime in 1979, 1980, and 1984, respectively. They are one of the remnant dry evergreen Afromontane forests in northwestern Ethiopia, if not in the country. Though it was not possible to get the size of the forests during their establishment, a later report indicated that Tara Gedam, Abebaye, and Fach forests had a total land area of 974.50 ha, 133.31 ha, and 823.19 ha, respectively (Anonymous 2004). The three forests are currently administered by Libokemkem District Agricultural Development Office. Within each protected forest are found churches and/or monasteries, which own a certain part of the respective forest and manage it, still under the umbrella of the protected area system.

Though a few preliminary surveys have been done by other scholars in the past, there has been neither particular nor detailed study on the vegetation ecology and socio-economic aspects of the forests so far. There is a need to generate ecological and socio-economic information, which would be relevant for conservation, management and sustainable utilization of the forests. Therefore, the present study was undertaken with the following objectives to: (a) investigate the floristic composition and diversity, population structure, regeneration status and socio-economic importance of the forests; (b) assess the forest management systems; (c) identify the anthropogenic factors affecting the forests.

9.2 Materials and Methods

9.2.1 Description of the Study Areas

The study was conducted in Tara Gedam, Abebaye and Fach forests located close to Addis Zemen town and northeast of Lake Tana, northwestern Ethiopia (Fig. 9.1). The study areas are found in Libokemkem district of SGAZ, within the Amhara National Regional State (ANRS). Geographically, Libokemkem district lies from 11°58′1.5″–12°22′6.7″ N and 37°33′25.4″–37°58′16.5″ E. The altitude of the district ranges from 1800 to 2850 m above sea level (m asl). The district shares borders with West Belessa district in the north, Ebinat district in the east, Fogera district in the south, and Lake Tana in the west. The capital of the district is Addis Zemen, which is located about 645 km northwest of the national capital Addis Ababa and 82 km northeast of the regional capital Bahir Dar on the Addis Ababa–Gondar highway. It is located at 12°06′59″–12°07′25″N and 37°46′14″–37°47′02″ E, and has an altitude of 1964 m asl. Tara Gedam and Abebaye forests are located very close to Addis Zemen, while Fach forest is located very close to the small town of Ambo Meda, which is located about 18 km east of Addis Zemen. Tara Gedam and Fach terrains consist of chains of rugged mountains and hills, steep and gentle slopes, valleys, and flat lands whereas Abebaye is a gentle hill. The altitude of Tara Gedam ranges from 2142 to 2484 m asl, Abebaye from 1921 to 2072 m asl, and Fach from 2025 to 2390 m asl.

Libokemkem district is characterized by moderate climate, locally known as *woina dega*. It has a unimodal rainfall distribution in which the main rainy season is from. June to August. The main dry season extends from December to March.

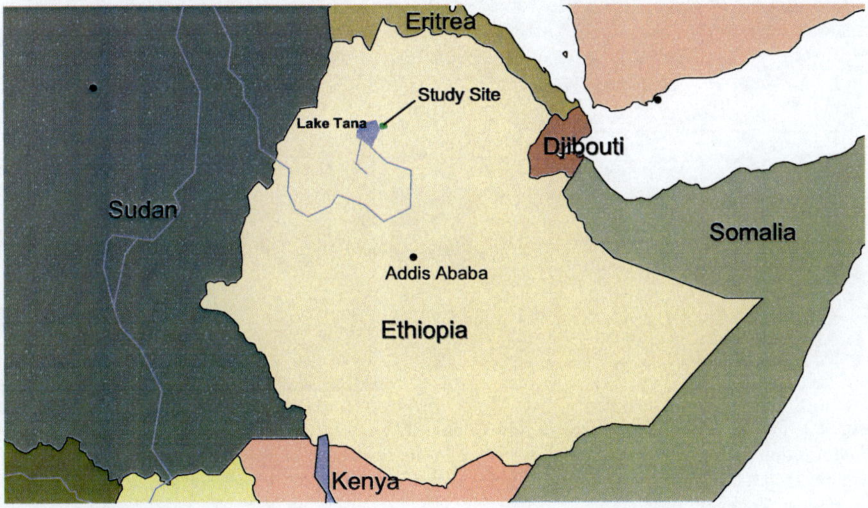

Fig. 9.1 Map of Ethiopia showing location of the study areas

The mean annual temperature of the district ranges from 18 to 25 °C, and the mean annual maximum and minimum temperatures are 27.9 °C and 11.1 °C, respectively. The mean annual rainfall ranges from 900 to 1200 mm.

Tara Gedam, Abebaye, and Fach forests are the main remnant forests in the district. The vegetation of Tara Gedam and Fach consists of a mosaic of forests, woodlands, bushlands, shrublands, and mixed/enrichment plantations (scattered plantation stands of exotic species among the stands of natural vegetation), while the vegetation of Abebaye consists of bushlands and mixed plantations. The forests contain a number of indigenous trees and shrubs interspersed with climbers and herbs. Partial views of the forests are shown in Fig. 9.2.

The study areas are generally poor in wildlife due to pronounced deforestation and other human interferences. The wild animals currently inhabiting the forests include mammals (apes, monkeys, bushbuck, klipspringer, bush duiker, wild pig, common fox, leopard, hyena, serval, Genet cat, black leopard, caracal, *mitmit*, *honey badger*, rock hyrax, rabbit, porcupine, rodents), a variety of birds (e.g., francolin, guinea fowl), and certain reptiles (e.g., Nile monitor, snake, phyton).

Fig. 9.2 Partial views of the forests: Tara Gedam (*Olea europaea* subsp. *cuspidata* and *Scolopia theifolia* dominated) (left top); Abebaye (*Acacia lahai* and *Acacia pilispina* dominated; note the firebreak) (left bottom); Fach (shrub and small-sized tree dominated) (right top and bottom) (photos by Haileab Zegeye)

Libokemkem district comprises a total of 32 *kebeles* (the smallest administrative units in Ethiopia) and a land area of 1082 km^2 (CSA 2007a). According to the 2007 Population and Housing Census of Ethiopia, the human population of the district was 198435, of whom 100987 were males and 97448 females (CSA 2007b). The economy of the local people is predominantly based on subsistence agriculture. The local people are engaged in crop cultivation and livestock rearing—thus, a mixed farming system. The people are involved in the collection of forest products, particularly fuelwood, construction material, timber, and charcoal for domestic consumption and/or income generation through selling. Traditional beekeeping is a minor source of income used by few farmers.

The churches and monasteries found in the study areas are: Tara Gedam (Tara Gedam Monastery with Mariam Church and Wostatewos Church, Abune Endrias Church, Abune Tekle Haymanot Church, Kidane Mihret Monastery, Yequala Yohannes Pre-Monastery, Addis Zemen Michael Church); Abebaye (Hanna Church, Kidane Mihret Church); Fach (Estifanos Church, Ambo Meda Michael Church, Shammo Medhane Alem Monastery Church). Mariam Church, Hanna Church, Ambo Meda Michael Church, and Shammo Medhane Alem Monastery Church are shown in Fig. 9.3.

Fig. 9.3 Mariam Church (at Tara Gedam Monastery) (left top); Hanna Church (Abebaye) (left bottom); Ambo Meda Michael Church (Fach) (right top); Shammo Medhane Alem Monastery Church (Fach) (right bottom) (photos by Haileab Zegeye)

9.2.2 Methods

9.2.2.1 Vegetation Sampling and Data Collection

Systematic sampling, in which plots are laid at regular intervals along line transects, was used to collect vegetation data following Sutherland (1996). Line transects were laid across the forests. Plots, measuring 20 m × 20 m (400 m^2) each, were established along the line transects approximately at 100 m intervals. A total of 64 plots (21 from Tara Gedam, 9 from Abebaye, 34 from Fach) were sampled from the forests. Herbaceous species were sampled using a 2 m × 2 m (4 m^2) subplot laid within each main plot where the herbaceous vegetation is assumed to be representative. In each plot, all vascular plant species were listed by their local and/or scientific names. The percentage cover-abundance of each species, defined here as the proportion of area in a plot covered by each species, was visually estimated and rated using the 1–9 modified Braun-Blanquet scale (van der Maarel 1979). Trees, shrubs, and woody climbers were counted. Diameter at Breast Height (DBH) of trees and shrubs with DBH ≥ 2 cm was measured with a diameter tape. Individuals of trees and shrubs with DBH < 2 cm were counted by species as seedlings. Altitude and geographic position were measured with a Magellan GPS 315 and a percentage slope with a Suunto clinometer. Plant species that occur outside the sample plots were also recorded to produce a more complete list of the plants in the study areas.

Plant specimens were collected, pressed, dried, and taken to the Department of Biology, Debre Tabor University (DTU). Identification of most plant species was made in the field; unidentified plant specimens were taken to the National Herbarium (ETH), Addis Ababa University for identification. Plant nomenclature in this paper follows the Flora of Ethiopia and Eritrea.

9.2.2.2 Socio-Economic Survey

A general survey consisting of field observations, key-informant interviews, and Focus Group Discussions (FGDs) was carried out following Martin (1995). Field observations were made on the biophysical features of the study areas, land use types, extent of deforestation and soil erosion, indigenous resource management systems such as agroforestry and soil conservation, and human impacts on the forests. Twenty five key informants (10 from Tara Gedam, 5 from Abebaye, 10 from Fach), which have been suggested by the forest guards, were selected and interviewed using the open-ended questionnaire prepared in advance for this purpose. For undertaking the FGDs, a group consisting of the representatives of communities around the forests, elderly, youth, clergymen, representatives of governmental institutions (Libokemkem Woreda Administrative Office, Libokemkem Woreda Agricultural Development Office), forest guards, and coordinators of seedling nurseries was established. At the beginning of the discussion, the aim of the survey was explained to the participants by the investigator so as to ensure openness in the discussion. The actual FGD was conducted using relevant discussion points or questions, mainly the socio-economic importance of the forests, the forest management systems, constraints/challenges in the management of the forests, and the human activities affecting the forests. The ideas and suggestions forwarded by the group participants were recorded.

9.2.3 *Data Analysis*

The diversity and evenness of woody species was determined using the Shannon–Wiener Diversity Index (H') and Evenness or Equitability Index (E) (Krebs 1989). The similarity in species composition between the forests was computed using Jaccard's Similarity Coefficient (Sj) (Krebs 1989). The density, percentage frequency, and dominance/basal area of woody species were calculated. The Importance Value Index (IVI) for each woody species was computed following Lamprecht (1989). The DBH data of trees and shrubs were categorized into 9 classes, and the DBH class distributions were presented using histograms. Socio-economic data (frequencies) were summarized using the 2007 Microsoft Office Excel software and were expressed in descriptive statistics (percentages).

9.3 Results and Discussion

9.3.1 *Floristic Composition*

A total of 263 vascular plant species belonging to 198 genera and 79 families were identified from Tara Gedam, Abebaye, and Fach forests (Table 9.1). From the total plant families, ferns were represented by 1 family (1.27%), gymnosperms by 1 family (1.27%), and angiosperms by 77 families (97.47%) with dicots by 69 families (87.34%) and monocots by 8 families (10.13%). The family with the highest number of species was Fabaceae (29 species, 11.03% of all species), followed by Asteraceae

Table 9.1 Family, genera, species and endemic richness of Tara Gedam, Abebaye, Fach and all forests

Forest	Richness			
	Family	Genera	Species	Endemic
Tara Gedam	62	112	136	8
Abebaye	53	104	114	3
Fach	76	183	230	17
All	79	198	263	17

(24 species, 9.13%), Poaceae (13 species, 4.94%), and Euphorbiaceae and Lamiaceae (10 species each). From the total species, 53 (20.15%) were trees, 68 (25.86%) trees/shrubs, 47 (17.87%) shrubs, 14 (5.32%) woody climbers, 11 (4.18%) herbaceous climbers and 70 (26.62%) herbs. Seventeen species (6.46%) were endemic to Ethiopia. Twenty eight trees, shrubs and treelike herbs (10.65%) were cultivated plants (most are exotic and some indigenous).

9.3.2 Similarity in Species Composition Between the Forests

The similarity values in species composition between the forests ranged from 0.34 to 0.52 (Table 9.2). The highest similarity (Sj = 0.52) in species composition was

Table 9.2 Similarity values in species composition among the three forests

Forest	Tara Gedam	Abebaye	Fach
Tara Gedam	1	0.40	0.52
Abebaye		1	0.34
Fach			1

between Tara Gedam (TG) and Fach (FC) forests, and this may be attributed to their altitudinal similarity (TG from 2142 to 2484 m asl, FC from 2025 to 2390 m asl). The lowest similarity (Sj = 0.34) in species composition was between Abebaye (AB) and Fach forests, and this may be attributed to their altitudinal differences (AB from 1921 to 2072 m asl). One similarity value was just above 0.5 (minimum is 0.0 and maximum 1.0) and the remaining two similarity values were below 0.5, indicating that there is low similarity among the forests and each forest has its own characteristic species. Thus, all of the forests are important as there are species restricted to each forest, and conserving the forest ecosystems where such species occur is of paramount importance.

9.3.3 Diversity and Evenness of Woody Species

Fach forest had the highest diversity and evenness (3.53 and 0.72, respectively), followed by Tara Gedam forest (2.98 and 0.65, respectively) and Abebaye forest (1.31 and 0.31, respectively) (Table 9.3). [For the data on woody species in Tara Gedam and Abebaye forests presented in this paper, you can also see Zegeye et al. 2011]. The high diversity (Fach and Tara Gedam forests) is attributed to habitat diversity and low human disturbances (partly slopy terrain limits human exploitation and livestock grazing/browsing), whereas the low diversity (Abebaye forest) is attributed to low habitat diversity and high human disturbances (partly there are less topographic barriers to human exploitation and livestock grazing/browsing). The evenness values

Table 9.3 Diversity (H') and evenness (E) of woody species in the three forests

Forest	H'	E
Tara Gedam	2.98	0.65
Abebaye	1.31	0.31
Fach	3.53	0.72

0.72 (Fach forest) and 0.65 (Tara Gedam forest) showed that there is a more or less balanced distribution of individuals among the different species. The high evenness may be attributed to species characteristics, suitable site conditions, and low human disturbances. On the other hand, the evenness value 0.31 (Abebaye forest) indicated that there is unbalanced representation of individuals of different species. The low evenness can be attributed to species characteristics for adaptation, poor conditions for regeneration, and high human disturbances (Wassie et al. 2005; Zegeye et al. 2011). The diversity and evenness values imply the need to conserve the forests from both floristic diversity and human disturbance perspectives.

9.3.4 Importance Value Index (IVI) of Woody Species

In Tara Gedam forest, the total density (inclusive of seedlings) of woody species was 3001 individuals ha^{-1}. The species with the highest density was *Albizia schimperiana* (252 individuals ha^{-1}), followed by *Calpurnia aurea* (245), *Olea europaea* subsp. *cuspidata* (204) and *Dodonaea angustifolia* (155). The species with the highest frequency was *A. schimperiana* (81%), followed by *Carissa spinarum* (71%), *C. aurea* and *O. europaea* subsp. *cuspidata* (62% each), and *Schrebera alata* and *Vernonia myriantha* (57% each). In Abebaye forest, the total density of woody species was 2850 individuals ha^{-1}. The species with the highest density was *Acacia lahai* (486 individuals ha^{-1}), followed by *Acacia pilispina* (336), *Acanthus polystachius* (261) and *Maytenus serrata* (253). The species with the highest frequency were *A. pilispina, Capparis tomentosa* and *Croton macrostachyus* (100% each), followed by *Acanthus polystachius* and *M. serrata* (89% each), and *A. lahai, Pterolobium stellatum, C. spinarum* and *Jasminum grandiflorum* subsp. *floribundum* (78% each). In Fach forest, the total density of woody species was 4938 individuals ha^{-1}. The species with the highest density was *D. angustifolia* (605 individuals ha^{-1}), followed by *M. serrata* (482), *C. aurea* (481) and *Euclea racemosa* subsp. *schimperi* (259). The species with the highest frequency was *C. aurea* (88%), followed by *D. angustifolia* and *P. stellatum* (82% each), and *C. spinarum* and *Rhus vulgaris* (79% each).

The total density of woody species in Fach forest was higher than that of both Tara Gedam and Abebaye forests but the total basal area of woody species was lower than that of both Tara Gedam and Abebaye forests. This is because most species having high densities were shrubs and small-sized trees, and big-sized trees were absent or represented by very few numbers of individuals. The inadequate number of big-sized trees indicates that Fach forest is still at the stage of secondary development.

Density and frequency of the woody species varied considerably among the species. The variation in density and frequency between species may be attributed to habitat differences, habitat preferences among the species, conditions for regeneration, and degree of human exploitation (Tesfaye and Teketay 2005; Zegeye et al. 2011).

In Tara Gedam forest, the total basal area of woody species was 115.36 m^2ha^{-1}. The species with the highest basal area was *O. europaea* subsp. *cuspidata* (47.06 m^2ha^{-1}), followed by *Schefflera abyssinica* (11.83), *A. schimperiana* (10.58) and *Ekebergia capensis* (9.36). In Abebaye forest, the total basal area of woody species was 49.45 m^2ha^{-1}. The species with the highest basal area was *Ficus vasta* (16.77 m^2ha^{-1}), followed by *A. lahai* (8.01), *C. macrostachyus* (4.40) and *A. pilispina* (4.3). In Fach forest, the total basal area of woody species was 19.17 m^2ha^{-1}. The species with the highest basal area was *Combretum molle* (3.62 m^2ha^{-1}), followed by *O. europaea* subsp. *cuspidata* (3.12), *F. vasta* (2.11) and *A. schimperiana* (0.85).

In Tara Gedam forest, the species with the highest IVI was *O. europaea* subsp. *cuspidata* (51.05%), followed by *A. schimperiana* (22.13%), *E. capensis* (13.82%) and *C. aurea* (12.58%). Species with low IVI values included *Clerodendrum myricoides* (0.35%), *Protea gaguedi* and *Ficus thonningii* (0.40% each), and *F. sycomorus* (0.41%). In Abebaye forest, the species with the highest IVI was *A. lahai* (36.88%), followed by *F. vasta* (35.14%), *A. pilispina* (25.17%) and *C. macrostachyus* (15.81%). Species with low IVI values included *Dombeya torrida* (0.72%), *Bridelia micrantha* (0.75%), *Maesa lanceolata* (0.76%) and *Euphorbia tirucalli* (0.91%). In Fach forest, the species with the highest IVI was *C. molle* (25.26%), followed by *O. europaea* subsp. *cuspidata* (21.19%), *D. angustifolia* (17.80%) and *C. aurea* (15.05%). Species with low IVI values included *Rhus retinorrhea* (0.14%), *F. sycomorus* (0.15%), and *Maytenus undata*, *Gardenia ternifolia* and *Abutilon longicuspe* (0.16% each).

The IVI is an important parameter that reveals the ecological significance and/or dominance of species in a given ecosystem (Lamprecht 1989). Species with high IVI values are considered more important than those with low IVI values. The IVI values can also be used to prioritize species for conservation: species with high IVI values need less conservation efforts whereas those with low IVI values need high conservation efforts (Shibru and Balcha 2004; Zegeye et al. 2011). The results suggest that species with low IVI values should be prioritized for conservation.

9.3.5 Regeneration Status of Woody Species

The DBH class distributions of the species exhibited different patterns (Fig. 9.4), and showed that there are species with high number of individuals in the lower classes, some species in the middle classes and others in the higher classes. The patterns of DBH class distributions indicated the general trends of population dynamics and recruitment processes of the species. From the DBH class distributions of the species, two types of regeneration status were determined, i.e., good and poor regeneration.

Some species possessed high number of individuals in the lower DBH classes, particularly in the first class (DBH < 2 cm), and this suggests that they have good regeneration potential. Most of the species, however, possessed low number of individuals in the lower DBH classes, particularly in the first class, and this suggests that the species are in poor regeneration status (demonstrated hampered natural regeneration).

In Tara Gedam forest, the species with good regeneration potential were *A. schimperiana, C. aurea, E. capensis, S. alata, Rhus glutinosa* and *Juniperus procera.* The species with poor regeneration status were *O. europaea* subsp. *cuspidata,*

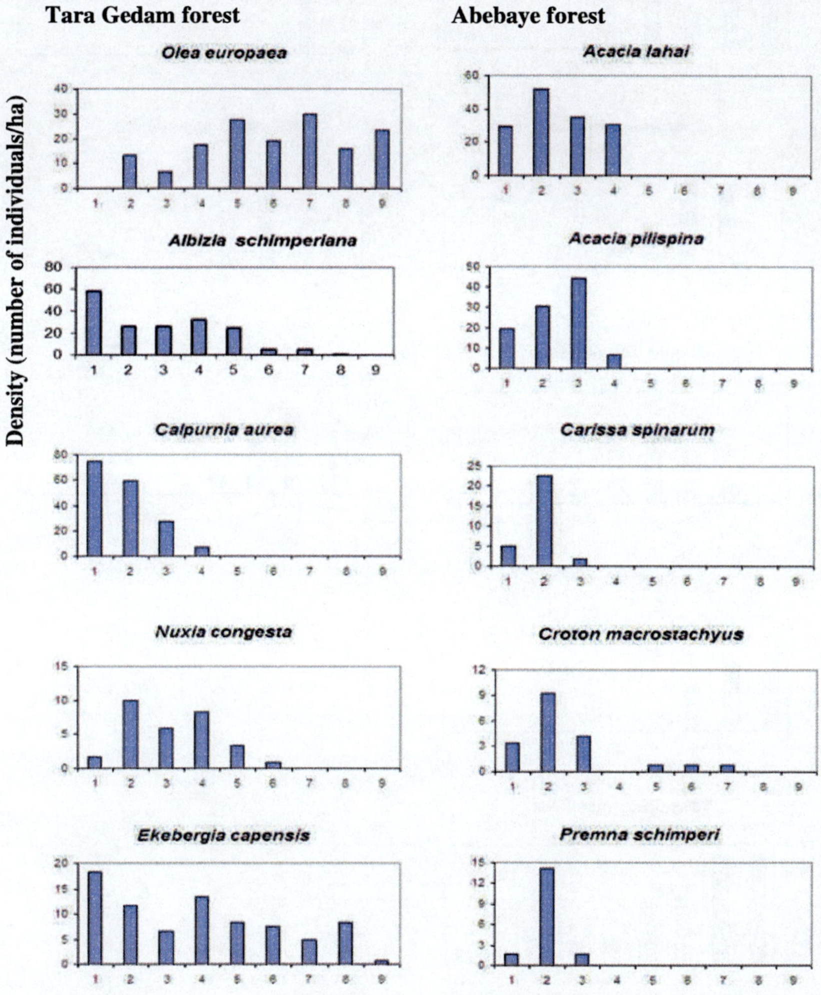

Fig. 9.4 DBH class distributions of some selected trees and shrubs and all woody species in Tara Gedam, Abebaye, and Fach forests (DBH classes: 1 = < 2 cm; 2 = 2–10 cm; 3 = 10–20 cm; 4 = 20–30 cm; 5 = 30–40 cm; 6 = 40–50 cm; 7 = 50–60 cm; 8 = 60–70 cm; 9 = >70 cm)

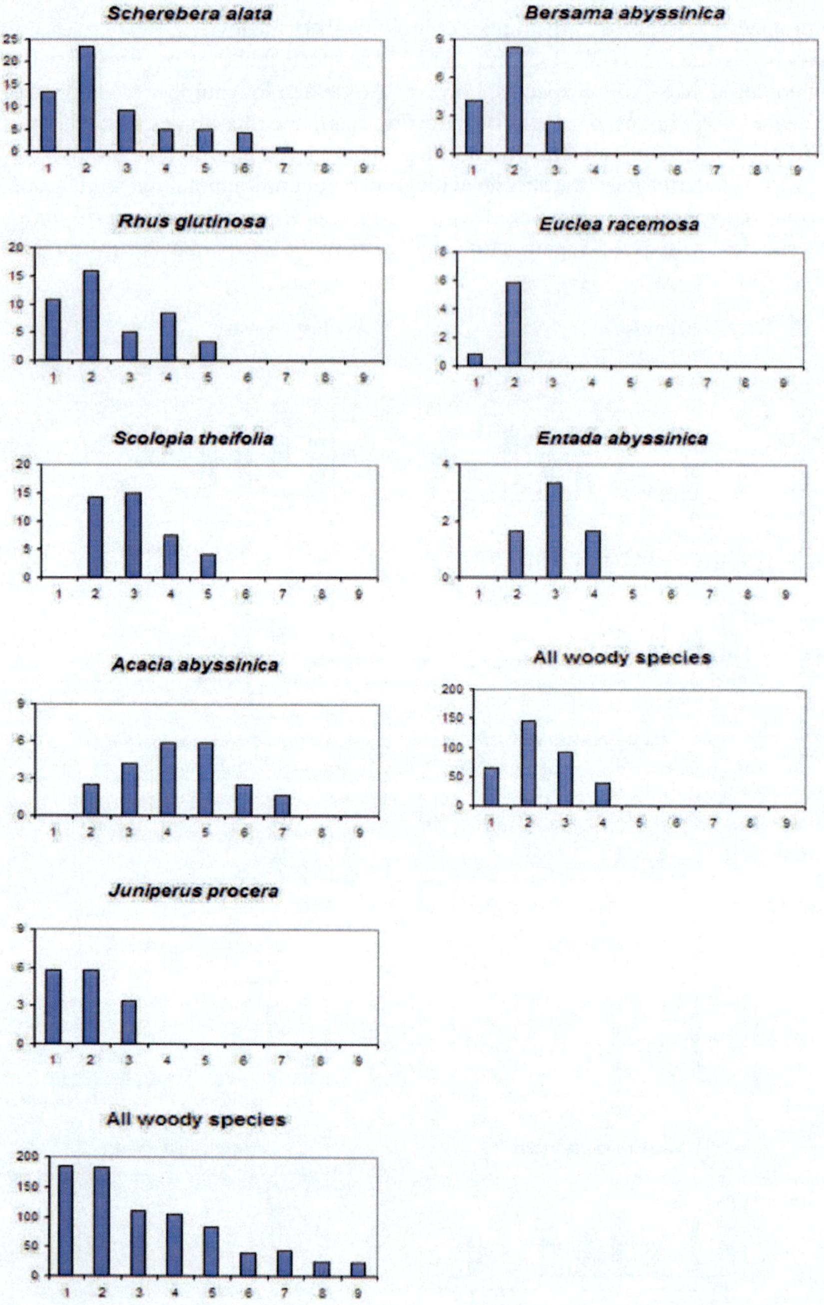

Fig. 9.4 (continued)

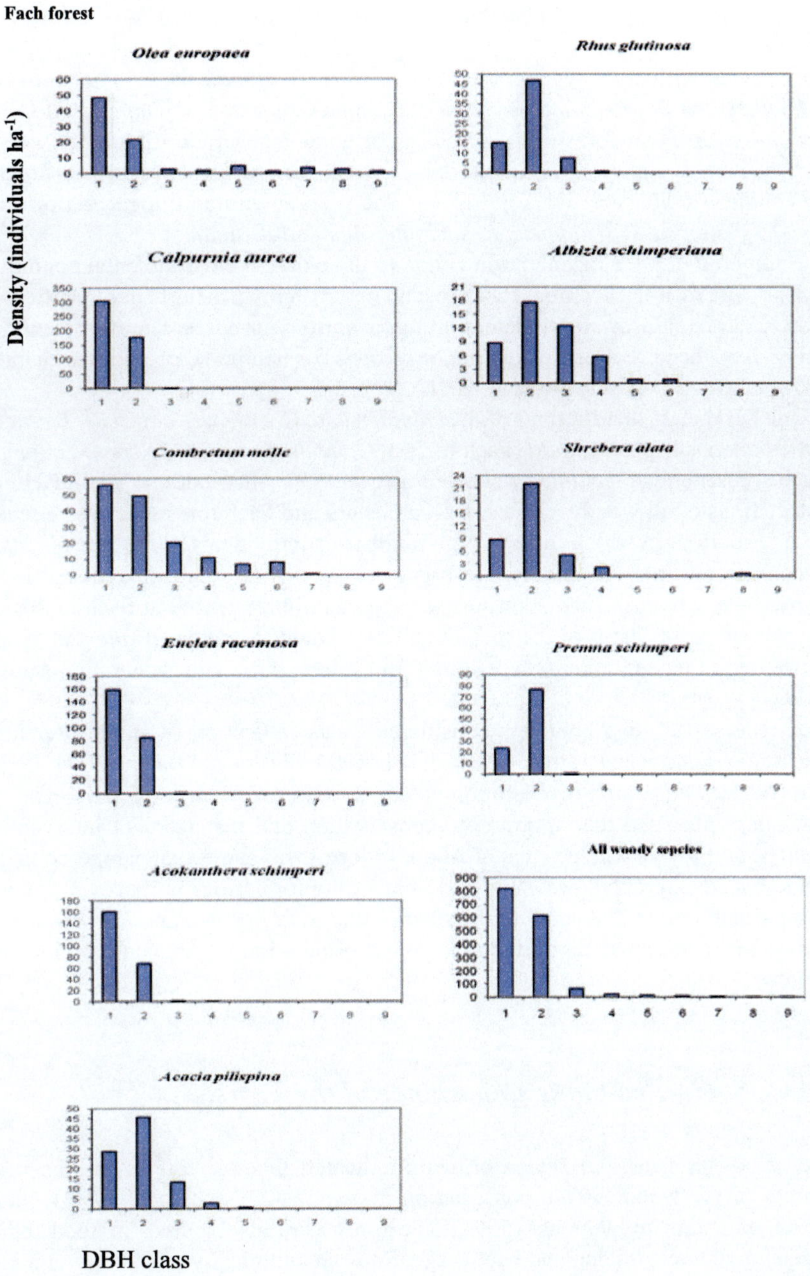

Fig. 9.4 (continued)

Nuxia congesta, Acacia abyssinica and *Scolopia theifolia.* In Abebaye forest, the species with a relatively good regeneration potential were *A. lahai* and *A. pilispina.* The species with poor regeneration status were *C. spinarum, C. macrostachyus, Premna schimperi, Bersama abyssinica, E. racemosa* subsp. *schimperi* and *Entada abyssinica.* In Fach forest, the species with good regeneration potential were *O. europaea* subsp. *cuspidata, C. molle, C. aurea, E. racemosa* subsp. *schimperi, Acokanthera schimperi* and *A. pilispina.* The species with poor regeneration status were *R. glutinosa, P. schimperi, A. schimperiana* and *S. alata.*

Hampered or poor regeneration is due to unfavorable environmental conditions, such as soil erosion and climate change, and mainly human disturbances, particularly livestock grazing/browsing and tree cutting for various purposes. The aforementioned factors have been reported as the major reasons for hampered or poor regeneration (Zegeye et al. 2011; Teketay et al. 2016).

The DBH class distributions of *A. schimperiana, C. aurea, E. capensis, J. procera, O. europaea* subsp. *cuspidata* (Fach forest), *C. molle, E. racemosa* subsp. *schimperi* (Fach forest) and *A. schimperi* showed a reverse "J" distribution. The DBH class distributions of all woody species in Tara Gedam and Fach forests showed a reverse "J" distribution, in which there is high number of individuals in the first class with a decrease towards the middle and higher classes. It is interesting to see that *O. europaea* subsp. *cuspidata* showed good regeneration potential at Fach, unlike the poor regeneration status of the species at Tara Gedam. *A. schimperiana* and *S. alata* showed good regeneration potential at Tara Gedam, but poor regeneration status at Fach.

A reverse "J" distribution is considered as an indication of stable population structure or good regeneration status (Bekele 1994; Shibru and Balcha 2004; Zegeye et al. 2006). The natural regeneration potential of Tara Gedam and Fach forests was promising, provided that appropriate conservation and management interventions could be employed, whereas that of Abebaye forest was somewhat hampered mainly due to livestock grazing/browsing and tree cutting for various purposes. But here it is important to note that some of the species like *A. abyssinica* and *P. schimperi* are in poor regeneration status and thus require due attention from a conservation point of view.

9.3.6 Socio-Economic Importance of the Forests

The responses from the key informants indicated that the forests are the major sources of fuelwood (90%), construction material (80%), charcoal (45%), timber (75%), and farm implements (55%). The forests are also sources of food (edible fruits), medicines, fodder, and bee forage. Some economically most important plant species in the forests and their uses are shown in Table 9.4. From the total species identified from Tara Gedam, Abebaye, and Fach forests, 147 (55.89%) had already been identified by Fichtl and Adi (1994) for their great values as bee forage. Forest management has created employment opportunities for the forest guards and workers

Table 9.4 Some economically most important woody sepeies in the forests and their uses (indicated by "x")

No.	Scientific name	Uses									
		FW	CM	CC	T	FI	F	MH	ML	FD	BF
1	*Acacia abyssinica*	x	x	x		x				x	x
2	*Acacia lahai*	x	x	x		x				x	x
3	*Acacia pilispina*	x	x	x		x				x	x
4	*Albizia schimperiana*	x	x		x	x					
5	*Apodytes dimidiata*		x		x						x
6	*Brucea antidysenterica*							x	x		
7	*Calpurnia aurea*	x	x			x				x	
8	*Carissa spinarum*	x					x			x	x
9	*Combretum molle*	x	x	x		x				x	x
10	*Cordia africana*		x	x	x		x	x		x	x
11	*Croton macrostachyus*		x	x		x			x		x
12	*Dichrostachys cinerea*	x		x				x		x	
13	*Dodonaea angustifolia*					x				x	x
14	*Dombeya torrida*	x								x	x
15	*Ficus sur*					x	x			x	
16	*Ficus vasta*					x	x			x	
17	*Gymnema sylvestre*							x	x		
18	*Justicia schimperiana*		x					x	x		
19	*Olea europaea* subsp. *cuspidata*		x	x		x				x	
20	*Rhus glutinosa*	x	x	x		x				x	
21	*Rhus vulgaris*	x	x	x			x			x	
22	*Schrebera alata*		x			x				x	
23	*Syzygium guineense*		x	x	x		x				x
24	*Vernonia amygdalina*							x	x		x
25	*Ximenia americana*						x	x			

Key FW-fuelwood, CM-construction material, CC-charcoal, T-timber, FI-farm implements, F-food, MH-medicine for humans, ML-medicine for livestock, FD-fodder, BF-bee forage

of seedling nurseries. Moreover, the forests have potential values for apiculture and tourism/ecotourism.

9.3.7 *Conservation and Management of the Forests*

The conservation and management of Tara Gedam, Abebaye, and Fach forests is a tripartite venture: the conservation efforts of local communities, religious institutions (churches and monasteries), and governmental and non-governmental institutions. Moreover, indigenous (sacred grove) and modern (protected area system) conservation methods have been integrated to conserve and manage the forests: a vital synergy. The integration of the conservation methods and the integration of the relevant stakeholders are crucial for conserving biodiversity.

9.3.7.1 Role of Local Communities

The socio-economic survey showed that the local communities have a good traditional agroforestry system, which provides various forest products and thereby reduces the pressure on the natural forests. The maintenance of the sacred groves is attributed to the strong religious belief and respect of the followers to the church, which is considered the house of God. Cutting trees from the sacred groves is taboo. If a person cuts trees from the sacred sites, the followers of the church inform the case to the religious fathers, and the doer is condemned. The followers actively participate in the religious, conservation (e.g., tree planting in church/monastery yards) and development activities of the churches and monasteries. The majority of the Ethjopian people have respect and trust for the Ethiopian Orthodox Tewahedo Church (EOTC) and it is this spirit that supported the church to maintain forest resources till this generation (Wassie 2002). The local communities were involved in the conservation activities of the government like tree plantation programs and construction of terraces. The local communities were also instrumental in controlling forest fire incidences that have occurred in the forests at different times.

9.3.7.2 Role of Religious Institutions (Churches and Monasteries)

Although the main role of the churches and monasteries is to give religious service to the followers, they are involved in protecting the sacred groves, planting trees in church/monastery yards, and giving advice to the followers about the importance of conserving the sacred groves. The EOTC has a long history of conserving sacred groves, i.e., patches of natural vegetation conserved on sacred sites. The EOTC has over 30 million followers, 400,000 clergies, and 35,000 churches in Ethiopia (Wassie 2002). Churches and monasteries have played a great role in the conservation of sacred groves in particular and forest resources of the country in general (Wassie et al. 2005; Zegeye et al. 2006).

9.3.7.3 Role of Governmental and Non-Governmental Institutions

As Tara Gedam, Abebaye and Fach forests are protected areas, they are managed by governmental institutions, particularly Libokemkem District Agricultural Development Office. Non-Governmental Organizations (NGOs) support local communities and governmental institutions in conservation and development activities. Governmental institutions enforce legal protection of the forests. As such, a total of 32 forest guards (23 for Tara Gedam and Abebaye forests, 9 for Fach forest) have been employed on a contractual basis to protect the forests from human and livestock interferences. The forest guards are trying their best to control tree felling and livestock grazing/browsing and enforce the existing forest law. Indeed, the existence of the forests is largely attributed to the remarkable efforts of the forest guards. Though it is very limited, enrichment tree planting has been carried out in the forests. Firebreaks or fire control gaps have been established in Abebaye and Fach forests by removing woody vegetation so as to control the spread of fire incidence and thus reduce its impacts. There is a model seedling nursery at Addis Zemen and a minor one near Ambo Meda. Moreover, the German Technical Cooperation (GTZ), now the German Agency for International Cooperation (GIZ), has a model seedling nursery at Addis Zemen. These nurseries produce seedlings of exotic and indigenous trees and distribute them to users (model farmers, schools, youth associations, churches and monasteries, institutions). Governmental and non-governmental institutions are promoting tree planting (reforestation, afforestation, agroforestry) in the study areas, and this reduces the pressure on the natural forests. In fact, rigorous tree planting is needed in the study areas and beyond.

9.3.8 Threats to the Forests

Tara Gedam, Abebaye, and Fach forests are protected areas and contain sacred places, but they are dwindling due to livestock grazing/browsing, tree cutting for various purposes, farmland expansion, human settlements, urbanization, fire incidences, exotic species plantations, and soil erosion. The forests will diminish in the near future unless appropriate and immediate measures are taken. The loss of the forests will lead to loss of biodiversity, particularly the endemic plant species. This calls for strengthening the conservation and management of the forests. A proper forest management plan should be developed and implemented to reverse or at least stabilize the present trend in the forests.

9.4 Conclusions and Recommendations

Tara Gedam, Abebaye and Fach forests possess high plant diversity and endemism. Woody species having low IVI values, such as *R. retinorrhea* and *F. sycomorus*, and those exhibiting poor regeneration status (as indicated by their DBH class distributions), such as *A. abyssinica* and *P. schimperi*, need high priority for conservation. The local communities are highly dependent on the forests for fuelwood, construction material, charcoal, timber, and farm implements, as well as food, medicines, fodder, and bee forage. Moreover, the forests have potential values for apiculture and tourism/ecotourism. The forests have been maintained to the present-day through the conservation efforts of local communities, religious institutions (churches and monasteries), and governmental institutions. At present, however, the forests are dwindling due to livestock grazing/browsing, tree cutting for various purposes, farmland expansion, human settlements, urbanization, and fire incidences. Effective conservation and management interventions are urgently needed to ensure the long-term maintenance of the forest. ecosystems, and benefit the local communities through sustainable utilization of the forests.

Therefore, in order to ensure the long-term maintenance of the forests, the following recommendations are forwarded:

- Employ in situ and ex situ conservation methods for the conservation of woody species having low IVI values and poor regeneration status;
- Develop appropriate forest management plan to enhance the conservation, development and sustainable utilization of the forests;
- Promote tree planting (reforestation, afforestation, agroforestry) in the study areas with emphasis on multipurpose indigenous and suitable exotic tree species to reduce the pressure on the natural forests;
- Allow sustainable utilization of the forests by the local communities (for example, collection of dried/dead trees, grasses—through cut-and-carry system, medicinal herbs or parts of trees and shrubs for medicines, and tree seeds; beekeeping in the forests) so as to develop a sense of ownership in the local people;
- Establish fire prevention and control system with the necessary facilities;
- Promote apiculture and tourism/ecotourism in the study areas;
- Provide the local communities with alternative sources of energy (hydroelectric, solar, biogas) and energy-saving stoves to reduce the dependency on the forests for fuelwood and charcoal;
- Promote agricultural and forestry extension services;
- Provide technical and financial support for the churches and monasteries to enhance their role in biodiversity conservation;
- Carry out further research on the forests, particularly ethnobotany, reproductive biology of the major woody species, soil seed banks, forest pathology, and carbon stock potentials.

Acknowledgments I would like to gratefully acknowledge the University of Gondar (UoG) for providing the grant needed for the research work on Tara Gedam and Abebaye forests and Debre Tabor University (DTU) for providing the grant needed for the research work on Fach forest. My heartfelt thanks go to the field assistants, especially Mr. Wasihun Belay and Mr. Wudu Sharew, for their help during the data collection. I also would like to thank Mr. Melaku Wondafrash and Mr. Wege Abebe, staff of the National Herbarium (ETH), Addis Ababa University for the identification of some plant specimens, and Dr. Mulugeta Kebede, staff of Department of Biology, Hawassa University for the production of map of the study areas.

References

Anonymous (2004) Summary report on the Tara Gedam forest demarcation

Bekele T (1994) Studies on remnant Afromontane forests on the central plateau of Shewa, Ethiopia. Ph.D. dissertation, Uppsala University. Comprehensive summaries of Uppsala dissertations from the Faculty of Science and Technology 23. Acta Universitatis Upsaliensis, Uppsala

CSA (2007a) Population and housing census atlas of Ethiopia. Central Statistical Agency (CSA), Addis Ababa

CSA (2007b) Population and housing census of Ethiopia. Statistical report: results at Country Level. Central Statistical Agency (CSA), Addis Ababa

Edwards S, Kelbessa E (1999) Indicators to determine the level of threat to tree species. In: Edwards S, Demissie A, Bekele T, Haase G (eds) Proceedings of the national forest genetic resources conservation strategy development workshop. IBCR and GTZ, Addis Ababa, pp 101–133

EFAP (1994) Ethiopian forestry action program (EFAP). Ministry of Natural Resources Conservation and Environmental Protection (MoNRCEP), Addis Ababa

Fichtl R, Adi A (1994) Honeybee flora of Ethiopia. Margraf Verlag, Weikersheim

Kelbessa E, Demissew S (2014) Diversity of vascular plant taxa of the flora of Ethiopia and Eritrea. Ethiop J Biol Sci 13(Suppl.):37–45

Krebs CJ (1989) Ecological methodology. Harper Collins Publishers, New York

Lamprecht H (1989) Silviculture in the tropics: tropical forest ecosystems and their tree species—possibilities and methods for their long-term utilization. Federal Republic of Germany, Eschborn

Martin GJ (1995) Ethnobotany: a methods manual. Chapman and Hall, London

Shibru S, Balcha G (2004) Composition, structure and regeneration status of woody species in Dindin natural forest, Southeast Ethiopia: an implication for conservation. Ethiop J Biol Sci 3(1):15–35

Sutherland WJ (ed) (1996) Ecological census techniques: a handbook. Cambridge University Press, Cambridge

Teketay D, Geeves G, Kopong I, Mojeremane W, Sethebe B, Smith S (2016) Diversity, stand structure and regeneration status of woody species, and spatial cover of herbaceous species in Mokolodi Nature Reserve, southeastern Botswana. Int J Biodivers Conserv 8(8):180–195

Tesfaye G, Teketay D (2005) The influence of logging on natural regeneration of woody species in Harenna montane forest. Ethiopia. Ethiop J Biol Sci 4(1):59–73

van der Maarel E (1979) Transformation of cover-abundance values in phytosociology and its effects on community similarity. Vegetatio 39:97–114

Wassie A (2002) Opportunities, constraints and prospects of the Ethiopian Orthodox Tewahido Churches in conserving forest resources: the case of churches in South Gondar, northern Ethiopia. M.Sc. thesis, Swedish University of Agricultural Sciences

Wassie A, Teketay D, Powell N (2005) Church forests in North Gonder Administrative Zone, northern Ethiopia. Forests Trees Livelihoods 15:349–373

Zegeye H (2017) Major drivers and consequences of deforestation in Ethiopia: implications for forest conservation. Asian J Sci Tech 8(8):5166–5175

Zegeye H, Teketay D, Kelbessa E (2006) Diversity, regeneration status and socio-economic importance of the vegetation in the islands of Lake Ziway, South-Central Ethiopia. Flora 201:483–498

Zegeye H, Teketay D, Kelbessa E (2011) Diversity and regeneration status of woody species in Tara Gedam and Abebaye forests, northwestern Ethiopia. J Forest Res 22(3):315–328

Chapter 10
Woody Vegetation Composition and Structure of Church Forests in Southeast of Lake Tana, Northwest Ethiopia

Ferehiwot Mequanint, Alemayehu Wassie, Shimeles Aynalem, Enyew Adgo, Jan Nyssen, Amaury Frankl, Diederik Strubbe, Bart Muys, and Luc Lens

Abstract Understanding woody plant species composition and structure is fundamental to design and optimize the needed conservation measures for Ethiopian church forests. The aim of this study was to describe the composition, structure, and regeneration status of woody species in church forests in southeast of Lake Tana, Ethiopia. Data were collected from twenty-four church forests. Four plots (20 m × 20 m) were established in each church forest. Plots were located in four cardinal directions (north, east, west, and south) at different distances from the forest center. Four subplots (5 m × 5 m) were established in each plot to assess seedlings and canopy cover. In each plot, all woody plants were identified and counted, and diameter at breast height (DBH) was measured. Species and family importance values were computed to characterize the species composition. Additionally, population structural

F. Mequanint (✉) · S. Aynalem
Department of Wildlife and Ecotourism Management, Bahir Dar University, Bahir Dar, Ethiopia
e-mail: ferehiwot.admassie@ugent.be

F. Mequanint · D. Strubbe · L. Lens
Terrestrial Ecology Unit, Ghent University, Ghent, Belgium
e-mail: diederik.strubbe@ugent.be

L. Lens
e-mail: luc.lens@ugent.be

A. Wassie
Organization for Rehabilitation and Development in Amhara, Bahir Dar, Ethiopia

E. Adgo
Department of Natural Resource Management, Bahir Dar University, Bahir Dar, Ethiopia

J. Nyssen · A. Frankl
Department of Geography, Ghent University, Ghent, Belgium
e-mail: jan.nyssen@ugent.be

A. Frankl
e-mail: amaury.frankl@ugent.be

B. Muys
Division Forest, Nature and Landscape, KU Leuven, Leuven, Belgium
e-mail: bart.muys@kuleuven.be

M. Kindu et al. (eds.), *State of the Art in Ethiopian Church Forests and Restoration Options*, https://doi.org/10.1007/978-3-030-86626-6_10

features were analyzed through the variation of tree size classes. Species richness (SR), Pilou evenness (J'), and Shannon–Wiener index (H') were used to determine species diversity. A total of 115 woody species representing 53 families and 97 genera were found. Of these, 62% were trees, 36% shrubs, 1.89% climber, and 0.06% reed species. Species richness differed among forests, ranging between 16 and 38 species. Fabaceae, Sapotaceae, and Rubiaceae were the dominant families with a high family importance values of 41, 28, and 22, respectively. The church forests have relatively high indices of species diversity (SR $= 26 \pm 1.25$), ($J' = 0.75 \pm 0.02$), and ($H' = 2.42 \pm 0.07$), indicating that they play a major role in the conservation of woody species. However, a relatively high densities of *Eucalyptus* spp. ranging from 13 to 1925 individuals ha^{-1} were recorded, and these exotic tree species, thus, form a potential threat to the conservation of native species. The diameter class distribution of some selected keystone and dominant species formed four main shape types, of which the irregular-shaped pattern was most predominant, which suggests missing cohorts and regeneration problems for most species. Higher densities of *Eucalyptus* plantations were recorded in more recently established than old church forests. Therefore, effective measures should be taken to address the major pressures, such as plantation of exotic species that negatively affect the species composition and vegetation structure of these church forests, which, in turn, affect their ecosystem functions and services.

Keywords Biodiversity · Conservation · Fragmentation · Native species · Exotic species · Sacred grove

10.1 Introduction

Forests play a significant role in providing multiple ecosystem services, such as climate change mitigation. However, due to anthropogenic and natural factors, forest cover has drastically declined worldwide (Contreras-Hermosilla 2000). Similarly, Ethiopia has suffered drastic historical deforestation. Logan (1946) reported that only 5% of the Ethiopian highlands were forested in 1946, suggesting that deforestation started a long time ago. The forest cover, particularly in the highlands of northern Ethiopia, has continued to decline (Darbyshire et al. 2003; Nyssen et al. 2004). For example, in Lake Tana Basin, during the mid-twentieth century, about 20% of the area was covered with woody vegetation, while only about 10% remained by 2014–16 (Frankl et al. 2019). The rapid human population growth is the main driving force that is responsible for the loss of forests (Bishaw 2001). Agricultural expansion, urbanization, free grazing, and unsustainable development activities are the major pressures that intensify deforestation and forest degradation (Hailu et al. 2015). Consequently, the forest coverage is fragmented to small patches, particularly in the central and northern highlands of Ethiopia. In most parts of the northern highlands, patchy remnants of natural forests are found almost only surrounding churches (Gashaw et al. 2015).

Ethiopian church forests provide essential ecosystem services and support diverse plant and animal species, including endangered indigenous tree species (Aerts et al. 2006; Bongers et al. 2006; Reynolds et al. 2017; Morgan et al. 2018). Furthermore, church forests serve as a habitat for many insects and animals, such as birds and mammals (Wassie et al. 2005a, b; Scull et al. 2017). Several studies (Nyssen et al. 2004; Aerts et al. 2006, 2016; Wassie et al. 2009a, b; Wassie et al. 2010; Aynekulu et al. 2011; Woldemedhin and Teketay 2016; Abiyot et al. 2017; Morgan et al. 2018) have investigated the species composition and tree community structure of Ethiopian church forests, their role for biodiversity conservation, and the associated conservation challenges. For example, in Amhara region, Woldemedhin and Teketay (2016) documented 56 woody species from nine church forests in West Gojjam, while Morgan et al. (2018) and Wassie et al. (2010) reported 47 from 11 church forests and 168 woody species from 28 church forests in the northern part of the region, respectively. The recorded species were mainly indigenous trees, indicating the high conservation value of Ethiopian church forests. Reynolds et al. (2015) reported that there are more than 8000 church forests across the Amhara National Regional State (NRS); however, if also small church forests are included, the current number of church forests in the NRS could be much higher than this report. Therefore, there is no doubt that church forests in the NRS are playing a significant role in woody species conservation.

Many indigenous trees and shrubs, which are exterminated in some localities, are still found in the compounds of churches (Wassie et al. 2005a). Protection of a large network of small church forests can be more effective for biodiversity conservation compared with the conservation of a few large patches of an equivalent area (Bhagwat and Rutte 2006; Dudley et al. 2009; Hokkanen et al. 2009; Aerts et al. 2016). This is due to the fact that the large network of small church forests covers a wide variety of habitat, located in different agroecological zones and managed by the respective parish. The significant role of small forest patches, such as church forests to the conservation of overall species diversity and structure, was also described by Teketay et al. (2018). However, pressures, such as free grazing, fuelwood harvesting, woody species removal to construct church buildings or expand burial sites, have been negatively affecting the species composition, community structure, and natural regeneration rate of church forests in northern Ethiopia (Reynolds et al. 2017; Wassie et al. 2009a, 2010). Furthermore, plantation of fast-growing and ornamental exotic species inside the church forests, such as *Cupressus lusitanica*, *Citrus* spp., *Eucalyptus* spp., *Grevillea robusta*, *Jacaranda mimosifolia*, *Melia azedarach*, *Pinus* spp., is strongly affecting species composition and community structure of church forests (Aerts et al. 2016; Cardelu's et al. 2019).

In general, these pressures could not only alter the ecological processes and functions of the church forests ecosystem, but also affect the ecosystem services that these church forests provide to the local community. Thus, to combat these pressures and maintain the existing ecosystem services, church forests should be supported in different ways. While stonewall construction around church forests can be a very effective management strategies to overcome the effect of free grazing (Wassie et al. 2009a; Woods et al. 2017), a proper understanding of species composition, structure, and regeneration status of church forests is required to draft long-term management plans.

In the southeast of Lake Tana located in the Amhara NRS, there is little information about the species composition and conservation status in the church forests, except for four church forests (i.e., Emashenekure Giworegis, Gebesiwit Mariyam, Wej Aregawi, and Zahara Mikael) that were assessed by Wassie et al. (2010). Furthermore, almost a decade has passed since the assessment made by Wassie et al. (2010), and, hence, the species composition and structure of the church forests have likely changed over time. To fill this important knowledge gap, we studied the composition, structure, and the regeneration status of woody species, in a network of 24 church forests located in the southeast of Lake Tana.

10.2 Materials and Methods

10.2.1 Description of the Study Area

The Lake Tana Basin has an area of 15,089 km^2 with an elevation ranging from 1782 to 4109 m above and located in the northwest highlands of Ethiopia (Lemma et al. 2018). This study was conducted in the lower elevations in southeast of Lake Tana between Bahir Dar and Debre Tabor (1794–2204 m). The main land use and land cover types are natural forests, bushland, plantation forest, urban, village, cultivated land, waterbody, wetland, woodlands, and grassland (Song et al. 2018). Cultivated land (62%) is the major land use type, followed by water bodies (20%). However, the forest cover, including plantation and woodlands, is only 5% (Song et al. 2018). This area experiences a strong seasonal rainfall regime (about 70–90% of the total rainfall occurs during June–September), but a very even monthly temperature regime (Peel et al. 2007). The annual average rainfall varies between 1250 and 1500 mm, and the mean annual temperature varies between 15.3 and 19.6 °C (Lemma et al. 2018). A total of 24 church forests located between 37° 27′ E–37° 55′ E and 11° 39′ N–11° 56′ N were selected in the southeast of Lake Tana (Table 10.1; Fig. 10.1). Church forest selection was based on accessibility, distance between the consecutive church forests (a minimum of 4 km) and variability of the surrounding matrix in the radius of 3 km. The surface area of the church forests ranged from 2 to 13 ha, and the churches in the forests were established between 340 and 2010 (EOTC, no date).

Table 10.1 List of the 24 church forests of this study with their names, abbreviations, districts, and numbers on the study area map

Church name	Abbreviation	Year of establishment	District	Number on study area map (Fig. 10.1)
Aba Gerima Mariyam	AG	1458	Bahir Dar Zuriya	2
Deber Kusekuam	DK	1008	Fogera	18
Delemo Tekelehayemanot	DT	1682	Fogera	13
Emashenekure Giworegis	ES	1660	Dera	7
Fisa Mikael	FM	1537	Dera	5
Gebesiwit Mariyam	Gma	1250	Dera	10
Hager Selam Mariyam	HSM	1363	Fogera	19
Kirekus	K	1983	Dera	12
Kulela Mesekel	KM	1422	Dera	4
Kudese Minas	Kmi	2004	Bahir Dar Zuriya	3
Meneguzer Eyesus	ME	1682	Fogera	17
Qere Giweregis	QG	1657	Fogera	16
Qere Mikael	QM	1682	Fogera	15
Robit Bat	RB	1361	Bahir Dar Zuriya	1
Seneko Medaniyalem	SeM	1682	Fogera	23
Siraba Mariyam	SiM	1563	Fogera	24
Sheleku Medaniyalem	SM	1270	Fogera	20
Shena Tekelehayemanot	ST	1682	Fogera	14
Tiwaz Abo	TA	2010	Fogera	21
Wenechet	W	340	Dera	9
Wej Aregawi	WA	1883	Fogera	22
Weyebela Kidanemehert	WK	1422	Dera	8
Zajor Mikael	ZA	1422	Dera	6
Zahara Mikael	ZM	343	Dera	11

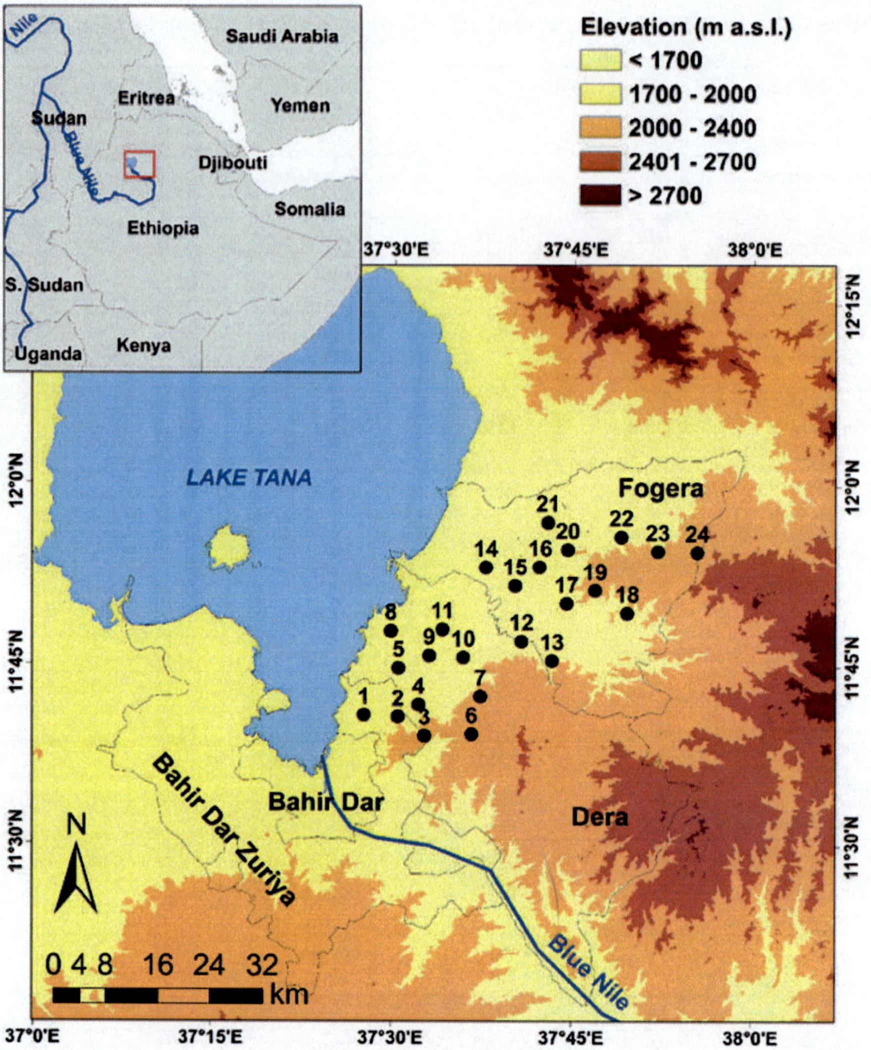

Fig. 10.1 Map of southeast of Lake Tana. Black dots and the respective numbers indicate the location and serial number of the selected 24 church forests, respectively

10.2.2 Sampling and Data Collection Methods

10.2.2.1 Vegetation Composition and Structure Survey

In each church forest, four sampling plots were systematically established in four cardinal directions (north, east, west, and south), but at different distances along the axis from the church buildings to edge of the compound (Fig. 10.2). The starting

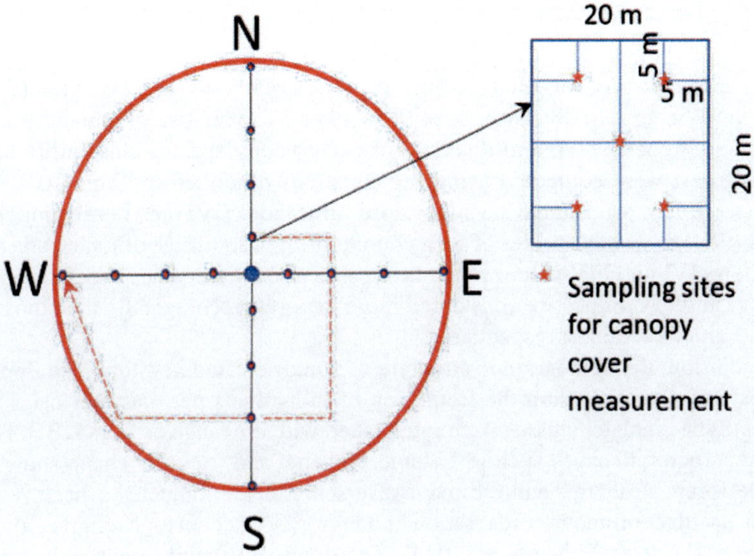

Fig. 10.2 Diagram showing sampling plot selection

direction was selected randomly and continued in a clockwise direction. A total of 96 plots with a size of 20 m × 20 m were established to identify mature and saplings individuals of the woody species. In each of the sampling plots, all woody species were identified, counted, and measured. Diameter at breast height (DBH) was measured using a diameter tape. When the height of the plant was greater than 1.6 m, DBH was measured at 1.3 m above ground level, but for smaller plants greater than 1 m, the diameter was measured at 10 cm above the ground. For trees containing buttressed stems, the DBH was measured above the buttress, and for multiple stems, all stems were counted and measured (Wassie et al. 2010). Woody plants less than 1 m height were considered as seedlings, and to score their presence and abundance, 5 m × 5 m plots were established at each corner of the large sampling plots. Identification was done to species level using identification keys (Bekele-Tesemma 2007). Species that were difficult to identify were identified in the National Herbarium Museum at Addis Ababa University. Plant nomenclature used in this article is based on the published guidelines of the Flora of Ethiopia and Eritrea, Vol. 1–8 (1989–2009) (Edwards et al. 1995, 1997; Edwards 1997; Hedberg et al. 2003). The recently updated nomenclature for *Acacia* spp. was also used (Kyalangalilwa et al. 2013). Lastly, the percent of canopy cover was measured using a convex mirror densiometer at four points within each of the main plots, and the average recording was used for each sampling plot. Maximum plant height within a church forest was taken as maximum forest height.

10.2.3 Data Analyses

The species diversity was analyzed using species richness (SR), evenness (J'), and Shannon–Wiener (H′) diversity index (Peet 1974). The relative dominance, density, frequency (Appendix 1), and diversity of each woody species and family at each church forest were computed using the equations described in Table 10.2. These values were used to calculate the importance value index (IVI) and family importance values (FIV) of woody species of each church forest in southeast of Lake Tana (Table 10.2). The IVI and FIV values range between 0 and 300. Species and families with high IVI and FIV values are considered more important ecologically than those with low IVI and FIV values, respectively.

In addition, the demographic structure of some selected keystone and dominant species was assessed using the frequency of individuals per diameter class (1–11 classes). The diameter classes were established with the range of 5 cm DBH. Demographic structural shapes such as I shape, J shape, and irregular shape show unfavorable forest structure, while broken reversed J shape indicates a healthy forest structure with continuous regeneration and ingrowth of cohorts (Alelign et al. 2007; Zegeye et al. 2011; Tadele et al. 2014). The dominant species are species that are very well represented in the church forests, whereas the keystone species are those which have social and ecological importance to local people (Wassie et al. 2009b; Gebeyehu et al. 2019): *Juniperus procera, Olea europaea, Podocarpus falcatus, Prunus africana, Ekebergia capensis* and *Mimusops kummel.*

Table 10.2 Structural measurements and formulae used for their calculation (Mueller and Ellenberg 1975; Mori et al. 1983)

Structural measurement	Equation (%)
Relative dominance	Total basal area of a species/total basal area of all species × 100
Relative density	Number of individuals of a species/total number of individuals × 100
Relative frequency	Frequency of a species/sum of all species frequencies × 100
Relative diversity	Number of a species in a family/total number of species × 100
Importance value index (IVI)	Relative dominance + relative density + relative frequency
Family importance value (FIV)	Relative dominance + relative density + relative diversity

10.3 Results

10.3.1 Species Richness, Evenness, and Diversity

A total of 115 woody species belonging to 53 families and 97 genera were identified in the 24 church forests studied, and the number of woody species ranged from 16 to 38 species per church forest (Tables 10.3 and 10.4). Of these species, 62% (n = 6268) were trees, 36% (n = 3571) were shrubs, 1.89% (n = 190) were climbers, and 0.06% (n = 6) were reed. The Kudese Minas (Kmi, 3 in Fig. 10.1) church forest had the highest number of species and genus, while Deber Kusekuam (DK, 18 in Fig. 10.1) church forest contained the lowest number of species and genus (Table 10.3).

The Kudese Minas (Kmi, 3 in Fig. 10.1) church forest had the highest number of families, while Qere Giweregis (QG, 16 in Fig. 10.1) church forest contained the lowest number of families (Table 10.3). The families with the highest number of species were Fabaceae (n = 17) followed by both Rubiaceae (n = 8) and Euphorbiaceae (n = 8). Oleaceae was represented by five species, whereas the remaining families were represented by $\leq$ 4 species (Table 10.4).

The overall evenness ($J' = 0.75 \pm 0.02$) and Shannon diversity index ($H' = 2.42 \pm 0.07$) of woody species were recorded in all church forests (Table 10.5). Both Delemo Tekelehayemanot (DT, 13 in Fig. 10.1) and Gebesiwit Mariya (Gma, 10 in Fig. 10.1) church forests had higher evenness than others, while church forest Kudese Minas (Kmi, 3 in Fig. 10.1) had higher Shannon diversity than other church forests. The lowest evenness and diversity were recorded at Wenechet (W, 9 in Fig. 10.1) church forest (Table 10.5).

10.3.2 Stand Structure of the Church Forests

Species and family importance ranged from 2 to 118% and 3% to 117%, respectively. The most ecologically important woody species (IVI) and families (FIV) differed between church forests (Tables 10.6 and 10.7). For example, the top three woody species in Aba Gerima Mariyam (AG, 2 in Fig. 10.1) church forest were *Millettia ferruginea* (IVI = 46%), *Ficus thonningii* (IVI = 39%), and *Grevillea robusta* (IVI = 25%), while *Mimusops kummel* (IVI = 97%), *Diospyros abyssinica* (IVI = 22%), and *Ocimum lamiifolium* (IVI = 21%) were the top three woody species in Zahara Mikael (ZM, 11 in Fig. 10.1) church forest (Table 10.6). Similarly, Fabaceae (63%), Moraceae (40%), and Proteaceae (24%) had the highest FIV values at Aba Gerima Mariyam (AG, 2 in Fig. 10.1) church forest while, Sapotaceae (93%), Rubiaceae (39%), and Fabaceae (26%) had the highest FIV values at Zahara Mikael (ZM, 11 in Fig. 10.1) church forest (Table 10.7).

The canopy cover of woody tree species at each church forest ranged from 37 to 83% with a mean of 59%. Kudese Minas (Kmi, 3 in Fig. 10.1) and Robit Bat (RB, 1

Table 10.3 Name of church forests, numbers of families, genera, species, canopy cover, and maximum height of the 24 church forests in the southeast part of Lake Tana (see full names of the churches in Table 10.1 and church forest numbers in Fig. 10.1)

Name	Family	Genera	Species richness	Canopy cover (%)	Maximum height (m)
Church name					
AG (2)[a]	24	30	30	58	32
DK (18)	12	16	16	67	30
DT (13)	17	21	21	42	25
ES (7)	21	25	27	54	27
FM (5)	20	28	29	54	30
Gma (10)	14	19	21	62	22
HSM (19)	15	18	19	67	30
K (12)	19	34	37	42	30
KM (4)	19	25	26	62	19
Kmi (3)	29	38	38	83	30
ME (17)	16	17	17	46	30
QG (16)	11	17	17	67	25
QM (15)	15	17	17	58	30
RB (1)	16	22	22	83	32
SeM (23)	21	27	29	42	27.3
SiM (24)	19	25	27	67	37
SM (20)	21	30	31	79	20
ST (14)	19	21	23	42	30
TA (21)	14	22	22	37	20
W (9)	19	24	24	62	35
WA (22)	22	29	30	62	25
WK (8)	22	31	31	58	42
ZA (6)	19	27	28	71	40
ZM (11)	21	30	30	62	30
Statistics					
Mean	19	25	26	59	29.09
Standard error	0.82	1.20	1.25	2.64	1.19
Minimum	11	16	16	37	19
Maximum	29	38	38	83	42
Statistics					
Total	53	97	115		

[a] Numbers in brackets indicate number of churches in Map 1

Table 10.4 Number of genera and species within families arranged alphabetically

Family name	Number of genera	Number of species
Acanthaceae	2	3
Anacardiaceae	2	4
Apocynaceae	2	2
Aquifoliaceae	1	1
Araliaceae	1	1
Arecaceae	1	1
Asteraceae	2	3
Bignoniaceae	1	1
Boraginaceae	2	2
Cactaceae	1	1
Capparidaceae	2	2
Casuarinaceae	1	1
Celastraceae	2	2
Combretaceae	1	1
Cupressaceae	2	2
Dracaenaceae	1	1
Ebenaceae	2	2
Euphorbiaceae	6	8
Fabaceae	13	17
Flacourtiaceae	2	2
Icacinaceae	1	1
Iridaceae	1	1
Lamiaceae	3	3
Lauraceae	1	1
Loganiaceae	1	1
Malvaceae	2	2
Meliaceae	3	3
Melianthaceae	1	1
Moraceae	1	4
Myricaceae	1	1
Myrtaceae	3	4
Olacaceae	1	1
Oleaceae	3	5
Phytolaccaceae	1	1
Pittosporaceae	1	1
Poaceae	1	1
Podocarpaceae	1	1

(continued)

Table 10.4 (continued)

Family name	Number of genera	Number of species
Proteaceae	1	1
Ranunculaceae	1	1
Rhamnaceae	1	1
Rhizophoraceae	1	1
Rosaceae	1	1
Rubiaceae	7	8
Rutaceae	3	4
Santalaceae	1	1
Sapindaceae	1	1
Sapotaceae	1	1
Simaroubaceae	1	1
Solanaceae	1	1
Sterculiaceae	1	1
Tiliaceae	1	1
Ulmaceae	1	1
Urticaceae	1	1

in Fig. 10.1) church forests had the highest canopy cover, while Tiwaz Abo (TA, 21 in Fig. 10.1) church forest had the lowest canopy cover. More than half of the church forests had a canopy cover greater than 59% (Table 10.3).

The diameter class distribution of some selected keystone and dominant species revealed four types of demographic structures. The four types of structures were broken reversed J shape, I shape, J shape, and irregular shape. The broken reversed J-shaped pattern was composed of a high number of individuals in the lowest diameter classes and progressively declining numbers in the highest diameter classes with an almost complete absence in the highest diameter classes. This pattern was, for example, exhibited by *Juniperus procera* at Delemo Tekelehayemanot (DT, 13 in Fig. 10.1), *Mimusops kummel* at Hager Selam Mariyam (HSM, 19 in Fig. 10.1) and Weyebela Kidanemehert (WK, 8 in Fig. 10.1), *Prunus africana* at Kudese Minas (Kmi, 3 in Fig. 10.1), and *Ekebergia capensis* at Kudese Minas (Kmi, 3 in Fig. 10.1) church forests.

The I-shaped pattern was formed when the numbers of individuals of a species were only presented in one of the eleven classes. *Juniperus procera* at Fisa Mikael (FM, 5 in Fig. 10.1), Hager Selam Mariyam (HSM, 19 in Fig. 10.1), Meneguzer Eyesus (ME, 17 in Fig. 10.1), Wenechet (W, 9 in Fig. 10.1) church forests, and *Mimusops kummel* at Emashenekure Giworegis (ES, 7 in Fig. 10.1) church forest were some of the species that showed I-shaped pattern.

J shape was composed of small numbers of individuals at the lowest classes and gradually increasing numbers at the highest diameter classes, and this pattern was represented by *Juniperus procera* at Wej Aregawi (WA, 22 in Fig. 10.1), *Olea*

Table 10.5 Mean, standard error, minimum, and maximum diversity values of woody species encountered in the 24 church forests in southeast of Lake Tana (see full names of the churches in Table 10.1)

Church forest name	Species richness	Shannon diversity (H′)	Pilou evenness (Jv)
Church name			
AG	30	2.9	0.85
DK	16	2.21	0.8
DT	21	2.78	0.91
ES	27	2.2	0.67
FM	29	2.36	0.7
Gma	21	2.77	0.91
HSM	19	2.62	0.89
K	37	2.76	0.77
KM	26	2.48	0.76
Kmi	38	3.08	0.85
ME	17	2.36	0.83
QG	17	2.19	0.77
QM	17	1.78	0.63
RB	22	2.36	0.76
SeM	29	2.49	0.74
SiM	27	2.35	0.71
SM	31	2.6	0.76
ST	23	2.27	0.72
TA	22	2.01	0.65
W	24	1.64	0.52
WA	30	1.97	0.58
WK	31	2.79	0.81
ZA	28	2.4	0.71
ZM	30	2.67	0.78
Mean	26	2.42	0.75
Standard error	1.25	0.07	0.02
Minimum	16	1.64	0.52
Maximum	38	3.08	0.91

europaea at Deber Kusekuam (DK, 18 in Fig. 10.1), and Gebesiwit Mariyam (Gma, 10 in Fig. 10.1) church forests. The irregular shape was formed when there was a complete absence of individuals in some diameter classes and a fair representation of individuals in other classes. *Juniperus procera* exhibited irregular-shaped pattern at Emashenekure Giworegis (ES, 7 in Fig. 10.1), Gebesiwit Mariyam (Gma, 10 in Fig. 10.1), Kirekus (K, 12 in Fig. 10.1), Kulela Mesekel (KM, 4 in Fig. 10.1), Siraba Mariyam (SiM, 24 in Fig. 10.1), Shena Tekelehayemanot (ST, 14 in Fig. 10.1), Zajor Mikael (ZA, 6 in Fig. 10.1), and Zahara Mikael (ZM, 11 in Fig. 10.1) church

Table 10.6 Importance value index (IVI) of the top three species encountered in the 24 church forests. Species are listed in rows and church forests in columns (see full names of churches in Table 10.1)

Species	AG	DK	DT	ES	FM	Gma	HSM	K	KM	Kmi	ME	QG
Acokanthera schimperi	–	–	–	–	46	–	–	–	–	–	–	–
Albizia schimperiana	–	–	24	–	–	–	–	–	–	32	–	–
Calpurnia aurea	–	–	–	–	–	–	–	–	–	–	–	–
Carissa spinarum	–	–	–	–	–	–	–	36	–	–	–	–
Celtis africana	–	–	–	–	–	–	–	–	–	–	32	–
Clausena anisata	–	33	–	–	–	–	–	–	–	–	–	–
Coffea arabica	–	–	–	–	–	–	–	–	–	–	–	–
Cordia africana	–	–	21	–	–	–	–	–	91	–	–	–
Croton macrostachyus	–	–	–	–	–	19	–	–	–	43	–	47
Diospyros abyssinica	–	–	–	–	41	23	–	–	–	–	–	–
Dodonaea angustifolia	–	–	–	–	–	–	–	23	–	–	–	–
Dracaena steudneri	–	–	–	–	–	–	30	–	–	–	–	–
Erythrina abyssinica	–	–	–	–	–	–	–	–	–	–	–	–
Eucalyptus camaldulensis	–	–	–	–	–	–	–	66	–	–	–	–
Euphorbia tirucalli	–	–	–	–	–	–	–	–	41	–	–	54
Ficus thonningii	39	–	64	–	–	–	52	–	–	–	–	–
Ficus vasta	–	–	–	–	–	–	–	–	–	–	–	–
Flueggea virosa	–	–	–	–	–	–	–	–	–	–	–	–
Grevillea robusta	25	–	–	–	–	–	–	–	–	–	–	–
Grewia ferruginea	–	–	–	–	–	–	–	–	23	–	–	–
Juniperus procera	–	–	–	45	–	–	–	–	–	–	–	–
Maytenus arbutifolia	–	–	–	–	–	–	–	–	–	–	–	–
Millettia ferruginea	46	46	–	–	–	–	–	–	–	25	–	44
Mimusops kummel	–	–	–	–	32	87	–	–	–	–	87	–
Ocimum lamiifolium	–	–	–	–	–	–	–	–	–	–	–	–
Olea capensis	–	–	–	–	–	–	–	–	–	–	–	–
Olea europaea	–	73	–	–	–	–	–	–	–	–	–	–
Pavetta abyssinica	–	–	–	–	–	–	–	–	–	–	43	–
Ritchiea albersii	–	–	–	–	–	–	–	–	–	–	–	–
Rothmannia urcelliformis	–	–	–	48	–	–	–	–	–	–	–	–
Teclea nobilis	–	–	–	36	–	–	54	–	–	–	–	–
Vernonia myriantha	–	–	–	–	–	–	–	–	–	–	–	–
Species	QM	RB	SeM	SiM	SM	ST	TA	W	WA	WK	ZA	ZM
Acokanthera schimperi	–	–	–	–	–	–	–	–	42	–	–	–
Albizia schimperiana	–	–	–	–	–	–	–	–	–	28	–	–

(continued)

Table 10.6 (continued)

Species	QM	RB	SeM	SiM	SM	ST	TA	W	WA	WK	ZA	ZM
Calpurnia aurea	35	–	–	45	–	–	–	–	–	–	–	–
Carissa spinarum	–	–	–	–	–	–	–	–	–	–	–	–
Celtis africana	–	–	–	52	–	–	–	–	–	–	–	–
Clausena anisata	–	–	–	–	–	–	–	–	–	–	–	–
Coffea arabica	–	40	–	–	–	–	–	54	–	–	–	–
Cordia africana	–	–	–	–	38	25	–	–	–	–	–	–
Croton macrostachyus	–	–	20	–	37	–	–	–	–	25	–	–
Diospyros abyssinica	–	–	–	–	–	–	–	–	–	–	–	22
Dodonaea angustifolia	–	–	–	–	–	–	–	–	–	–	–	–
Dracaena steudneri	–	–	–	–	–	–	–	–	–	–	–	–
Erythrina abyssinica	–	–	–	–	–	–	17	–	–	–	–	–
Eucalyptus camaldulensis	59	–	–	–	–	118	109	–	–	–	–	–
Euphorbia tirucalli	–	–	–	–	–	–	–	–	–	–	–	–
Ficus thonningii	46	–	–	–	–	–	–	–	–	–	–	–
Ficus vasta	–	–	73	–	–	–	–	39	52	–	–	–
Flueggea virosa	–	–	–	–	–	–	20	–	–	–	–	–
Grevillea robusta	–	–	–	–	–	–	–	–	–	–	–	–
Grewia ferruginea	–	–	–	–	–	25	–	–	–	–	–	–
Juniperus procera	–	–	–	–	–	–	–	–	37	–	80	–
Maytenus arbutifolia	–	–	–	–	33	–	–	–	–	–	–	–
Millettia ferruginea	–	27	–	–	–	–	–	–	–	35	–	–
Mimusops kummel	–	85	–	–	–	–	–	–	–	–	–	97
Ocimum lamiifolium	–	–	–	–	–	–	–	–	–	–	–	21
Olea capensis	–	–	–	30	–	–	–	–	–	–	–	–
Olea europaea	–	–	–	–	–	–	–	–	–	–	–	–
Pavetta abyssinica	–	–	–	–	–	–	–	–	–	–	–	–
Ritchiea albersii	–	–	–	–	–	–	–	39	–	–	–	–
Rothmannia urcelliformis	–	–	–	–	–	–	–	–	–	–	28	–
Teclea nobilis	–	–	–	–	–	–	–	–	–	–	31	–
Vernonia myriantha	–	–	32	–	–	–	–	–	–	–	–	–

forests. Generally, of the total 57 species assessed, 28 and 21 revealed irregular-shaped and I-shaped structures, respectively, while only five and three showed broken reversed J-shaped and J-shaped structures, respectively. However, a single species exhibited different diameter class distribution patterns across the different church forests. For example, the diameter class distribution for *Mimusops kummel* was I

Table 10.7 List of families having woody species with the top three importance value indexes in the 24 church forests in southeast of Lake Tana (see full names of the churches in Table 10.1)

Family	AG	DK	DT	ES	FM	Gma	HSM	K	KM	Kmi	ME	QG
Apocynaceae	–	–	–	–	50	–	–	35	–	20	–	–
Asteraceae	–	–	–	–	–	–	–	–	–	–	–	–
Boraginaceae	–	–	–	–	–	–	–	–	88	–	–	–
Capparidaceae	–	–	–	–	–	–	–	–	–	–	–	–
Cupressaceae	–	–	21	56	–	–	–	–	–	–	–	–
Ebenaceae	–	–	–	–	40	27	–	–	–	–	–	–
Euphorbiaceae	–	–	–	–	–	–	–	–	63	46	–	110
Fabaceae	63	66	57	–	40	–	–	26	26	60	–	98
Lamiaceae	–	–	–	–	–	–	–	–	–	–	–	16
Meliaceae	–	–	–	–	–	–	–	–	–	–	–	–
Moraceae	40	–	63	38	–	27	54	–	–	–	–	–
Myrtaceae	–	–	–	–	–	–	–	69	–	–	–	–
Oleaceae	–	68	–	–	–	–	–	–	–	–	–	–
Proteaceae	24	–	–	–	–	–	–	–	–	–	–	–
Rubiaceae	–	–	–	58	–	–	47	–	–	–	34	–
Rutaceae	–	59	–	–	–	–	51	–	–	–	–	–
Sapotaceae	–	–	–	–	–	84	–	–	–	–	83	–
Tiliaceae	–	–	–	–	–	–	–	–	–	–	–	–
Ulmaceae	–	–	–	–	–	–	–	–	–	–	28	–
Family	QM	RB	SeM	SiM	SM	ST	TA	W	WA	WK	ZA	ZM
Apocynaceae	–	–	–	–	–	–	–	–	47	–	–	–
Asteraceae	–	–	31	–	–	–	–	–	–	–	–	–
Boraginaceae	–	–	–	–	38	–	–	–	–	–	–	–
Capparidaceae	–	–	–	–	–	–	–	41	–	–	–	–
Cupressaceae	–	–	–	–	–	–	–	–	34	–	80	–
Ebenaceae	–	–	–	–	–	–	–	–	–	–	–	–
Euphorbiaceae	–	–	–	–	50	–	29	–	–	–	–	–
Fabaceae	74	34	44	71	34	23	55	–	–	68	–	26
Lamiaceae	–	–	–	–	–	–	–	–	–	–	–	–
Meliaceae	–	–	–	–	–	–	–	–	–	–	32	–
Moraceae	44	–	80	–	–	–	–	41	63	–	–	–
Myrtaceae	57	–	–	–	–	117	105	–	–	–	–	–
Oleaceae	–	–	–	40	–	–	–	–	–	–	–	–
Proteaceae	–	–	–	–	–	–	–	–	–	–	–	–
Rubiaceae	–	59	–	–	–	–	–	54	–	30	44	39

(continued)

Table 10.7 (continued)

Family	QM	RB	SeM	SiM	SM	ST	TA	W	WA	WK	ZA	ZM
Rutaceae	–	–	–	–	–	–	–	–	–	43	–	–
Sapotaceae	–	83	–	–	–	–	–	–	–	–	–	93
Tiliaceae	–	–	–	–	–	23	–	–	–	–	–	–
Ulmaceae	–	–	–	48	–	–	–	–	–	–	–	–

Table 10.8 List of church forests and the total number of species recorded with the numbers and proportions of species with important value index of <5% in the 24 church forests in southeast of Lake Tana (see full names of the churches in Table 10.1)

Church name	Total number of species	Number of species (IVI < 5%)	Proportion of species (IVI < 5%) (in %)
ZA	29	17	59
KM	26	15	58
W	24	13	54
Kmi	38	20	53
SM	31	16	52
K	37	19	51
WA	30	15	50
FM	29	14	48
WK	31	14	45
ZM	30	13	43
SiM	27	11	41
ST	23	9	39
ES	27	10	37
AG	30	11	37
TA	22	7	32
SeM	29	9	31
RB	22	6	27
DK	16	4	25
Gma	21	5	24
QG	17	4	24
QM	17	4	24
HSM	19	4	21
ME	17	3	18
DT	21	3	14

shaped at Emashenekure Giworegis (ES, 7 in Fig. 10.1), broken reversed J shaped at Hager Selam Mariyam (HSM, 19 in Fig. 10.1), and irregular shaped at Wenechet (W, 9 in Fig. 10.1) church forests (Fig. 10.3).

10.3.3 Species Richness and Stand Structure of Exotic Species

Of the total 115 woody species, 18 were exotic species (one climber, three shrubs, and 14 trees). Exotic species occurred at 19 church forests with the range from one to four species per church forest. *Eucalyptus camaldulensis* was recorded at nine church forests (38%), followed by *Cupressus lusitanica* at seven church forests (29%), and *Grevillea robusta* at six church forests (25%). The remaining exotic species occurred at ≤ 3 church forests. Relatively high densities of *Eucalyptus* spp. ranging from 13 to 1925 individuals ha^{-1} were recorded. Of the exotic species, *Eucalyptus camaldulensis* had the highest IVI in Kirekus (K, 12 in Fig. 1, IVI = 66%), Qere Mikael (QM, 15 in Fig. 1, IVI = 59%), Shena Tekelehayemanot (ST, 14 in Fig. 1, IVI = 118%), and Tiwaz Abo (TA, 21 in Fig. 1, IVI = 109%) church forests. *Grevillea robusta* (IVI = 25%) and *Cupressus lusitanica* (IVI = 21%) exhibited the highest IVI values in Aba Gerima Mariyam (AG, 2 in Fig. 10.1) and in Kirekus (K, 12 in Fig. 10.1) church forests, respectively.

10.3.4 Species Richness of Seedlings

The seedlings of a total of 62 species (57 indigenous and five exotic) were identified, in the 24 church forests. Of these, 80%, 16%, and 4% were seedlings of trees, shrubs, and climbers, respectively. Species richness of seedlings varied among the 24 church forests, ranging from 2 to 31 species per church forest with a mean and standard error of 18 ± 1.61. *Ruta chalepensis* was the only species found at the seedling stage. Zahara Mikael (ZM, 11 in Fig. 10.1) church forest had the highest total number of seedlings (391,200 ha^{-1}), while the lowest number of seedlings (200 ha^{-1}) was recorded in Delemo Tekelhayemanot (DT, 13 in Fig. 10.1) church forest (Fig. 10.4). Among the total woody species recorded in the church forests, *Diospyros abyssinica* had the highest number of seedlings (38%), followed by *Mimusops kummel* (13%). In general, thirty-three species (29 indigenous and four exotic) exhibited low densities of seedlings 100–1700 ha^{-1} (53%), whereas four (three indigenous and one exotic), seven (all indigenous), and eighteen (all indigenous) species had 2100–3700 (7%), 4200–5900 (11%) and $\geq$7600 (29%) individuals ha^{-1}, respectively.

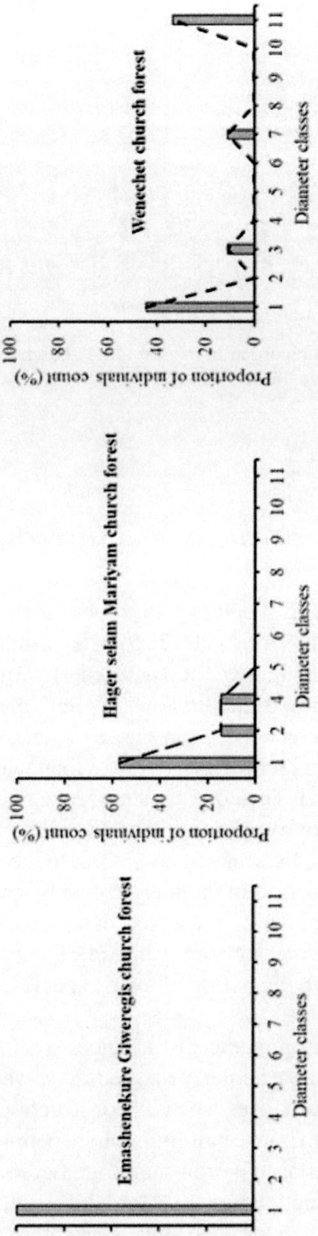

Fig. 10.3 Demographic structure of the dominant species *Mimusops kummel* in different church forests. Diameter classes: 1 = 0–5; 2 = 5–10; 3 = 10–15; 4 = 15–20; 5 = 20–25; 6 = 25–30; 7 = 30–35; 8 = 35–40; 9 = 40–45; 10 = 45–50; 11 ≥ 50 cm DBH

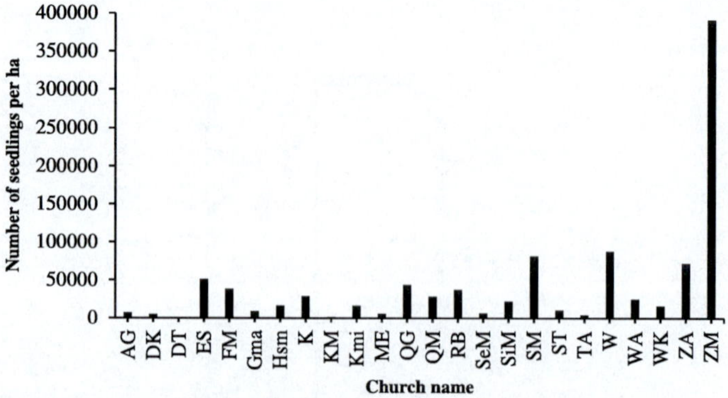

Fig. 10.4 Total number of seedlings recorded in each studied church forest in southeast of Lake Tana (see full names of the churches in Table 10.1)

10.4 Discussion

10.4.1 Species Richness, Evenness, and Diversity

In the present study, a total of 115 woody species were identified in the 24 church forests. Of the 24 church forests, four church forests (Emashenekure Giworegis, Gebesiwit Mariyam, Wej Aregawi, and Zahara Mikael) were already assessed by Wassie et al. (2010). The assessment methods were relatively similar to this study. However, except for Zahara Mikael (ZM, 11 in Fig. 10.1) church forest, the numbers of families and woody species we recorded were lower than what Wassie et al. (2010) found ten years earlier (Fig. 10.5). The continued presence of relatively high species richness at Zahara Mikael church forest was likely due to construction of stone wall since 2014, and this could avoid disturbance and increase the probability of recruitment of new seedlings. The reduction in number of woody species in the remaining three church forests could be a sign of increased disturbance from anthropogenic pressures. Among the different pressures, expansion of graveyard, second church buildings inside the church forest, plantation of exotic species, and free grazing were the major factors.

The negative effects of anthropogenic disturbances, including free grazing on woody species of church forests, particularly on seedlings, were reported by Wassie et al. (2009a). The level of disturbances, such as construction of additional church buildings, grave houses, and small house buildings where people organize themselves into associations to celebrate a chosen patron saint together *(mehabirs)* and grazing, has continuously increased in the study area (Cardelús et al. 2017; Orlowska and Klepeis 2018). Although, some species have always been rare, the presence of many species with IVI < 5% per church forest suggests that the disturbance is still ongoing. However, the presence of old remnant trees (i.e., rarely recorded trees with low IVI values) and those which are economically and ecologically important species, such as

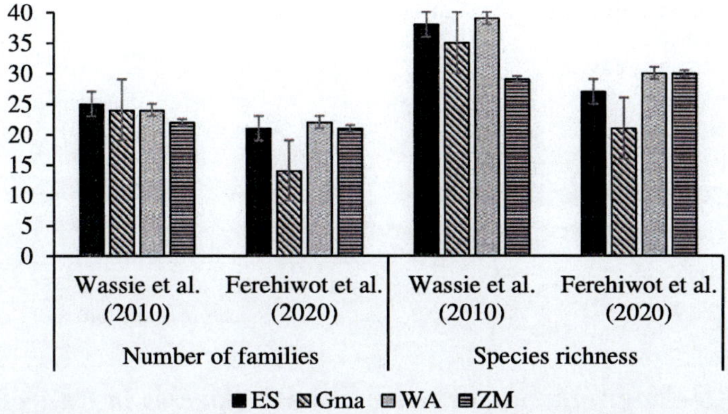

Fig. 10.5 Comparison of the number of families and species between results from Wassie et al. (2010) and the current study. The four church forests were Emashenekure Giworegis (ES, 7 in Fig. 10.1), Gebesiwit Mariyam (Gma, 10 in Fig. 10.1), Wej Aregawi (WA, 22 in Fig. 10.1), and Zahara Mikael (ZM, 11 in Fig. 10.1)

Juniperus procera, Olea europaea, Prunus africana, Podocarpus falcatus, Ekebergia capensis, Mimusops kummel, and *Cordia africana* in the compounds of churches, highlights the importance of church forests in southeast of Lake Tana for indigenous tree species and wildlife conservation similar to other sacred forests in the world (Bhagwat and Rutte 2006).

The Pilou evenness value ranges from 0 to 1, and values close to 1 indicate even representation of individuals of the occurring species in the area (Help et al. 1998). Except for Wenechet (W, 9 in Fig. 10.1) (J = 0.52) and Wej Aregawi (WA, 22 in Fig. 10.1) (J = 0.58) church forests, all church forests in the present study had evenness values (0.63 to 0.91), indicating even representation of the individuals of the occurring species in each church forest. Moreover, except for Qere Mikael (H' = 1.78), Wenechet (H' = 1.64), and Wej Aregawi (H' = 1.97) church forests, all church forests had Shannon diversity greater than two, which indicates medium to high species diversity (Giliba et al. 2011). The highest and lowest Shannon diversity values were recorded in Kudese Minas (Kmi, 3 in Fig. 10.1) and Wenechet (W, 9 in Fig. 10.1) church forests, respectively. In the present study, when more church forests were included, the number of newly described species increased (Fig. 10.6). Therefore, since we studied only 24 church forests out of many available church forests in the larger region of Lake Tana, more species richness and diversity could be recorded when more church forests are studied, owing to increases in the environmental heterogeneity. A similar explanation was also given by Wassie (2007).

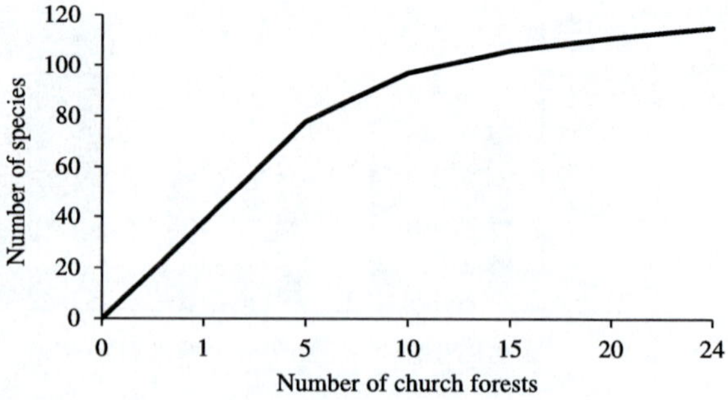

Fig. 10.6 The number of species as a function of the number of church forests inventoried

10.4.2 Demographic Structure of Woody Species in Church Forests

Investigating the IVI and FIV values are significant to understand the ecological importance of the species and families, respectively. The IVI and FIV values are good indicators to understand the current condition of the church forests since they provide important insights into the basal area, abundance, frequency, and relative diversity of the species in a particular forest area. In this study, fleshy-fruited tree species had the highest IVI in most church forests. The fruits of these species are indehiscent and consumed by a variety of birds and mammals. As these animals, particularly birds, move from one church forest to another before defecating the seeds, they could further promote seed dispersal between church forests. Therefore, this might be the probable reason why the fleshy-fruited tree species had high IVI values in different church forests. Furthermore, the highest IVI values of these species were due to their high frequency, abundance, and basal area values. The importance of woody species in a given area can be better explained by their basal area than simple stem count (Lamprecht 1989; Bekele 1994; Abyot et al. 2014; Meragiaw et al. 2018). The species that had the highest IVI values had large basal areas and, hence, could play a significant contribution to biodiversity conservation by providing habitat and food for frugivore birds and mammals as well as bee forage. However, several woody species at each church forest had low IVI values, indicating that most species had small basal areas, small numbers of individuals, and are generally rare in the church forests. For example, ≥50% of the woody species in seven church forests had IVI < 5%. The different anthropogenic disturbances in the study area could be the main reasons for the presence of many species with low IVI values, and this could likely affect their ecological significance. The rarity of these species could also be caused by other factors, such as their poor dispersal ability and competition for nutrients or other resources (Hubbell et al. 2001; Engelbrecht et al. 2007). Thus, woody species with a low IVI values require high conservation priority to maintain their composition and diversity.

The Fabaceae family had the highest FIV at five church forests, Sapotaceae at four church forests, and Rubiaceae at two church forests. Teketay et al. (2018) also reported that Fabaceae had the highest FIV value. The main reason why this family had the highest FIV value is likely due to its wide range of ecological adaptations. Although it was represented by only one species and small numbers of individuals, Sapotaceae had the second-highest FIV value in the study area. This is due to the fact that most individuals of this species had large basal area. Species with a large basal area (even though the density of the species is low) had more structural complexity and provide a lot of habitat for various species and also serve as shade for animals and human beings. Hence, Sapotaceae species could have significant ecological importance. They could also provide vital social values, as large trees are culturally and spiritually important in Ethiopian (Orlowska and Klepeis 2018).

The difference in canopy cover among the church forests was due to the relatively higher dominance of big old canopy tree species in some church forests than others. For example, the high canopy covers at Kudese Minas (Kmi, 3 in Fig. 10.1) and Robit Bat (RB, 1 in Fig. 10.1) church forests were due to the relative dominance of big old canopy trees such as *Albizia schimperiana, Croton macrostachyus, Millettia ferruginea,* and *Mimusops kummel.* However, Tiwaz Abo (TA) church forest had small canopy cover due to the relative dominance of *Eucalyptus camaldulensis* individuals, kept small by frequent harvesting or coppicing. According to Sabine and Miehe (1994), forests with > 80% canopy cover are considered as closed forests. Therefore, Kudese Minas (Kmi, 3 in Fig. 10.1) and Robit Bat (RB, 1 in Fig. 10.1) church forests are closed forests. However, in the present study, the majority of church forests had canopy cover <80% with a mean canopy cover of 59%, which implies that most church forests in the study area are open forests. The presence of low canopy cover in the church forests is an indication of the degradation of the primary forests to a shrubland. This could be due to anthropogenic disturbances, such as selective cutting of large canopy trees or natural disturbance due to the death of large canopy trees without any replacement. Although the death of large canopy trees in a natural system can be filled in by young trees, this was not the case in the studied church forests, probably, due to the succession being blocked by grazing or other disturbances, such as soil degradation. On the other hand, canopy cover is an indicator of the microclimate conditions that determine the species composition and structure. For example, temperature decreases with increasing canopy cover, with the greatest cooling when canopy cover exceeds 40% (Ziter et al. 2019), while light, soil water, and airflow exchange increases with decreasing canopy cover (Muscolo et al. 2014). Therefore, understanding the canopy cover is significant to conserve the old growth forests.

Based on the diameter class distribution analyses of the selected species, four types of demographic structure were revealed. Of the total 57 species structure, 49%, 37%, 9%, and 4% showed irregular-shaped, I-shaped, broken reversed J-shaped, and J-shaped patterns, respectively. The irregular-shaped, I-shaped, and J-shaped patterns represent abnormal demographic structure due to the removal of woody species at different diameter classes. The underlying reasons for such kind of demographic structure could be related to overgrazing at young stages of plants and removal of vegetation for burial activities. Celebration of some spiritual activities inside the church forests could have a negative effect on the recruitment of seedlings due to trampling. Additionally, activities such as selective cutting of trees for construction of church buildings, poor reproduction of old trees, and loss of seeds to predators could likely cause the abnormal population structure (Abyot et al. 2014). However, broken reversed J-shaped demographic structures represent relatively a healthy forest.

10.4.3 Stand Structure of Exotic Species

Eucalyptus camaldulensis had the highest IVI value in 17% of church forests in the study area. This species is mainly planted inside or around the church forests matrix for different purposes, such as construction materials, firewood, timber, ornamental, and also windbreak (Bekele-Tesemma 2007). Similarly, Burkhard et al. (2012) reported that this species was planted in the agricultural matrix around the church forests due to its multiple benefits for local communities. *Eucalyptus camaldulensis* is a strong competitor within the native tree community due to its fast growth and resilience to disturbances. Thus, the relatively high dominance and density of *Eucalyptus* spp. in the study area could negatively affect the ecological diversity and structure of the indigenous woody species. As a backlash to its strong wood provisioning performance, it can cause threats to the ecological conditions due to decreased understory cover which lead to increased soil loss rates (Nyssen et al. 2004). It also has a high water consumption rate compared with the other forest communities (Bekele-Tesemma 2007) and also reduces the regeneration potential of native tree species compared with other tree canopies (Thijs et al. 2014a). Plantations of *Eucalyptus* spp. were higher in recently established church forests compared with the old church forests within the study area (field observation). For example, the relative density of *Eucalyptus camaldulensis* was high (41%) in Tiwaz Abo (TA, 21 in Fig. 10.1), which was established in 2010 and 0.3% in Meneguzer Eyesus (ME, 17 in Fig. 10.1), which was established in 1682. This might be because most of the recently established churches are constructed at most degraded areas, and the local people prefer to plant exotic and fast-growing species than indigenous plants. Therefore, considering the ecological threat of exotic species plantation, in general, and *Eucalyptus camaldulensis* plantation, in particular, emphasis should be given for re-introduction of indigenous species in recently established church forests.

10.4.4 The Regeneration Status of Church Forests in Southeast of Lake Tana

The regeneration status of each species can be determined by looking at the pattern of their seedlings, saplings, and matured stage (Malik and Bhatt 2016). To sustain their future regeneration status, woody species should successfully complete their life cycle. When a species is present in both mature and sapling or seedling stages, it indicates the good regenerating status of the species, whereas when the species is present either in mature stage or sapling or seedling stages only, it may show an extinction debt or a colonization credit, respectively (Thijs et al. 2014b). In the present study, most woody species showed poor regeneration status, which exhibit potential extinction debt. Of the total 115 woody species, seedlings were recorded for 62 woody species with 33 species represented only by 100–1700 seedlings ha^{-1}. However, Zahara Mikael (ZM, 11 in Fig. 10.1) church forest had a large number of seedlings (391,200 ha^{-1}). The possible reason for this is because Zahara Mikael (ZM, 11 in Fig. 10.1) has been fenced with stone walls since 2014, and, thus, this could avoid grazing, creation of new paths and clearings, which are the major factors for seedling regeneration.

Of the species that had the highest IVI values, only few species at few church forests showed good regeneration status. For example, *Mimusops kummel* had higher numbers of seedlings and saplings than mature trees in Gebesiwit Mariyam (Gma, 10 in Fig. 10.1), Meneguzer Eyesus (ME, 17 in Fig. 10.1), Robit Bat (RB, 1 in Fig. 10.1), and Zahara Mikael (ZM, 11 in Fig. 10.1) church forests, suggesting that the species has a good regeneration status. However, one or two of the growth stages were missed or lower numbers of seedlings and saplings than the number of mature trees were recorded in many woody species, such as *Ficus vasta* in Gebesiwit Mariyam (Gma, 10 in Fig. 10.1), Kulela Mesekel (KM, 4 in Fig. 10.1), Seneko Medaniyalem (SeM, 23 in Fig. 10.1), Tiwaz Abo (TA, 21 in Fig. 10.1), Wenechet (W, 9 in Fig. 10.1), and Wej Aregawi (WA, 22 in Fig. 10.1) church forests, which indicate their poor regeneration status.

The presence of low numbers of seedlings could pose a threat by reducing the viable population size, which, in turn, affects the ecological function of church forests. The poor regeneration status of woody species was also corroborated by our demographic analysis. Of the selected species for structural analyses, only 5% exhibited the broken reversed J-shaped structure, which presents relatively a healthy forest with good natural regeneration and recruitment potential while the rest (95%) of the species showed abnormal population structures that confirmed the poor regeneration status of the species. The poor regeneration status of woody species in church forests in southeast of Lake Tana is primarily attributed to anthropogenic pressures, such as free grazing, vegetation removal for different purposes, and plantation of fast-growing exotic species. However, climate change and other natural factors, such as physical and chemical soil degradation, could also be considered responsible for the poor regeneration status of the woody species in the study area. Effects of anthropogenic and natural disturbance on the regeneration status of forests have been reported by previous studies (Dai et al. 2002; Tesfaye et al. 2002; Wassie et al. 2009b; Abyot et al. 2014; Meragiaw et al. 2018; Maua et al. 2020).

10.5 Conclusions

Church forests in the study area are important for regional biodiversity conservation since they are the only remnant forests and the last option to harbor woody species and their associated fauna and flora. This study characterized the species and structural composition of woody species at 24 church forests in southeast of Lake Tana. The diversity indices revealed that most of the church forests contained high species diversity. The ecologically and culturally most important woody species, such as *Juniperus procera, Olea europaea, Ficus vasta,* and families in the church forests, which were identified by means of their IVI and FIV values, have to be prioritized for conservation. Furthermore, conservation activities are also urgently required for the rare indigenous species and families that had low IVI and FIV values in all forests. For these species, the church forests are often the only safe haven of survival, and action is required to prevent their regional extermination.

The demographic structure for majority of the species showed I-shaped, irregular-shaped, and J-shaped patterns, suggesting poor regeneration status. Human influence, such as plantation of exotic species, affects the structural composition and regeneration status of church forests. Additionally, there were several species at different church forests that had lower numbers of seedlings and saplings than mature trees, which confirm the low regeneration and recruitment potential of the forests. Woody species with low IVI values and those with poor regeneration status should be prioritized for conservation. Since other patch-level and landscape-level factors could possibly affect species and structural composition of church forests, due attention should be given to such factors in future studies. Additionally, the causes of poor regeneration status of woody species and the possible options to increase their natural regeneration should be studied.

Acknowledgements This study received financial support from VLIR-UOS, Belgium, through the VLIR-IUC Interuniversity cooperation with Bahir Dar University, Ethiopia (BDU-IUC). We are grateful to all EOTC priests, monks, students, and local communities for access to the church forests, technical support and field assistance.

Appendix 1: Species Name, Family, and the Relative Frequency of Occurrence in 24 Church Forests Southeast of Lake Tana, Ethiopia

Species name	Family	Relative frequency of occurrence (%)
Capparis tomentosa L	Capparidaceae	92
Justicia schimperiana (Hochst.ex Nees) T. Anders	Acanthaceae	92

(continued)

(continued)

Species name	Family	Relative frequency of occurrence (%)
Cordia africana Lam	Boraginaceae	83
Croton macrostachyus Del	Euphorbiaceae	83
Grewia ferruginea Hochst. ex A. Rich	Tiliaceae	83
Millettia ferruginea (Hochst.) Bak	Fabaceae	79
Maytenus arbutifolia (A.Rich.) Wilczek	Celastraceae	71
Mimusops kummel A. DC	Sapotaceae	71
Albizia schimperiana Oliv	Fabaceae	67
Calpurnia aurea (Ait.) Benth	Fabaceae	63
Carissa spinarum L	Apocynaceae	63
Celtis africana Burm.f	Ulmaceae	63
Pavetta abyssinica Fresen	Rubiaceae	63
Juniperus procera Hochst. ex Endl	Cupressaceae	58
Teclea nobilis Del	Rutaceae	58
Ficus thonningii Blume	Moraceae	54
Premna schimperia Engl	Lamiaceae	54
Clausena anisata (Willd.) Benth	Rutaceae	50
Vernonia myriantha Hook.f	Asteraceae	50
Acanthus sennii Chiov	Acanthaceae	42
Ocimum lamiifolium Hochst. ex Benth	Lamiaceae	42
Olea europaea L	Oleaceae	42
Coffea arabica L	Rubiaceae	38
Dracaena steudneri Engl	Dracaenaceae	38
Eucalyptus camaldulensis Dehnh	Myrtaceae	38
Diospyros abyssinica (Hiern) F.White	Ebenaceae	33
Vernonia amygdalina Del	Asteraceae	33
Cupressus lusitanica Mill	Cupressaceae	29
Entada abyssinica Steud.ex.A.Rich	Fabaceae	29
Ficus vasta Forssk	Moraceae	29
Ricinus communis L	Euphorbiaceae	29
Ritchiea albersii Gilg	Capparidaceae	29
Bersama abyssinica Fresen	Melianthaceae	25
Euclea racemosa Murr	Ebenaceae	25
Grevillea robusta R.Br	Proteaceae	25
Rhus quartiniana A.Rich	Anacardiaceae	25

(continued)

(continued)

Species name	Family	Relative frequency of occurrence (%)
Dovyalis abyssinica (A.Rich.) Warb	Flacourtiaceae	21
Flueggea virosa (Willd.) Voigt	Euphorbiaceae	21
Osyris quadripartita Decn	Santalaceae	21
Podocarpus falcatus (Thunb.) R. B. ex. Mirb	Podocarpaceae	21
Rhus vulgaris Meikle	Anacardiaceae	21
Rothmannia urcelliformis (Hiern) Robyns	Rubiaceae	21
Schrebera alata (Hochst.) Welw	Oleaceae	21
Senna singueana (Del.) Lock	Fabaceae	21
Acokanthera schimperi (A. DC.) Schweinf	Apocynaceae	17
Arundo donax L	Poaceae	17
Ekebergia capensis Sparrm	Meliaceae	17
Euphorbia tirucalli L	Euphorbiaceae	17
Ilex mitis (L.) Radlk	Aquifoliaceae	17
Acanthus pubescens (Oliv.) Engl	Acanthaceae	13
Brucea antidysenterica J.F.Mill	Simaroubaceae	13
Casuarina equisetifolia L	Casuarinaceae	13
Clerodendrum myricoides (Hochst.) Vatka	Lamiaceae	13
Dodonaea angustifolia L.f	Sapindaceae	13
Ehretia cymosa Thonn	Boraginaceae	13
Erythrina abyssinica Lam.ex DC	Fabaceae	13
Flacourtia indica (Burm.f) Merr	Flacourtiaceae	13
Gardenia fiorii Chiov	Rubiaceae	13
Jasminum abyssinicum Hochst. ex DC	Oleaceae	13
Jasminum grandiflorum L	Oleaceae	13
Phytolacca dodecandra L'H'erit	Phytolaccaceae	13
Pittosporum viridiflorum Sims	Pittosporaceae	13
Prunus africana (Hook. f.) kalkm	Rosaceae	13
Apodytes dimidiata E. Mey ex. Arn	Icacinaceae	8
Azadirachta indica A.Juss	Meliaceae	8
Citrus aurantifolia (Christm.)	Rutaceae	8
Citrus aurantium L	Rutaceae	8
Combretum molle R.Br.ex G.Don	Combretaceae	8
Dombeya torrida (J.F.Gmel.) P.Bamps	Sterculiaceae	8

(continued)

(continued)

Species name	Family	Relative frequency of occurrence (%)
Euphorbia abyssinica Gmel	Euphorbiaceae	8
Galiniera saxifraga (Hochst.) Bridson	Rubiaceae	8
Olea capensis L	Oleaceae	8
Opuntia ficus-indica (L.) Miller	Cactaceae	8
Piliostigma thonningii (Schumach.) Milne-Redh	Fabaceae	8
Rhamnus prinoides L' Herit	Rhamnaceae	8
Rhus glutinosa A.Rich	Anacardiaceae	8
Schefflera abyssinica (Hochst. ex A. Rich.) Harms	Araliaceae	8
Senna didymobotrya (Fresen.) Irwin and Barneby	Fabaceae	8
Vachellia abyssinica (Hochst. ex. Benth.) Kyal. and Boatwr.	Fabaceae	8
Abutilon figarianum Webb	Malvaceae	4
Albizia anthelmintica (A. Rich.) Brogn	Fabaceae	4
Bridelia micrantha (Hochst.) Baill	Euphorbiaceae	4
Buddleja polystachya Fresen	Loganiaceae	4
Cassipourea malosana (Baker) Alston	Rhizophoraceae	4
Clematis hirsuta Perr. and Guill	Ranunculaceae	4
Croton dichogamus Pax	Euphorbiaceae	4
Delonix regia (Boj.ex Hook.) Raf	Fabaceae	4
Dichrostachys cinerea (L.) Wight and Arn	Fabaceae	4
Dolichos sericeus E. Mey	Fabaceae	4
Eucalyptus saligna Smith	Myrtaceae	4
Ficus ingens (Miq.) Miq	Moraceae	4
Ficus sycomorus L	Moraceae	4
Gladiolus psittacinus Hook. F	Iridaceae	4
Gossypium arboreum L	Malvaceae	4
Hippocratea africana (Willd.) Loes	Celastraceae	4
Indigofera arrecta Hochst. ex A. Rich	Fabaceae	4
Lepidotrichilia volkensii (Giirke) Leroy	Meliaceae	4
Mangifera indica L	Anacardiaceae	4
Myrica salicifolia A. Rich	Myricaceae	4

(continued)

(continued)

Species name	Family	Relative frequency of occurrence (%)
Oxyanthus speciosus Dc.	Rubiaceae	4
Persea americana Mill	Lauraceae	4
Phoenix reclinata Jacq	Arecaceae	4
Psidium guajava L	Myrtaceae	4
Sapium ellipticum (krauss) pax	Euphorbiaceae	4
Senna petersiana (Bolle) Lock	Fabaceae	4
Sesbania sesban (L.) Merr	Fabaceae	4
Solanecio gigas (Vatke) C. Jeffrey	Asteraceae	4
Solanum giganteum Jacq	Solanaceae	4
Stereospermum kunthianum Cham	Bignoniaceae	4
Syzygium guineense (Willd.) DC	Myrtaceae	4
Urera hypselodendron (A. Rich.) Wedd	Urticaceae	4
Vachellia lahai (Steud. and Hochst. ex. Benth.) Kyal. and Boatwr.	Fabaceae	4
Vangueria apiculata K. Schum	Rubiaceae	4
Vangueria madagascariensis Gmel	Rubiaceae	4
Ximenia americana L	Olacaceae	4

References

Abiyot B, Sebsebe D, Zerihun W, Motuma D (2017) Woody species composition and structure of Kuandisha afromontane forest fragment in northwestern Ethiopia. J for Res 28(2):343–355

Abyot D, Teshome S, Ensermu K, Abiyou T (2014) Diversity, structure and regeneration status of the woodland and riverine vegetation of Sire Beggo in Gololcha District Eastern Ethiopia. Momona Ethiop J Sci 6(1):70–96

Aerts R, Van Overtveld K, Haile M, Hermy M, Deckers J, Muys B (2006) Species composition and diversity of small Afromontane forest fragments in northern Ethiopia. Plant Ecol 187(1):127–142

Aerts R, Van Overtveld K, November E, Wassie A, Abiyu A, Demissew S, Daye DD, Giday K, Haile M, TewoldeBerhan S, Teketay D, Teklehaimanot Z, Binggeli P, Deckers J, Friis I, Gratzer G, Hermy M, Heyn M, Honnay O, Paris M, Sterck FJ, Muys B, Bongers F, Healey JR (2016) Conservation of the Ethiopian church forests: threats, opportunities and implications for their management. Sci Total Environ 551–552:404–414

Alelign A, Teketay D, Yemshaw Y, Edwards S (2007) Diversity and status of regeneration of woody plants on the peninsula of Zegie, Northwestern Ethiopia. Trop Ecol 48(1):37–49

Aynekulu E, Denich M, Tsegaye D, Aerts R, Neuwirth B, Boehmer HJ (2011) Dieback affects forest structure in a dry Afromontane forest in Northern Ethiopia. J Arid Environ 75(5):499–503

Bekele T (1994) Phytosociology and ecology of a humid Afromontane forest on the Central Plateau of Ethiopia. J Veg Sci 5(1):87–98

Bekele-Tesemma A (2007) Useful trees and shrubs of Ethiopia: Identification, Propagation and Management for 17 Agroclimatic Zones. In: Bo T, Ensermu K, Sebsibe D, Maundu P (eds) RELMA in ICRAF Project, Nairobi Kenya

Bhagwat SA, Rutte C (2006) Sacred groves: potential for biodiversity management. Front Ecol Env 4(10):519–524

Bishaw B (2001) Deforestation and land degredation in the Ethiopian highlands: a strategy for physical recovery. Northeast Afr Stud 8(1):7–25

Bongers F, Wassie A, Sterck F, Bekele T, Teketay D (2006) Ecological restoration and church forests in Northern Ethiopia. J. Dry Lands 1(1):35–44

Burkhard B, Kroll F, Nedkov S, Müller F (2012) Mapping ecosystem service supply, demand and budgets. Ecol Indic 21:17–29

Cardelús CL, Scull P, Wassie Eshete A, Woods CL, Klepeis P, Kent E, Orlowska I (2017) Shadow conservation and the persistence of sacred church forests in Northern Ethiopia. Biotropica 49(5):726–733

Cardelu's C, Woods C, Bitew MA, Dexter S, Scull P, Tsegay B (2019) Human disturbance impacts the integrity of sacred church forests, Ethiopia. PLoS One 14(3):1–14

Contreras-Hermosilla (2000) The underlying causes of forest decline the CGIAR system. In CIFOR Occas. Paper, pp 1–25

Dai N, Kenji S, Akiko S (2002) Seedling establishment of deciduous trees in various topographic positions. J Veg Sci 13:35–44

Darbyshire I, Lamb H, Umer M (2003) Forest clearance and regrowth in northern Ethiopia during the last 3000 years. Holocene 13(4):537–546

Dudley N, Higgins-Zogib L, Mansourian S (2009) The links between protected areas, faiths, and sacred natural sites. Conserv Biol 23(3):568–577

Edwards S, Tadesse M, Hedberg I (1995) Flora of Ethiopia and Eritrea Canellaceae to Euphorbiaceae. In: Sue E, Mesfin T, Inga H (eds) Natl. Herb. Ethiop. The National Herbarium Addis Ababa University, Addis Ababa, and Department of Systematic Botany, Uppsala University,Uppsala, Sweden

Edwards S, Demissew S, Hedberg I (1997) Flora of Ethiopia and Eritrea Hydrocharitaceae to Arecaceae. In: Sue E, Sebsebe D, Inga H (eds) The National Herbarium Addis Ababa University, Addis Ababa, and Department of Systematic Botany, Uppsala University,Uppsala, Sweden

Edwards S (1997) Flora of Ethiopia. In: Inga H, Sue E (eds) The national herbarium Addis Ababa University, Addis Ababa, and Department of Systematic Botany, Uppsala University,Uppsala, Sweden

Engelbrecht BMJ, Comita LS, Condit R, Kursar TA, Tyree MT (2007) Drought sensitivity shapes species distribution patterns in tropical forests. Nature 447

EOTC (no date) Metsahef Senkesar and Areganone holy books

Frankl A, Nyssen J, Adgo E, Wassie A, Scull P (2019) Can woody vegetation in valley bottoms protect from gully erosion? Insights using remote sensing data (1938–2016) from subhumid NW Ethiopia. Reg Environ Change 19(7):2055–2068

Gashaw T, Terefe H, Soromessa T, Ahmed S (2015) Riparian areas rehabilitation and restoration: an overview. Point J Agric Biotechnol Res 1(2):55–63

Gebeyehu G, Soromessa T, Bekele T, Teketay D (2019) Species composition, stand structure, and regeneration status of tree species in dry Afromontane forests of Awi Zone, Northwestern Ethiopia. Ecosyst Health Sustain 5(1):199–215

Giliba RA, Boon EK, Kayombo CJ, Musamba EB (2011) Species composition, richness and diversity in Miombo Woodland of Bereku Forest Reserve Tanzania. Biodiversity 2(1):1–7

Hailu BT, Maeda EE, Heiskanen J, Pellikka P (2015) Reconstructing pre-agricultural expansion vegetation cover of Ethiopia. Appl Geogr 62:357–365

Hedberg I, Edwards S, Nemomissa S (2003) Flora of Ethiopia and Eritrea Apiaceae to Dipsaceae. In: Inga H, Sue E, Sileshi N (eds) The National Herbarium Addis Ababa University, Addis Ababa, and Department of Systematic Botany, Uppsala University,Uppsala, Sweden

Help CHR, Herman PMJ, Soetaert K (1998) Indices of diversity and evenness. Oceanis 24(4):61–87

Hokkanen PJ, Kouki J, Komonen A (2009) Nestedness, SLOSS and Conservation Networks of Boreal Herb-Rich Forests Nestedness, SLOSS and conservation networks of boreal herb-rich forests. Appl Veg Sci 12(3):295–303

Hubbell SP, Ahumada JA, Condit R, Foster RB (2001) Local neighborhood effects on long-term survival of individual trees in a neotropical forest. Ecol Res 16:859–875

Kyalangalilwa B, Boatwright JS, Daru BH, Maurin O, Bank MVD (2013) Phylogenetic position and revised classification of Acacia s. l. (Fabaceae: Mimosoideae) in Africa, including new combinations in Vachellia and Senegalia. Bot J Linn Soc 172:500–523

Lamprecht H (1989) Silviculture in the Thopics. In: Tropical forestry ecosystems. Their Tree species possibilities methods their long-term utilization. Technical Cooperation Federal Republic of Germany

Lemma H, Admasu T, Dessie M, Fentie D, Deckers J, Frankl A, Poesen J, Adgo E, Nyssen J (2018) Revisiting lake sediment budgets: how the calculation of lake lifetime is strongly data and method dependent. Earth Surf Process Landforms 43(3):593–607

Logan WEM (1946) An introduction to the forests of central and southern Ethiopia. Imp For Inst 24:65

Malik ZA, Bhatt AB (2016) Regeneration status of tree species and survival of their seedlings in kedarnath wildlife sanctuary and its adjoining areas in Western Himalaya. Trop Ecol 54(4):677–690

Maua JO, MugatsiaTsingalia H, Cheboiwo J, Odee D (2020) Population structure and regeneration status of woody species in a remnant tropical forest: a case study of South Nandi forest Kenya. Glob Ecol Conserv 21:e00820. https://doi.org/10.1016/j.gecco.2019.e00820

Meragiaw M, Woldu Z, Martinsen V, Singh BR (2018) Woody species composition and diversity of riparian vegetation along the Walga River Southwestern Ethiopia. PLoS ONE 13(10):1–18

Morgan R, KarimAly K, Asfaw Z (2018) Human ecology of sacred space: Church forests in the highlands of Northwestern Ethiopia. Enviromental Conserv 45(3):291–300

Mori SA, Boom Baian M, de Carvalho AM, dos Santos TS (1983) Sothern Bahian moist forests. New York Bot Gard 49(2):155–232

Mueller DD, Ellenberg H (1975) The count-plot method and plotless sampling techniques. In: Aims methods vegetation ecology, pp 158–159

Muscolo A, Bagnato S, Sidari M, Mercurio R (2014) A review of the roles of forest canopy gaps. J for Res 25(4):725–736

Nyssen J, Poesen J, Moeyersons J, Deckers J, Haile M, Lang A (2004) Human impact on the environment in the Ethiopian and Eritrean highlands—a state of the art. Earth-Science Rev 64(3–4):273–320

Orlowska I, Klepeis P (2018) Ethiopian church forests: a socio-religious conservation model under change. J. East. African Stud 12(4):674–695

Peel MC, Finlayson BL, Mcmahon TA, Peel MC, Finlayson BL, Updated TAM (2007) Updated world map of the Köppen-Geiger climate classification To cite this version: HAL Id: hal-00298818 Updated world map of the K oppen-Geiger climate classification. Hydrol Earth Syst Sci Discuss 4(2):439–473

Peet RK (1974) The measurement of species diversity. Annu Rev Ecol Syst 5(1):285–307

Reynolds T, Collins CD, Wassie A, Liang J, Briggs W, Lowman M, Sisay TS, Adamu E (2017) Sacred natural sites as mensurative fragmentation experiments in long-inhabited multifunctional landscapes. Ecography (cop.) 40:144–157

Reynolds T, Sisay ST, Wassie A, Lowman M (2015) Sacred natural sites provide ecological libraries for landscape restoration and institutional models for biodiversity conservation. Policy Br 2015 U.N. Glob Sustain Dev Report, pp 1–4. Available at: https://sustainabledevelopment.un.org/content/documents/614059-Sacrednaturalsitesprovideecologicallibrariesforlandscaperestorationandinstitutionalmodelsforbiodi.pdf

Sabine G, Miehe G (1994) Ericaceous forests and heathlands in the Bale Mountains of South Ethiopia: ecology and man's impact. T. Warnke Verlag, Hamburg

Scull P, Cardelús CL, Klepeis P, Woods CL, Frankl A, Nyssen J (2017) The resilience of Ethiopian Church forests: interpreting aerial photographs. Land Degradation Dev 1938–2015458:450–458

Song C, Nigatu L, Beneye Y, Abdulahi A, Zhang L (2018) Mapping the vegetation of the Lake Tana basin, Ethiopia, using Google Earth images, pp 2033–2041

Tadele D, Lulekal E, Damtie D, Assefa A (2014) Floristic diversity and regeneration status of woody plants in Zengena Forest, a remnant montane forest patch in northwestern Ethiopia. J For Res 25(2):329–336

Teketay D, Kashe K, Madome J, Kabelo M, Neelo J, Mmusi M (2018) Enhancement of diversity , stand structure and regeneration of woody species through area exclosure: the case of a mopane woodland in northern Botswana. Ecol Process 7(5). https://doi.org/10.1186/s13717-018-0116-x

Tesfaye G, Teketay D, Fetene M (2002) Regeneration of fourteen tree species in Harenna forest, southeastern Ethiopia. Flora 197:461–474

Thijs KW, Aerts R, Van de Moortele P, Musila W, Gulinck H, Muys B (2014a) Contrasting cloud forest restoration potential between plantations of different exotic tree species. Restor Ecol 22(4):472–479

Thijs KW, Aerts R, Musila W, Siljander M, Matthysen E, Lens L, Pellikka P, Gulinck H, Muys B (2014b) Potential tree species extinction, colonization and recruitment in Afromontane forest relicts. Basic Appl Ecol 15(4):288–296

Wassie A, Teketay D, Powell N (2005) Church forests provide clues to restoring ecosystems in the degraded highlands of Northern Ethiopia. Ecol Restor 23(2):115–144

Wassie A, Teketay D, Powell N (2005) Church forests in north gonder administrative zone, northern Ethiopia. For Trees Livelihoods 15:349–373

Wassie A, Sterck FJ, Teketay D, Bongers F (2009) Effects of livestock exclusion on tree regeneration in church forests of Ethiopia. For Ecol Manage 257(3):765–772

Wassie A, Sterck FJ, Teketay D, Bongers F (2009) Tree regeneration in church forests of Ethiopia: effects of microsites and management. Biotropica 41(1):110–119

Wassie A, Sterck FJ, Bongers F (2010) Species and structural diversity of church forests in a fragmented Ethiopian Highland landscape. J Veg Sci 21(5):938–948

Wassie (2007) Ethiopian church forests opportunities and challenges for restauration. Doctoral Dissertation, Wageningen University Wageningen, Netherlands. https://doi.org/10.1017/S0014479702003046

Woldemedhin T, Teketay D (2016) Forest conservation tradition of the Ethiopian Orthodox Tewahdo Church: a case study in West Gojjam Zone, north-western Ethiopia. Symb Bot Ups 38(16):57–73

Woods CL, Cardelús CL, Scull P, Wassie A, Baez M, Klepeis P (2017) Stone walls and sacred forest conservation in Ethiopia. Biodivers Conserv 26(1):209–221

Zegeye H, Teketay D, Kelbessa E (2011) Diversity and regeneration status of woody species in Tara Gedam and Abebaye forests, northwestern Ethiopia. J for Res 22(3):315–328

Ziter CD, Pedersen EJ, Kucharik CJ, Turner MG (2019) Scale-dependent interactions between tree canopy cover and impervious surfaces reduce daytime urban heat during summer. Proc Natl Acad Sci USA 116(15):7575–7580

Chapter 11
Woody Species Composition, Diversity, Structure and Uses of Selected Church Forests in Central Ethiopia

Eguale Tadesse Kifle, Asia Khamzina, Yowhan Son, Florent Noulèkoun, Abdu Abdelkadir, and Wubalem Tadesse

Abstract The objectives of this study were to investigate woody species diversity, species richness, and identify endangered plant species of three church forests. The churches were located across three agro-ecologies. Transect lines and plots were established for woody species inventory. All woody species with a diameter at breast height of ≥5 cm were considered. Species richness, α-diversity, β-diversity, and Sørensen similarity were calculated. Results revealed the presence of 34, 17, and 27 woody species in Assela Teklehymanot (Church in highland agroecology), Etisa Teklehymanot (Chuch in mid-lowland agroecology), and Saramba Kidanemhret (Church in the lowland agroecology), respectively. The importance value index (IVI) indicated native tree species *Podocarpus falcatus* (IV = 164.59), *Croton macrostachyus* (IV = 80.28), and the exotic *Eucalyptus globulus* (IV = 199.06) are important in church 1, church 2, and church 3, respectively. Moreover, 9 IUCN Red List of Threatened Species recorded in the study. The conservation and additional enrichment planting with the indigenous woody species found in the respective church forests and from similar other areas by performing species-site matching is recommended.

Keywords Conservation · Diversity index · Endangered species · Species richness

E. T. Kifle (✉) · A. Khamzina · Y. Son · F. Noulèkoun
Department of Environmental Science and Ecological Engineering, Korea University, 145 Anam-ro, Seongbuk-gu, Seoul 02841, Korea

A. Khamzina
e-mail: asia_khamzina@korea.ac.kr

Y. Son
e-mail: yson@korea.ac.kr

F. Noulèkoun
e-mail: bigflo@korea.ac.kr

E. T. Kifle · A. Abdelkadir · W. Tadesse
Ethiopian Environment and Forest Research Institute (EEFRI), 24536 Addis Ababa, Ethiopia

M. Kindu et al. (eds.), *State of the Art in Ethiopian Church Forests and Restoration Options*, https://doi.org/10.1007/978-3-030-86626-6_11

11.1 Introduction

Deforestation has been taking place since agriculture has the main livelihood source for humanity (Hance and Jeremy 2008). In Ethiopia, the main pressure on the forest remains to be land clearing for crop and cattle farming (Adam and David 2008). Deforestation of forest ecosystems has also been traced to economic incentives that make forest conversion appear more profitable than forest conservation in Ethiopia (Ref). In Ethiopia, deforestation is an ongoing issue that is causing extinction, changes to climatic conditions, desertification, and displacement of indigenous people (Rojahn 2006). The overall effect of such a loss and widespread forest degradation is a decline in environmental goods and services, including climate stabilization reduction, human well-being, and loss of biodiversity in general (Lamb et al. 2013).

Biodiversity is an attribute of an area and specifically refers to the variety within and among living organisms and is the total variety of life on earth comprising genes, species, ecosystems, and the ecological process as a whole (DeLong 1996). The diversity of woody species is central to total forest biodiversity since woody species deliver resources and habitats for almost all other species obtainable in the forest (Huange 2003). Ethiopia is an important regional center for biological diversity due to its wide range of altitude, diverse topography with high and rugged mountains, mountain ridges, flat-topped plateaus, deep gorges, notched stream gorges, and undulating plains (Teketay 2002; Zegeye et al. 2011). Moreover, the presence of a number of relic isolated forest trees, even on farmlands or patches of forests around churchyards and religious burial grounds in the country indicate the occurrence of well-diversified forests (Bekele 1993).

Church forests, befalling around more than 30,000 churches in Ethiopia, persevere vital biodiversity in a rather degraded landscape (Bongers et al. 2006; Cardelús et al. 2012). The value of these church forests in continuing to offer ecosystem services and serve as 'stepping stones' for restoration will depend on their long-term sustainability. Recent records show that populations of the tree species in the church forests are small, and decreasing in extent over time (Wassie 2002; Bingelli et al. 2003; Wassie et al. 2005).

Ethiopian churches and monasteries have a long-standing tradition of preserving and conserving their forests, including the many native plants and animals within them (Wassie et al. 2005; Bongers et al. 2006). Additionally, sacred groves as a whole can be particularly useful for conserving biodiversity by covering a wider variety of habitats than would be covered by a single large area (Bongers et al. 2006).

Ethiopian Orthodoc Tewahido Church (EOTC) has a long history of conservation of forest resources that encircle the churches besides its religious activities (Wassie et al. 2009). Church gardens are serving as in situ conservation and hotspot sites for biodiversity resources, especially for indigenous trees and shrubs of Ethiopia. The positive attitude to the resources sheltered by the church and the acceptance of the church tradition is also an opportunity for forest ecosystem conservation and restoration (Wassie 2002).

The EOTC conservation trend bases on its strong attachment to the words of God. The Bible teaches about environmental protection to be undertaken by human beings. For instance, in Genesis 2:15 it is written: "The Lord God took the man and put him in the Garden of Eden to work it and keep it." It is also written by the prophet Ezekiel 34:18 "Is it not enough for you to feed on the good pasture, that you must tread down with your feet the rest of your pasture; and to drink of clear water, that you must muddy the rest of the water with your feet?" (Ezekiel 34:18).

Studies of Church forest have focused mostly on those in northern parts of the country, Amhara and Tiogray regions. Few studies have been done on church forests in central Ethiopia. Based on this, the species composition, structure, and diversity of woody species were examined in central Ethiopia church forests. The aim of the study was towards the management and conservation of the church forests. Consequently, the objectives of the study were: (1) to investigate and describe woody species composition, structure, and diversity and the species richness of the selected church forests, (2) to explore the uses of woody species, and (3) to identify threatened (endangered) woody species in the study areas for future conservation.

11.2 Materials and Methods

11.2.1 The Study Areas

Three Ethiopian Orthodox Tewahdo Church forests have been selected for this study. Site I is Assela Teklehaymanot, located 175 km from Addis Ababa (AA). It has 25 ha church forest. Its elevation ranges 2521–2581 m a.s.l., which belongs to Dega (cool highland) agro-climatic zone. Site 2 is Etisa Teklehaymanot found 75 km far from AA and has an area of 23 ha. The elevation ranges 1800–2301 m a.s.l, which belongs to weyandega (midland) agro-climatic zone. Site 3 is Saramba Kidanemhret, found 200 km from AA and has an area of 22 ha. The elevation ranged 2164–2251 m.a.s.l. showing midland agro-climatic zone similar to Site 2 (Fig. 11.1).

11.2.2 Woody Species Inventory

To collect data on woody species, transect lines were laid 100 m apart along an altitudinal gradient based on altitudinal variations. The distance between consecutive plots within transect lines was also 100 m. The size of a plot was 20 m × 20 m following Gillison (2001). The number of plots on a transect line was determined by its length. Information such as elevation and geographical position was recorded at the center of each plot using a Garmin GPS receiver.

Site 1 had three transects. Transects 1 and 2 had 4 plots each and transect 3 had 3 plots. Hence a total of 11 plots were established at Site 1. Site 2 had only two

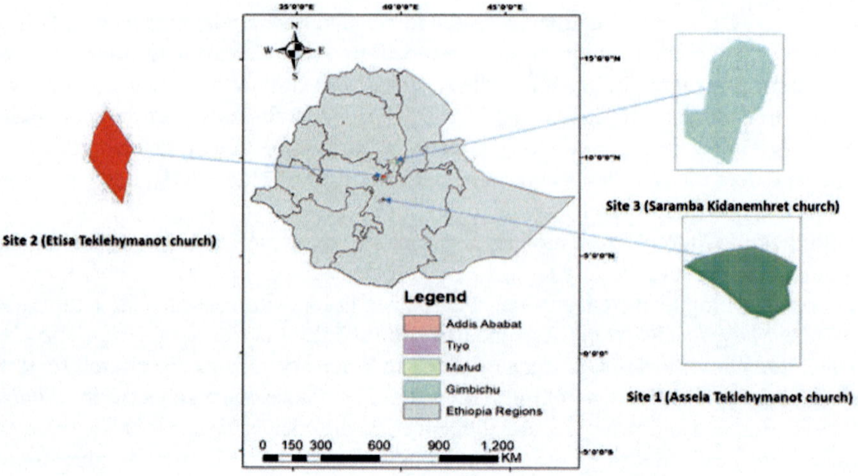

Fig. 11.1 Location of the study sites in central Ethiopia

transects. Transect 1 had 6 plots and transect 2 had only 1 plot. Therefore, the total number of plots on this site was 7. Site 3 had 2 transects. Transects 1 and 2 had 4 and 2 plots respectively. Hence, a total of 6 plots were established at Site 3.

Woody species encountered in the plots were surveyed and most identified in the field with the help of flora books on Ethiopia such as 'Flora of Ethiopia and Eritrea' (Edwards et al. 1995, 1997, 2000, 2003; Hedberg and Edwards, 1989; Hedberg et al. 2004, 2006), 'Useful Trees and Shrubs for Ethiopia' (Bekele et al. 2007). For the plant species difficult to identify in the field, voucher specimens were collected, pressed, and dried, and then taken to the herbarium laboratory of Addis Ababa University (Ethiopia) for identification.

All woody plants with a diameter at breast height, (DBH) $\geq$5 cm were measured using a caliper. For trees and shrubs that are forking at and below 1.3 m height, the DBH of each fork was measured and averaged.

11.2.3 Uses of Woody Species

A focus group discussion was undertaken after inventory of the woody species. The focus group discussants were provided with the list of woody species. Local names were used to grasp the indigenous knowledge of the local community. Finally, the group discussants have given the different uses of the woody species.

11.2.4 Data Analysis

The data were analyzed based on density, frequency, DBH, and basal area. The calculated indices were compared for the values of Shannon–Wiener, Simpson, and evenness indices among the three church forests. The analysis was performed using Generalized Linear Models (GLMs) with Gaussian distribution in SPSS software (v. 20) and Microsoft Excel 2010 for preliminary calculations.

The density and coverage area per ha were used as estimates of the relative standing yield of wood resources (Ludwig and Reynolds 1988; Kent and Coker 1992).

Relative density (RD), relative frequency (RF), and relative coverage (RC) were computed as:

$$RD = \frac{\text{Number of species (ns)}}{\text{Total number of individuals in a sample}} * 100\% \quad (11.1)$$

$$RF = \frac{\text{Frequency of species (nps/np)}}{\text{Total frequency of all species}} * 100\% \quad (11.2)$$

$$RC = \frac{\text{Coverage area of species (Cs)}}{\text{Total coverage area (C)}} * 100\% \quad (11.3)$$

Finally, the importance value (IV) was calculated as in Ludwig and Reynolds (1988):

$$IV = \frac{RD + RF + RC}{3} \quad (11.4)$$

where—***nps*** is the number of plots in which the species *S* appears and ***np*** is the total number of plots.

The church level wood resource was estimated in terms of the number of stems and basal area (coverage) (m^2) per ha for each church forest.

11.2.5 Diversity Indices

Because tree diversity can be measured at different levels, 3 biodiversity components were evaluated in this study in order to provide exhaustive biodiversity information: (i) α-diversity, as the mean species diversity of each study site; (ii) β-diversity, as both the difference in diversity among sites and among plots at each site; and (iii) effective species number, as the overall species richness at each study site (Cazzola Gatti et al. 2015). β-diversity was calculated by using Marczewski-Steinhaus index among plots of each study site (β-diversity within site) to evaluate internal small-scale diversity patterns.

Shannon diversity is the very widely used index for comparing diversity between various habitats (Clarke and Warwick 2001). The species diversity of the church forest in the selected forests was assessed using Shannon–Wiener diversity index (H'), Shannon evenness and Simpson's diversity index following Krebs (1989) and Magurran (2004).

β-diversity analysis was performed by computing the Marczewski-Steinhaus index to understand the dissimilarity between pairs of sites (i.e., β-diversity among sites). Because this measure considers only species presence and not species abundance data, it is not influenced by the dominance of certain species in the assemblage (Magurran 2013). Marczewski-Steinhaus index, the simplest of these diversity indices computes and compares the percentage similarity between two samples or communities (Marczewski and Steinhaus 1959, cited in Jakubowska 2011):

$$MS = \frac{c}{a + b - c} * 100 \tag{11.5}$$

where:

"*MS*" is the Marczewski-Steinhaus index, "*a*" is the total number of species in sample ≠ 1, "*b*" is the number of species in sample ≠ 2 and "*c*" is the number of species common to both samples.

11.3 Results

11.3.1 Woody Species Richness and Floristic Composition

The vegetation survey in the three church forests identified a total of 52 woody species (trees, shrubs, and lianas) belonging to 47 genera and 36 families. There were 34, 17, and 27 woody species in Site 1, Site 2, and Site 3 church forests, respectively.

In Site 1 there were a total of 27 families and 32 genera. The most represented family was Myrsinaceae having 3 woody species and 2 genera followed by Rosaceae (2 species and 2 genera), Euphorbiaceae (2 species and 2 genera), Cupressaceae (2 species and 2 genera), Fabaceae (2 species and 2 genera) and Oleaceae (2 species and 1 genus). The remaining 21 families had 1 species each and a total of 21 genera. The identified species were 56% trees, 41% shrubs, and 3% liana.

The importance value index in Site 1 indicated *Podocarpus falcatus* (Thunb.) R.Br. ex Mirb as the most important woody species (IV = 146.13) followed by *Maytenus arbutifolia* (A. Rich.) (IV = 105.47), *Rhoicissus tridentate* (L.f.) Wild & R.B. Drumm. (IV = 73.85), *Olinia rochetiana* A. Jussieu (IV = 71.26), *Juniperus procera* Hochst. ex Endl. (IV = 69.67) and *Rosa abyssinica* R. Br. ex Lindl. (IV = 69.23) (Appendix, Table 11.5).

In Site 2 17 species with 16 genera and 15 families were identified. The most represented families were Cupressaceae (with 2 species and 2 genera) and Moraceae (with 2 species and 1genus). The remaining 13 families had 1 species each and a total of 13 genera. The species identified were 53% trees, 41% shrubs, and 6% liana.

The importance value in Site 2 indicated *Croton macrostachyus* Hochst. was the most important woody species (IV = 26.76) followed by *Ficus thonningii* Blume (IV = 26.10), *Juniperus procera* Hochst. ex Endl. (IV = 24.15), *Euclea divinorum* Hiern (IV = 23.05), *Allophylus abyssinicus* Hochst. Radlk. (IV = 21.12) (Appendix, Table 11.6).

In Site 3 the survey identified 20 families with 26 genera. The families represented most are Anacardiaceae (2 species and 1 genus), Asteraceae (2 species and 2 genera), Euphorbiaceae (2 species and 2 genera), Fabaceae (2 species and 2 genera), Myrsinaceae (2 species and 2 genera), and Oleaceae (2 species and 2 genera). The rest 14 families had 1 species each and 15 genera. The habit of the species surveyed was 70% shrubs and 30% trees.

The importance value in Site 3 indicated the planted exotic tree species *Eucalyptus globulus* Labil. was the most important woody species (IVs = 66.35) followed by indigenous species *Dodonea angustitolia* L. f. (IV = 35.66), *Olea europaea* L. (IV = 30.50), *Rhus vulgaris* Meikle (IV = 23.59), *Jasminum grandiflorum* L. (IV = 23.50) and *Carissa spinarum* L. (IV = 23.10) and *Premna schimperi* Engl. (IV = 22.78) (Appendix, Table 11.7).

11.3.2 Woody Species Comparison and Structure in the Church Forests

There were 5 woody species common to all three church forests. Those species were *Eucalyptus globulus* Labil., *Osyris quadripartite* Salzm. ex Decne., *Croton macrostachyus* Hochst., *Juniperus procera* Hochst. ex Endl. and *Olea europaea* L. A similarity analysis was also carried out to evaluate the resemblance between the church forests based on the presence of woody species. Evaluation was conducted using the Sørensen similarity index as described in Sørensen (1984) and Marczewski-Steinhaus index.

The similarity as calculated by Sørensen similarity index was 26.1% between Site 1 and Site 2, 29.9% between Site 1 and Site 3, and 26.7% between Site 2 and Site 3. Similarly, Marczewski-Steinhaus index was calculated to determine the similarity of woody species in the 3 church forests. As a result Site 1 and Site 2 church forests had 18.22%, Site 1 and Site 3 forests had 27% and Site 2 and Site 3 church forests had 22.22% similarities.

The density of trees and shrubs with DBH $\geq$5 cm in highland church forest (Site 1) was 1058 stems ha^{-1}. The density is very high in the lower diameter classes, especially from 5 to 20 cm. On the other hand, the density was low in the higher diameter classes above 40.1 cm. In Site 2 the total number of stems DBH $\geq$5 cm was 305 stems ha^{-1}. The highest density was similarly found in the lower DBH classes 5–20 cm. The density in the higher DBH classes (i.e., greater than 20.1 cm)

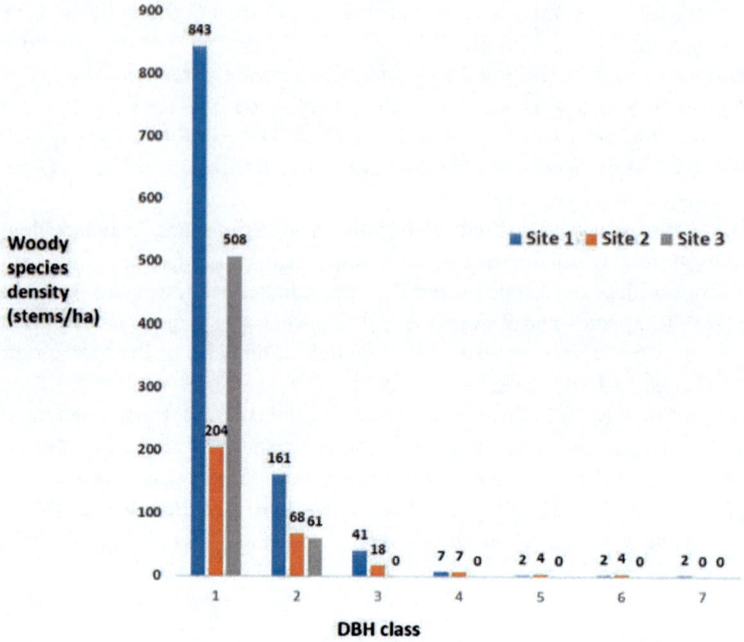

Fig. 11.2 Distribution of woody species density among DBH classes (1 = 5–10 cm, 2 = 10.1–20 cm, 3 = 20.1–40 cm, 4 = 40.1–60 cm, 5 = 60.1–80 cm, 6 = 80.1–100 cm, 7 ≥ 100.1 cm), in the study sites, central Ethiopia

was very small. In Site 3 the total density was concentrated only in the lower DBH classes (i.e., 5–20 cm) similar to Site 1. The density in the higher diameter classes (DBH >20.1 cm) was very small. In midland church forest (Site 3) the total density of woody species DBH ≥5 cm was 569 stems ha^{-1}(Fig. 11.2).

11.3.3 Woody Species Abundance and Diversity in the Church Forests

The results from the calculated Generalized Linear Models with Gaussian distribution revealed no significant variation in the mean Shannon–Wiener (H'), Simpson (D), and evenness indices among sites. The woody species richness and abundance of species in Site 1 was significantly higher than in Site 2 and similarly, in Site 3 was significantly higher than in Site 2 at p <0.05. Hence, Site 2 had a lower species richness and abundance (Table 11.1).

Table 11.1 Woody species richness, abundance, diversity, and evenness in the study church forests in central Ethiopia

Study site (forest)	Species richness	Mean species richness	Mean species abundance	Mean Shannon–Wiener diversity index	Mean Simpson diversity index	Mean species evenness (equitability)
Assela Teklehymanot (Site 1)	34	12.55 ± 1.41^{a}	106.55 ± 16.88^{c}	1.76 ± 0.08	0.74 ± 0.02	0.72 ± 0.03
Etisa Teklehymanot (Site 2)	17	4.71 ± 0.75^{b}	17.29 ± 12.38^{d}	1.19 ± 0.19	0.60 ± 0.07	0.81 ± 0.07
Saramba Kidanemhret (Site 3)	27	10.67 ± 1.86^{a}	71.79 ± 12.38^{c}	1.63 ± 0.32	0.67 ± 0.12	0.71 ± 0.12

Note Letters a, b, c, and d show the difference is significant at $p < 0.05$ i.e. similar letters indicate the difference is not significant at $p<0.05$

11.3.4 Uses of Woody Species

The uses of the species included cultural medicine for both humans and livestock, human food, fodder, construction wood, and making house utensils among others. More than 41% of the woody species were used for human medication, livestock medication, and/or both, revealing the supreme importance of plant-based medicine. Of these 34 plant species, 14 were used predominantly for medicinal purposes. The rest of the species (58%) found in the highland forest site were used for various purposes such as tying and fastening material, fodder, and construction wood (Table 11.2).

In Site 2, 35% of the woody species were used for human medicine, livestock medicine, and both for human and livestock medicine. The rest 65% of the woody species in this site are used for different purposes such as for fodder, construction, charcoal making, and fire wood (Table 11.3).

The survey of Site 3 has shown that more than 22% of the woody species were used as human medicine livestock medicine and both human and livestock medicine.

Table 11.2 Number of woody species richness used for various purposes in highland church forest (Site 1, Assela Teklehymanot), central Ethiopia

Use type	Number of woody species
Human medicine	6
Livestock medicine	5
Livestock and human medicine	3
For tying	3
Fodder	3
Construction	4
Food	3
House utensils	3
Fire wood	4

Table 11.3 Number of woody species richness used for various purposes in the midland mixed lowland church forest (Site 2, Etisa Teklehymanot), central Ethiopia

Use type	Number of woody species
Human medicine	3
Livestock medicine	1
Livestock and human medicine	2
Tying	2
Fodder	2
Construction	2
Food	1
House utensils	1
Fire wood	2
Charcoal making	1

Table 11.4 Number of woody species richness used for various purposes in the midland church forest (Site 3, Saramba Kidanemhret), central Ethiopia

Use type	Number of woody species
Human medicine	2
Livestock medicine	2
Livestock and human medicine	2
Tying	1
Fodder	7
Construction	1
Food	2
House utensils	2
Fire wood	3
Charcoal making	2
For washing utensils	1
Fencing	2

Fodder was obtained from 26% of the species and fire wood from 11% of the mentioned species. The rest 41% of the woody species were used for various uses such as for tying, construction, food, house utensils, and fencing (Table 11.4).

Of the 17 woody species found in this midland church forest site, 6 (35%) species were predominantly used for medicinal purposes. Of these species, 2 are used for human medicine, 2 are used for livestock medicine and 2 are used for both human and livestock medicine.

11.3.5 Threat and Pressure on the Forests

The loss of biodiversity is the world's known crisis as observed from day to day events of our surroundings. There is also tangible evidence from science and public media about the issue. The loss is still increasing as seen from the present study's social survey. The present study might be one example regarding the study of some species in the church forests that were surveyed through ethnobotanical and vegetation census.

Most of the problems mentioned particularly habitat loss and overexploitation are most prevalent in the surrounding areas of the church forests whereas church forests maintain some species including most endangered ones. The present study identified 14 woody and herbaceous species listed by "The IUCN Red List of Threatened Species" (https://www.iucnredlist.org/) in the three sites. The identified woody species for the conservation priority are: *Cupressus lusitanica*, *Juniperus procera*, *Prunus africana*, *Erica arborea*, *Pinus radiata*, *Ekebergia capensis*, *Embelia schimperi*, *Ferula communis*, and *Ficus carica*.

11.4 Discussions

11.4.1 *Woody Species Richness and Diversity*

The significant difference observed in species richness and abundance between Site 1 and Site 2 may be attributed to the elevation differences. In the present study, opposite trend was found as compared to a study by Ren et al. (2006) that stated the number of species will decrease with increasing altitude. Other factors such as physiographic factors may impact soil moisture, chemical properties, and also physical properties such as drainage, porosity, and other characteristics which influence the distribution and diversity of plant species (Enright et al. 2005).

The identified highest floristic composition of woody species in the present study was comparable to other areas with more or less similar elevations. The floristic composition of Site 1 (2521–2581 m a.s.l.) showed a different family and species representatives from a similar study in Arsi Negele, Ethiopia (1557–1650 m a.s.l.) (Kedir et al. 2015) because of their elevation differences.

Comparisons of the recorded species richness of the present church forests with other Afromontane forests revealed that the level of the diversity in the forests was low maybe because of the small area of the church forests that ranged from 22 to 25 ha. For instance, in Ethiopia, the Yayu forest had 220 (Woldemariam et al 2008), Chilimo forest 90 (Woldemariam et al. 2000), Dakata forest 202 (Teketay and Bekele 1995), Jibat forest 58 (Bekele 1994), and the Harrena forest 128 species (Tadesse and Nigatu 1996). The lesser number of woody species richness in the present study might also be because of the deforestation of indigenous tree shrub species and replacement of the indigenous species by exotic ones such as *Eucalyptus* species that have more economic value for immediate needs. Church administration might need the immediate incomes from the sale of exotic fast-growing species to pay salaries to the church workers as stated by the group discussants.

The findings of this study were higher than that found from burial sites in Tehuledre district, Ethiopia which had 28 species from the whole representative sites (Seid and Santini 2017) while the present study identified 52 different woody species from the 3 church forests. It was also higher when compared to a study in Tanzania (Sitati et al. 2016) that recorded a total of 26 woody species.

There were only 5 woody species common to the 3 church forests of the present study. In a similar study of church forests in Ethiopia by Bongers et al. (2006), the species similarity in studied forests that are found above 2300 m a.s.l., was very small (only 4 species) in common, and for those found below 2300 m a.s.l. 9 species in common.

On the other hand, the results obtained from the present 3 study sites indicated DBH class distribution was concentrated in the lower diameter class especially 5–10 cm followed by 10.1–20 cm diameter classes. The general trend was shown in Fig. 11.2 approached an inverted J-shape. According to studies by Senbeta (2006), Yeneger et al. (2008), and Fisaha (2013), this pattern of diameter classes indicated good potential for reproduction and enrollment of the forest. Therefore, the diameter class distribution in the present study showed a high potential for forest reproduction and staffing.

The basal area (coverage) calculated from the measured DBH also identified the highest coverage in Site 1 was *Podocarpus falcatus* (Thunb.) R.Br. ex Mirb. (6.66 m^2/ha) followed by the planted exotic species *Eucalyptus globulus* Labil. (1.11m^2/ha) and the other planted exotic species *Cupressus lusitanica* Mill. (0.75 m^2/ha). In Site 2 the highest basal area recorded for *Ficus thonningii* Blume (3.71 m^2/ha) followed by *Juniperus procera* Hochst. ex Endl. (2.04 m^2/ha) and *Croton macrostachyus* Hochst. (1.14 m^2/ha). In Site 3 the highest basal area was recorded for the planted exotic species *Eucalyptus globulus* Labil. (2.04 m^2/ha) followed by *Olea europaea* L. (0.46 m^2/ha) and *Myrsine africana* Linn. (0.17 m^2/ha). According to some studies (e.g., Bekele 1994) basal area is an important parameter for measuring the relative importance of plant species. Therefore, the woody plant species with a larger basal area in a forest was considered the most important species in that forest.

The woody species diversity in the present study was said to be lower (mean 1.56) as compared to similar studies in Ethiopia and elsewhere. For example in a study of church forests by Ayanaw (2016) the overall average Shannon–Wiener Diversity Index (H′) and the average evenness values for the entire forest were 2.88 and 0.79, respectively where the present study the highest Shannon diversity was 1.76 and the highest evenness index was 0.72 for Site 1 and less for the other 2 church forests. From a study in Tanzania (Sitati et al. 2016) the overall Shannon Wiener diversity index (H′) was 2.36 which was much higher than the present study's mean result (mean $H' = 1.56$).

11.4.2 Uses of Woody Species

The result of this study showed there are various uses of woody species. This is supported by a study in northern Ethiopia by Moravec et al. (2014), who explained that rural communities depend on plant resources for their domestic needs. Additionally, indigenous people all over the world are still using medicinal plants for treatment of various diseases (Akram et al. 2011; Ovesna et al. 2011).

In the present study, the representative life forms of the observed medicinal species are shrubs, trees, and climbers. Climbers have been mentioned in the list in Site 2 and Site 3. In Site 1 the number of medicinal shrubs and trees was equal while there are no climbers mentioned. A similar study by Moravec et al. (2014) recorded shrubs (21%) being used in traditional medicine in Tigray northern Ethiopia.

From the results of informal discussion and direct observation, there are no conservation activities to protect the different woody species that have varied uses. In contrast to the present study, studies in Mena Angetu district of western Ethiopia showed there was some conservation practice by the local people in cultivating 13.8 and 5.7% of the medicinal plants in home gardens by Tollosa and Moa (2018) and by Lulekal et al. (2008), respectively. Natural resources could be utilized best in a sustainable way if management activities are practiced (Yeneger et al. 2008). The requirement of appropriate action to do such valuable activities is irrefutable and so is the involvement of the full range of societies and stakeholders in the conservation, production, and management as well as the use of medicinal plants (Schippmann et al. 2002). Woody plants, particularly trees, provide by far the largest number of products harvested. The various tree-based products contribute to rural household welfare in a range of ways, providing food and non-food goods, inputs into income generating activities, and inputs into agricultural production.

11.4.3 Threatened Species in the Study Area

Threatened species found in the present study sites indicate the requirement of conservation of these species with special concern. According to Borokini (2013), the highest rates of biodiversity in the world are found in tropical regions' developing countries that lack resources to manage all these species. Additionally, the species have widely ranged population, which are as a result of widely ranged trends of human mistreatment, growth forms, and varying bio-geographical distribution Borokini (2013). Thus, setting priorities for conservation efforts for these species is imperative. Those species whose population have reduced drastically, with a slight range of bio-geographical distribution, rampant species, and species those who belong to monospecific genera need more concern.

Those species requiring urgent conservation efforts are said to be threatened or endangered (Borokini 2013). Based on this, the present study identified 9 different woody species listed as threatened. The identification of the species whether they are threatened or not in the present study was done using the website (www.iucnredlist.org). However, two of the 2 woody species (*Cupressus lusitanica* and *Pinus radiata*) identified as threatened in the present study are increasing in population in Ethiopia and hence their extinction is in doubt. The IUCN Red List is possibly the most useful worldwide list of species in danger of extinction (Lamoreux et al. 2003). Its usefulness is based on its reliance on a number of objective criteria (IUCN 2012).

In addition to extinction risk assessment, the Red List provides an excess of useful evidence on each species assessed, including dispersal-trends, threats, and management actions (Cardoso et al. 2016). Assessment of threatened species is done either as Critically Endangered (CE), Endangered (EN) or Vulnerable (VU). Extinct or non-threatened species are also assessed and listed (Cardoso et al. 2016). Hence, the identified plant species in the present study attract attention in the future conservation and management of the species.

11.5 Conclusion and Recommendations

Floristic composition and species richness in the church forests were optimum as compared to the surrounding degraded lands including cultivated lands but it was low as compared to other similar studies in Ethiopia and elsewhere. Highland site (Site 1) was comprised of big size trees and other shrubs and lianas than the mixed midland and lowland and midland sites. There was a higher similarity of woody species between Site 1 (highland) and Site 3 (midland) sites (29.9%). All the three church forests have less than 50% similarities indicating the need to enrich the church forests by diversified woody species (specifically indigenous ones). The diameter distribution of the 3 church forests showed the density of woody species was concentrated in the lower diameter classes indicating the high possibility of forest reproduction and enrollment. The woody species diversity of the present study was lower as compared to many studies in Ethiopia and elsewhere but better as compared to studies in some burial sites in Ethiopia. The uses of the woody species in the study area indicated the value of the trees, shrubs, and lianas for traditional medicine, food, fodder, and other multiple uses. Some 9 woody species have been found threatened in the study area that attracts future concern to their conservation. In the present study sites, there was a tendency to replace the indigenous species with exotic planted species such as *Eucalyptus globulus* Labil. and *Cupressus lusitanica* Mill. Which might be another threat for the conservation of natural indigenous forests. Therefore, Conservation of the present stand by fencing and local rules and religious sanctions, additional enrichment planting with the indigenous woody species found in the respective church forests, giving priorities for the IUCN red-listed species, planting other indigenous woody species by undertaking species-site matching and using the plots established for data collection as experimental plots for the indigenous species is recommended. Generally, the present remaining forests in the 3 church forests can also be used as a seed source for future management of the church forests.

Acknowledgments The study was supported financially by Ethiopian Environment and Forest Research Institute (EEFRI) and Korea International Cooperation Agency (KOICA), KOICA-Ethiopia project.

My greater gratitude also goes to my major advisor professor Yowhan Son and my co-advisor professor Asia Khamzina for their unprecedented help and encouragement to accomplish the writing of my research dissertation.

Appendix

See Tables 11.5, 11.6 and 11.7.

Table 11.5 Top 10 woody species in Site 1 (Highlan d, Asela Tekle Hymanot), central Ethiopia

Woody species	Family	Origin	Importance value (RCs + RFs + RDs)/3
Podocarpus falcatus (Thunb.) R.Br. ex Mirb	Podocarpaceae	Ind	146.13
Maytenus arbutifolia (A. Rich.)	Celastraceae	Ind	105.47
Rhoicissus tridentate (L.f.) Wild & R.B.Drumm	Vitaceae	Ind	73.85
Olinia rochetiana A.Jussieu	Oliniaceae	Ind	71.26
Juniperus procera Hochst. ex Endl	Cupressaceae	Ind	69.67
Rosa abyssinica R. Br. ex Lindl	Rosaceae	Ind	69.23
Dovyalis abyssinica (A. Rich.) Warb	Flacourtiaceae	Ind	65.00
Schefflera abyssinica A. Rich. Harms	Araliaceae	Ind	64.98
Osyris quadripartite Salzm. ex Decne	Santalaceae	Ind	64.45
Nuxia congesta R. Br. ex Fresen	Stilbaceae	Ind	58.36

Note 'Ind' stands for Indigenous and 'Exo' for exotic

Table 11.6 Top 10 woody species in Site 2 (mixed midland and Lowland, EtisaTekle Hymanot), central Ethiopia

Woody species	Family	Origion	Importance value (RCs + RFs + RDs)/3
Croton macrostachyus Hochst	Euphorbiaceae	Ind	26.76
Ficus thonningii Blume	Moraceae	Ind	26.10
Juniperus procera Hochst. ex Endl	Cupressaceae	Ind	24.15
Euclea divinorum Hiern	Ebenaceae	Ind	23.05
Allophylus abyssinicus Hochst. Radlk	Sapindaceae	Ind	21.12
Eucalyptus globulus Labil	Myrtaceae	Exo	16.57
Calpurnia aurea (Ait.) Benth	Fabaceae	Ind	12.43
Carissa spinarum L	Apocynaceae	Ind	11.88
Ficus sur Forssk	Moraceae	Ind	11.39
Olea europaea L	Oleaceae	Ind	10.80

Note 'Ind' stands for Indigenous and 'Exo' for exotic

Table 11.7 Top 10 woody species in Site 3 (midland, Saramba Kidanemhret church), central Ethiopia

Woody species	Family	Origin	Importance value (RCs + RFs + RDs)/3
Eucalyptus globulus Labil	Myrtaceae	Exo	66.35
Dodonea angustitolia L. f	Sapindaceae	Ind	35.66
Olea europaea L	Oleaceae	Ind	30.50
Rhus vulgaris Meikle	Anacardiaceae	Ind	23.59
Jasminum grandiflorum L	Oleaceae	Ind	23.50
Carissa spinarum L	Apocynaceae	Ind	23.10
Premna schimperi Engl	Lamiaceae	Ind	22.78
Osyris quadripartite Salzm. ex Decne	Santalaceae	Ind	17.39
Maytenus gracilipes Welw. ex Oliv	Celastraceae	Ind	17.23
Croton macrostachyus Hochst	Euphorbiaceae	Ind	12.41

Note 'Ind' stands for Indigenous and 'Exo' for exotic

References

Adam D (2008) Global deforestation figures questioned. The Guardian

Ayanaw (2016) Woody species diversity structure and regeneration status of Yemrehane Kirstos church forest of Lasta Woreda, north Wollo zone, Amhara region, Ethiopia. M.Sc. thesis (Unpublished). Addis Ababa University, Addis Ababa, Ethiopia

Bekele T (1993) Vegetation ecology of remnant afromontane forests on the central plateau of Shewa, Ethiopia. Acta Phytogeographica Suecica 79:1–59

Bekele T (1994) Studies on remnant afromontane forests on central plateau of Shewa. Ph.D. thesis Uppsala University, Uppsala, Sweden

Bekele T, Lemenih M, Tadesse T, Amaha A (2007) Determining forest carrying capacity in PFM/JFM sites: the case of Adaba-Dodolla forest. In: Kelbesa E, De Stoop C (eds.) Participatory forest management (PFM), Biodiversity and livelihoods in Africa. Proceedings of the international conference. FARM Africa/SOS Sahel, Addis Ababa, Ethiopia, pp 202–213

Bongers F, Wassie A, Sterck FJ, Bekele T, Teketay D (2006) Ecological restoration and church forests in northern Ethiopia. J Drylands 1:35–45

Borokini TI, Onefeli AO, Babalola FD (2013) Inventory analysis of Milicia excelsa (Welw. C.C. Berg.) in Ibadan (Ibadan Metropolis and University of Ibadan), Nigeria. J Plant Stud 2(1):97–109. https://doi.org/10.5539/jps.v2n1p97

Cardoso P, Stoev P, Georgiev T, Senderov V, Penev L (2016) Species conservation profiles compliant with the IUCN red list of threatened species. Biodiversity Data J 4:e10356. https://doi.org/10.3897/BDJ.4.e10356

Cazzolla Gatti R, Castaldi S, Lindsell JA, Coomes DA, Marchetti M, Maesano M, Di Paola A, Paparella F, Valentini R (2015) The impact of selective logging and clearcutting on forest structure, tree diversity and above-ground biomass of African tropical forests. Ecol Res 30(1):119–132. https://doi.org/10.1007/s11284-014-1217-3

Clarke KR, Warwick RM (2001) Changes in marine communities: an approach to statistical analysis and interpretation, 2nd edn. Plymouth, PRIMER-E

DeLong DC (1996) Defining biodiversity. Wildl Soc Bull 24:738–749

Edwards S, Mesfin T, Sebsebe D, Hedberg I (2000) Flora of Ethiopia and Eritrea, Part 1. Magnoliacea to Flacourtiacea. The National Herbarium, Addis Ababa University, Addis Ababa and Uppsala

Edwards S, Sebsebe D, Hedberg I (1997) Flora of ethiopia and eritrea, vol 6, Hydrocharitaceae to Arecaceae. The National Herbarium, Addis Ababa University, Addis Ababa and Uppsala

Edwards S, Tadesse M, Hedberg I (1995) Flora of ethiopia and eritrea, vol 2(2), Canellaceae to Euphorbiaceae. The National Herbarium, Addis Ababa University, Addis Ababa and Uppsala

Enright NJ, Miller BP, Akhtar R (2005) Desert vegetation and vegetation-environment relationships in Kirthar national park, Sindh, Pakistan. J Arid Environ 61:397–418

Fisaha G, Hundera K, Dalle G (2013) Woody plants' diversity, structural analysis and regeneration status of Wof-Washa Natural Forest, north-east Ethiopia. Afr J Ecol 51:599–608

Food and Agriculture Organization of The United Nations (FAO) (2007) State of the World's Forests, FAO, Forestry Department, p 144

Hance J (2008) Tropical deforestation is one of the worst crises since we came out of our caves. Mongabay.com/a place out of time: tropical rainforests and the perils they face. http://news.mongabay.com/2008/0515-hance_myers.html

Hedberg I, Edwards S (1989) Flora of Ethiopia, Pittosporaceae to Araliaceae. The National Herbarium, Addis Ababa University, Addis Ababa/Asmara and Uppsala

Hedberg I, Edwards S, Sileshi N (2003) Flora of Ethiopia and Eritrea, Part 1. Apiaceae to Dipsacaceae. The National Herbarium, Addis Ababa University, Addis Ababa and Uppsala

Hedberg I, Friis I, Edwards S (2004) Flora of ethiopia and eritrea, part 2. Asteraceae (Compositae). The national herbarium, Addis Ababa University, Addis Ababa and Uppsala

Hedberg I, Kelbesa E, Edwards S, Demisew S, Persson E (2006) Flora of Ethiopia and eritrea, Gentianaceae to Cyclocheilaceae. The National Herbarium, Addis Ababa University, Addis Ababa and Uppsala

Huange W, Pohjonen V, Johansson S, Nashanda M, Katigula MIL, Luukkanen O (2003) Species diversity, forest structure and species composition in Tanzanian tropical forests. For Ecol Manage 173:11–24. https://doi.org/10.1016/S0378-1127(01)00820-9

International Union for Conservation of Nature IUCN (2012) IUCN Red List Categories and Criteria: Version 3.1. IUCN, Gland, Switzerland and Cambridge, UK, 32 pp. ISBN 978-2-8317-1435-6

Jakubowska M (2011) Quality structure of the Noctuidae family in selected habitats of the agricultural landscape. J Plant Protect Res 51(2):166–170. https://doi.org/10.2478/v10045-011-0029-5

Kent M, Coker P (1992) Vegetation description and analysis A practical approach. BP Press, London, UK. p, p 363

Krebs CJ (1989) Ecological methodology. Harper Collins Publishers, New York, USA

Lamb D, Gilmour D (2013) Rehabilitation and restoration of degraded forests. IUCN and WWF, Gland, Switzerland

Lamoreux J, Akçakaya HR, Bennun L, Collar NJ, Boitani L, Brackett D, Bräutigam A, Brooks TM, da Fonseca GB, Mittermeier RA, Rylands AB, Gärdenfors U, Hilton-Taylor C, Mace G, Stein BA, Stuart S (2003) Value of the IUCN red list. Trends Ecol Evol 18(5):214–215. https://doi.org/10.1016/s0169-5347(03)00090-9

Ludwig JA, Reynolds JF (1988) Statistical ecology: a primer on methods and computing. Wiley, New York, p 337

Lulekal E, Kelbessa E, Bekele T, Yineger H (2008) An ethnobotanical study of medicinal plants in Mana Angetu District, southeastern Ethiopia. J Ethnobiol Ethnomed 4:10. https://doi.org/10.1186/1746-4269-4-10

Magurran AE (2004) Measuring biological diversity. Blackwell Science, Oxford, UK, p 256

Magurran AE (2013) Measuring biological diversity. Wiley, Oxford, UK, p 264. [Online] http://books.google.com/books?id=fIjsaxmL_S8C

Ren HB, NiuS K, Zhang LY, MaK P (2006) Distribution of vascular plant species richness along an elevational gradient in the Dongling mountains, Beijing China. J Integr Plant Biol 48(2):153–160

Rojahn A (2006) Incentive mechanisms for a sustainable use system of the montane rain forest in Ethiopia: PhD thesis. Economics Faculty, Christian Albrechts University of Kiel, Germany

Schippmann U, Leaman D, Cunningham AB (2002) Impact of cultivation and gathering of medicinal plants on biodiversity: global trends and issues. FAO, Rome, Italy

Seid MA, Santini G (2017) A survey on species diversity, abundance and community structure of woody plants in burial sites in Gobeya rural administrative of Tehuledere district, South-Wollo zone, Ethiopia. J Ecol Nat Environ 9(4):62–70. https://doi.org/10.5897/JENE2017.0634

Senbeta F (2006) Biodiversity and ecology of Afromontare rainforests with wild *Coffea arabica* L. populations in Ethiopia. Ecology and development series No. 38, Center for Development Research, University of Bonn, Bonn, Germany

Sitati N, Warinwa F, Maina M, Sumba DS (2016) Tree species diversity and dominance in Ketumbeine forest reserve, Tanzania. J Biodiv Manage Forest 5:3. https://doi.org/10.4172/2327-4417.1000161

Sørensen T (1984) A method of establishing groups of equal amplitude in plant sociology based on similarity of species content and its application to analyses of vegetation on Danish commons. Biol J Lin Soc 5(4):1–34

Tadesse M, Nigatu L (1996) An ecological and ethnobotanical study of wild or spontaneous coffee, *Coffea arabica* in Ethiopia. In: van der Maesen LGJ, van der Burgt XM, van Eedenbach de Rooy JM (eds) The biodiversity of African plants. Kluwer Academic Publishers, Dordrecht, The Netherlands, pp 227–294

Teketay D (2002) Germination responses of *Discopodium penninervium* to temperature and light. Flora 197:76–80

Teketay D, Bekele T (1995) Floristic composition of Wof-Washa natural forest, Central Ethiopia: implications for the conservation of biodiversity. Feddes Repertorium 106:127–147

Tolossa TJ, Megersa M (2018) Ethnobotanical study of medicinal plants used to treat human diseases in berbere district, bale zone of Oromia regional state, south east Ethiopia. Evidence Complem Alternative Med 16:1–33. https://doi.org/10.1155/2018/8602945

Woldemariam T, Borsch T, Denich M, Teketay D (2008) Floristic composition and environmental factors characterizing coffee forests in south-west Ethiopia. For Ecol Manage 255:2138–2150

Woldemariam T, Teketay D, Edwards S, Olsson M (2000) Woody plant and avian species diversity in a dry Afromontane forest on the central plateau of Ethiopia: biological indicators for conservation. Ethiopian J Nat Res 2:255–293

Yineger H, Kelbessa E, Bekele T (2008) Floristic composition and structure of the dry afromontane forest at Bale mountains National Park Ethiopia. SINET. Ethiopian J Sci 31:103–120

Zegeye H, Teketay D, Kelbessa E (2011) Diversity and regeneration status of woody species in Tara Gedam and Abebaye forests, northwestern Ethiopia. J Forest Res 22:315–328. https://doi.org/10.1007/s11676-011-0176-6

Part IV
Restoration Options to the Surrounding Landscapes of Ethiopian Church Forests

Chapter 12
Church Forests as Sources for Forest Reproductive Material of Native Species and Their Possible Role as Starting Points for the Restoration of Degraded Areas in Ethiopia

Bernd Stimm, Mengistie Kindu, and Thomas Knoke

Abstract One of the major bottlenecks in restoration of degraded land areas worldwide is the lack of adequate numbers of high quality forest reproductive material, i.e., forest seeds and planting stocks. The objectives of our contribution are to give an overview of initiatives in forest landscape restoration with an emphasis in Ethiopia. We try to point out the possible purpose of Church forests as sources for reproductive material and to highlight the potential role of new church nurseries in high quality planting stock production. Native tree species should be an essential part of the ambitious projects in greening the country. Unfortunately, adequate amounts of native seedlings are not available. We can fall back on many years of experience from research projects in Ecuador, Egypt, Namibia, and other countries as well. We finish with recommendations for a conceptual framework on tree seed procurement and its implementation for the purpose of successful restoration of Ethiopian forests and end up with a plea for the country's efforts on conservation of biodiversity and national culture, which may act as a regional revitalization and demonstration object for innovative environmental enhancement and job creation for the people.

Keywords FGR conservation · Seed procurement · Seed supply · Planting stock

B. Stimm (✉)
Ecosystem Dynamics & Forest Management in Mountain Landscapes, TUM School of Life Sciences Weihenstephan, Technical University of Munich, Freising, Germany
e-mail: bernd.stimm@tum.de

M. Kindu · T. Knoke
Institute of Forest Management, TUM School of Life Sciences Weihenstephan, Technical University of Munich, Freising, Germany
e-mail: mengistie@tum.de

T. Knoke
e-mail: knoke@forst.wzw.tum.de

M. Kindu et al. (eds.), *State of the Art in Ethiopian Church Forests and Restoration Options*, https://doi.org/10.1007/978-3-030-86626-6_12

12.1 Introduction

Ongoing land degradation is globally a very serious problem. A group of scientists (Willemen et al. 2020) engaged in the Intergovernmental Science-Policy Platform on Biodiversity and Ecosystem Services (IPBES) gives a statement on the assessment of land degradation and restoration, where they mention that evidence is provided that land degradation is avoidable, and in many instances reversible. But why has the issue failed to attract global attention in a similar way to climate change? The scientists mentioned five systemic reasons, and ten strategies to overcome the five systemic policy barriers and, thus, transform the effectiveness of land protection and restoration. The recommendations are focused mainly on socio-political stakeholders, i.e., decision makers, scientists, and citizens. Willemen et al. (2020) are well aware that land degradation is a typically local, visible, and immediate issue, and successful countermeasures must put emphasis also on practice-oriented implementation. Support could be provided through the United Nations (UN), which has announced 2021 as the start of the Decade on Ecosystem Restoration (Aronson et al. 2020). While the challenges are great, so are the opportunities. The landscape approach will be an important vehicle to provide countries with implementation packages tailored to a wide range of landscapes and facilitate scaling-up (Stanturf et al. 2019).

If we have a look at the present situation in Ethiopia, with a land area of about 113 million ha, a forest and tree cover (30%) extent (basis of 2018) of 12 million ha (including 1.8 million ha of primary forests), we have to recognize that the forest/tree cover share of the country is only 10.7% (Mongabay 2018). In the past two decades (between 2001 and 2018), there was a 370,127 ha tree cover loss (Mongabay 2018: Data Source: Tree cover loss: Hansen/UMD/Google/USGS/NASA via Global Forest Watch). Interestingly and praiseworthy, in 2016, Ethiopia has made a restoration commitment of 15 million ha in forest landscape restoration (FLR) under AFR100 and Bonn Challenge Programme (Pistorius et al. 2017; AFR100 2020; Bonn Challenge 2020). It is also worth to mention that the Ethiopian Prime Minister announced in May 2019 that the country has set a goal of planting 4 billion seedlings on 1.5 million ha across the country (40 trees per person) under the country's National Green Development Programme in a bid to combat climate change and environmental degradation (UNEP 2019). These programs, of course, are very ambitious, but necessary to halt degradation and deforestation in the country.

When people talk about restoration, one often has to learn that every person has a different understanding of the term. We prefer the following definition for restoration: A process of returning ecosystems or habitats to their original structure and species composition. Restoration requires detailed knowledge of the (original) species, ecosystem functions, and interacting processes involved (B. C. Min. For. 2008). If this definition is applied to the tree-planting campaign and forest landscape restoration activities recently started in Ethiopia, we are forced to have an eye on the sufficient and well-balanced integration of native woody species in these activities.

According to the Institute of Biodiversity Conservation (IBC) (2012), in Ethiopia, woody plants constitute about 1000 species, out of which 300 are tree species, which makes the conservation of forest genetic resources a priority area in biodiversity conservation in the country. In consulting to the Red List of Endemic Trees and Shrubs of Ethiopia and Eritrea (Vivero et al. 2005), we learn that there are 428 endemic and near endemic woody taxa in Ethiopia and Eritrea, of which 107 are trees and 321 are shrubs. In this report, IUCN Red List Categories and Criteria are given for 135 endemic taxa. These 135 threatened woody taxa (31 trees and 104 shrubs) represent 13% of the total woody plant flora estimated for the Flora of Ethiopia and Eritrea area (Vivero et al. 2005, referring to Teketay et al. 2000).

Governmental organizations, regional state forest enterprises, non-government organizations (NGOs), farmers, and communities throughout Ethiopia are involved with reforestation or other tree-planting activities. All of these groups can make important contributions to the restoration of degraded land in Ethiopia (Derero 2011).

Derero et al. (2011) are citing reports of four National Regional States (Oromia, Amhara, Southern Nations, Nationalities and Peoples (SNNPR) and Tigray), which claimed the annual planting of a total of over 3 billion tree seedlings in 2009 and 2010. The authors have calculated that planting of such a huge number of seedlings would cover 1.2 million ha of land if planted at a spacing of 2500 trees ha^{-1}. They reported data on tree seed request and supply at the Forestry Research Center (FRC) from 2007 to 2010 indicating that the Center was supplying, on average, 7278 kg of pure seeds annually in the stated period satisfying 78% of the request.

In consequence, planting over 4 billion seedlings, which for example, was aimed at by the country's National Green Development Program in 2019, would require a much higher amount of seeds.

12.2 Purpose and Objectives of Restoration in Ethiopia

Forest restoration in Ethiopia has a multi-functional approach, as newly restored forests can be of considerable economic, ecological, and social importance.

Important overall objectives can be: (1) food security for increasing population through protection of arable and newly restored lands from wind and combating desertification, (2) protection of human settlements from strong winds and sandstorms, (3) prevention of soil erosion from flooding and storms, (4) production of biofuel-crops as a renewable energy source, (5) wood production, (6) provision of other goods and services from forests, e.g., wild edible and medicinal plants, (7) absorption of carbon dioxide from the atmosphere, and (8) creating new job and qualification opportunities among young people.

Successful forest landscape restoration needs proper planning and implementation (Ministry of Environment, Forest and Climate Change 2018), and usually is based on land capability surveys. Regarding forest restoration, data on the natural distribution of native tree species is required. In this context, the identification of: (1) conservation areas, where rare species might be found, (2) areas with species of high timber value, and (3) possible sites, where mother seed trees of certain species can be found is a necessary prerequisite. In addition, potential sites for enrichment plantings in natural forests and the landscape should be identified (Bortoleto et al. 2016; Birhane et al. 2017; Kassa et al. 2017; Beltrán and Howe 2020).

Factors influencing the choice of species are policy decisions to establish FLR, environmental conditions, and availability of potential seed species (see Evans 1992). Selection criteria for the choice of tree species should include considerations on: (1) adaptedness to harsh conditions of the dry environment and climate, and adaptability to climate change and dynamic environments, (2) tolerance to basicity and salinity of irrigation water, (3) ability to provide high quality timber, NTFPs and services from the new forests, e.g., carbon sequestration, etc., and (4) preferably native species to support restoration of ecosystems and regional landscape, and selected exotic species.

12.3 Church Forests and the Bottleneck on the Way to Successful Restoration

Church forests are quite successful in conservation of their own forests and old tree resources, but obstacles, for instance, livestock grazing in the forests, are affecting tree regeneration and a healthy population structure of the forest stands (Bongers et al. 2006). Thus, it is interesting to experience, apart from mitigating the danger of senescence, whether Church forests can be motivated to contribute to sustainable provision of forest reproductive materials, preferably from native tree species, in terms of the planned restoration activities in the region.

Of course, there are challenges ahead, which must be overcome to reach the goals.

Decisions on political and institutional levels have to be made on the purposes of forest landscape restoration, including the establishment of productive plantations, hedgerows, and corridors. Choice of adequate species for the various objectives need to be made, and it is imperative that the supply of high quality seed material in sufficient amounts can be ensured (León Lobos et al. 2018). The latter builds the basis for the production of high quality planting stocks in nurseries (Bannister et al. 2018). Last, but not least, the application of good practice in tree planting has a strong impact on seedling establishment, survival, and recruitment, with other factors, some of them are outside human control, like drought or natural disasters (see UNEP 2019; Duguma et al. 2020).

Forest reproductive material (FRM) is the umbrella term for fruits, cones, seeds, cuttings and planting stocks used in the establishment of forests (EC 1999; FAO 2020). High quality forest reproductive material matching with the site where it is planted improves production, yield, and timber quality, but is also essential for health and stability.

Ethiopia has developed a national tree seed system (Derero et al. 2011), with the aim to manage suitable species and provenances, delineate seed (ecotype) zonation

(Bekele-Tesemma 2007), and identification of qualified seed stands and trees for seed collection.

We acknowledge that there are renowned and well-working tree seed centers in Amhara, Oromia, Tigray, and SNNP National Regional States and private seed suppliers. In addition, the Tree Seed Technology Coordination Unit (TSTCU) of the Ethiopian Environment and Forest Research Institute (EEFRI) is also mandated to procure quality tree seeds, develop referral tree seed sources and generate technologies and scientific information that can be applied to improve the seed system of the country (TSTCU-EEFRI 2020). Unfortunately, the seed centers, including TSTCU, have a limited capacity in terms of seed supply in the demanded quality and quantity. For instance, TSTCU has published a Tree Seed Catalog, which gives an insight into the profiles of seeds collected and managed by the unit (Adefa et al. 2019). At the moment, tree seeds are available from 73 different species, and half of the species are native while the remaining half are exotic.

As we have already learned from a publication of the Institute of Biodiversity Conservation (IBC) (2012), there are about 300 tree species in Ethiopia, which makes the conservation of forest genetic resources a priority area not only for biodiversity conservation but also for FLR in the country. Hence, there is a high potential for tree seed centers to intensify their activities because about 90% of the native tree species are not in the focus of professional tree seed conservation and management activities at the moment.

Apart from their role in biodiversity conservation, church forests are valuable seed sources (Wassie and Teketay 2006). There are about 35,000 churches (Wassie 2007; Lemineh and Bongers 2011), most of them with forests with an average size of 2–2.5 ha, and harboring 15–78 tree species per church, on average (Lemineh and Bongers 2011). Estimates of native species identified in church forests indicated more than 150 different tree species (Wassie 2007). This supports our impression that these forests could really act as repositories for forest reproductive material of native species and could tremendously help to support FLR.

12.4 Producing High Quality Seedlings in Nurseries Supported by Church Forests

Successful planting stock production starts with detailed planning and organization of forest restoration in the localities adjacent to church forests. People involved should have a sound basis in sustainable provision of high quality seeds, including good knowledge in seed biology and technology. The knowledge basis should also encompass topics of dormancy breaking and germination (see Schmidt 2000; Stimm et al. 2008; Kildisheva et al. 2020).

Commonly, a neglected issue is knowledge transfer of good practice in the raising of seedlings in nurseries and the steady improvement of bulk propagation techniques.

Observations in our research nursery in Ecuador have shown that after germination, seedlings of several tree species did not readily establish, but remained tiny and susceptible to getting moldy, even when subjected to adjusted light intensities. Various substrates (with and without fertilizer) were examined in a comprehensive study by Leischner (2005) in which she has been able to show that a mix of soil from the natural environment with compost in a ratio of 3:2 was very effective in promoting early growth. This observation suggested that inoculation with mycorrhiza could play an essential role in the production of young plants for reforestation. To evaluate the possibility of mycorrhizal inoculation an exploratory experiment has been set up, where native plant seedlings of two forest species, *Cedrela montana* Moritz ex Turcz. and *Heliocarpus americanus* L., growing under controlled environmental conditions, has been inoculated with arbuscular mycorrhizas (AM). An evaluation of the tree seedlings six months after sowing showed that the rate of mycorrhization and biomass production was best after addition of forest soil as inoculum source (Stimm et al. 2008). In an additional experiment, we have found that application of mycorrhizal roots improves growth of tropical tree seedlings in the nursery (Urgiles et al. 2009). Results from Ethiopia established that arbuscular mycorrhizas are predominant in the dry Afromontane forests. Therefore, AM should receive special attention in indigenous tree seedling production and restoration activities for the dry Afromontane ecosystems of the country (Wubet et al. 2003).

How should the ideal seedling look like? Which criteria must be realized? The main criteria of external and internal quality are choice of the right provenance or ecotype, age of planting stock, seedling size, root collar diameter, height/diameter-ratio as an indicator for a well-balanced biomass ratio of above versus below ground biomass, and a sufficient to good nutritional and water status (see also Ivetic and Novikov 2019). Criteria for planting success may include an evaluation of root growth potential, drought tolerance, and stress resistance. Extensive information on important seedling attributes is published in Grossnickle and MacDonald (2018).

Both, criteria of external and internal quality as well as the criteria for planting success, give comprehensive information to the nursery grower about plant quality, i.e., with special focus on the possibility of survival, healthiness, and good growth. The Target Plant Concept (TPC) provides a flexible framework that nursery managers and their clients can use to improve the survival and growth of these seedlings (Dumroese et al. 2016). A key tenet of the TPC is that emphasis is placed on how seedlings perform on the outplanting site. For this reason, a cooperative partnership should exist between the nursery manager and the client to determine the target plant based on site characteristics, and that information gathered from post-planting monitoring is used to improve subsequent plant materials. Dumroese et al. (2016) gave examples from the southeastern United States, Hawai 'i, and Lebanon on how the TPC process has improved performance of seedlings deployed for reforestation and forest restoration.

If seed production in a stand or individual tree is not sufficient or of low quality, vegetative propagation techniques, like cutting, grafting, and in vitro propagation can be applied and used in the production of planting stock, e.g., as a tool to rescue endangered populations, or specimens (Fig. 12.1).

Fig. 12.1 Cutting propagation is a tool for mass propagation but also for rescue of endangered populations. Left: Hybrid acacia (*Acacia mangium x Acacia auriculiformis*) propagation under intermittent mist in chambers in Sabah, Malaysia (Photo: Stimm 2009). Parent species are *Acacia mangium* Willd. and *Acacia auriculiformis* A. Cunn. ex Benth. Right: Cutting propagation of *Podocarpus oleifolius* D. Don ex Lamb. in Loja, Ecuador (Photo: Stimm 2007)

All the above mentioned features predetermine the success of planting in the field. Therefore the most important determinants for successful planting are summarized below.

- Use high quality planting stock.
- Planting stock should not have damages or pests and diseases.
- The optimum harvesting age of well-balanced target seedlings from the nursery is species dependent and related to ambient growth conditions, e.g., whether seedlings are fertilized or not. Seedlings raised in containers might not be older than 6 – 12 months when planted. If seedlings are produced from polybags, bag size must promote well developed roots without deformations. The crucial point is that in many cases seedlings have been kept in relatively small polybags or containers without air root pruning at the bottom, which results in root deformations (see Fig. 12.2).
- Seedling height at the time of outplanting may be about 50 (–80) cm. It is not appropriate to define only one size for seedlings from many different native species and for the various site conditions in the field.
- Roots must be kept fresh before planting and planting techniques must guarantee proper positioning of the seedling in the soil without deforming the roots (Fig. 12.3).

Fig. 12.2 Left: Girdling root deformation on 2 year old *Picea abies* (L.) H. Karst. seedling caused by extended production time in mini plug containers, Germany (Photo: Stimm 2015); Right: About 25 year old deformed root system on *Picea abies* (L.) H. Karst. from a high altitude reforestation site on Stillberg, Davos, Switzerland. The diameter of the "root-ball" is 26 cm (Photo: Stimm 1998)

Fig. 12.3 Toppling in drip-irrigated Aleppo pine (*Pinus halepensis* Mill.) plantations near Ismailia, Egypt, caused by underdeveloped and malformed root system (Photo: Stimm 2012)

- Plant handling must be done with care to avoid damages.

12.5 Church Forests and the Way Ahead—Perspectives and Challenges

A lot of knowledge about church forests has been collected and summarized (Bongers et al. 2006; Wassie et al. 2010; Cardelús et al. 2013; Reynolds et al. 2015, 2017; Klepeis et al. 2016; Mekonen et al. 2019). From knowing this knowledge, it becomes obvious that church forests are in situ biodiversity hotspots.

Figure 12.4 shows conceivable options in the interrelations between (genetic) variability, yield (productivity), and different units of genetic resources corresponding to different levels of improvement intensity. The more well-defined a genetic resource the smaller the circle. The level of variability is initially decreasing (natural forest stand → church forest) because of past use. In the proposed new phase, the establishment of church forest seed orchards facilitates the initiation of a network of church forest nurseries, which build up an important basis of forest landscape restoration in the country. One of the complementary ways involves the establishment of monoculture plantations of exotic and native species, and subsequently clonal plantations for the selection of the best individuals resulting in higher yields (modified after Keiding and Graudal 1996).

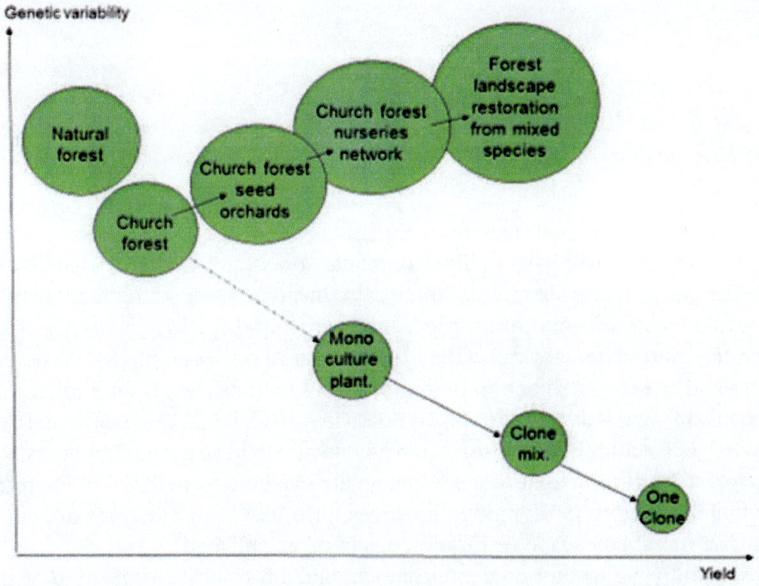

Fig. 12.4 Forest landscape restoration at the crossroads. Creating a balance between conservation, restoration, risk minimization, and productivity (modified after Keiding and Graudal 1996)

We can only speculate about the reasons for plantations with exotic tree species: one of them might be limited knowledge about characteristics and properties of native species (e.g., site requirements and growth). Knowledge about silvics, including reproductive biology, appropriate seed procurement, propagation methods and silvicultural treatment options have to be adequately retrieved, compiled, applied, and communicated by research organizations, universities, and NGOs. A good example of what could be done is given in the publication of Teketay (2011) on natural regeneration and management of *Podocarpus falcatus* (Thunb.) R. Br. ex Mirb. Complementary forest site classification can essentially support restoration planning and management. Since many soils in tropical areas are heavily degraded, investigations should be carried out on how soil biodiversity in tropical ecosystems can be facilitated.

We already mentioned that there is a factual lack of adequate reproductive material of native tree species and availability of exotics (e.g., Pines and Eucalypts) in nurseries. Restoration with native species and mixed forests with higher ecological and economic stability is not yet considered in restoration practices adequately. One option is to foster the establishment of mixed forests with native species, including tested enrichment plantings with native tree species in naturally regenerated stands (see Birru et al. 2011).

Because an overall objective of restoration with native species is a process of returning ecosystems or habitats to their original structure and species composition (after B. C. Min. For. 2008) such an approach gives the chance to combine both, biodiversity conservation and production.

12.6 Challenges for Policy, Stakeholders, and Implementation of Best Practices of Church Forests

It is mandatory to raise awareness for the urgent necessity of forest landscape restoration in the country (Ministry of Environment, Forest and Climate Change 2018). Within this context, it is also important to raise awareness for sound management of forest genetic resources and forest reproductive material (FRM), including effective conservation and sustainable use (IBC 2012, León Lobos et al. 2018). Participatory planning and consolidated action on regional and national levels on a trustful basis must be established between the stakeholders involved, e.g., the government should acknowledge the church for conservation and decide to have a clear boundary for the church forests to minimize further encroachment (Mekonen et al. 2019). Therefore, a conceptual framework for defining objectives, priorities, and functions of a national program on forest landscape restoration has to be established.

Subsequently, an institutional environment and a formal sector need to be developed: for government, public, church, and private organizations with specialized roles and legal recognition, including regulations for production, distribution, and use of reproductive material. Seed policies may distinguish between procedures to regulate production of native seed for restoration and procedures to regulate seed supply for exotic or improved plant material for the forestry industry (see the case of Brazil, de Urzedo et al. 2019). Mandatory seed quality testing in accredited laboratories could prevent formal recognition of local native seed suppliers and, therefore, simpler quality assurance might be appropriate. Governmental support is essential for structuring a consistent restoration market and technological innovation and for the desired success, already existing research organizations should be strengthened.

Churches are not only well known for proclaiming God's Word, but they fulfill the role of trustworthy cornerstones of societies. They are famous for their capabilities to spread knowledge and ethical values to people.

We see a great chance in the incorporation of Forest Genetic Resources (FGR) Conservation Programme (IBC 2012), Participatory Forest Management (PFM) (Kassa et al. 2017) and restoration activities in the beneficial work of churches in addition to their conservation efforts (Mekonen et al. 2019).

The application of new technologies can facilitate the identification of high quality seed sources by means of remote sensing and on ground work (e.g., population size and minimum number of mother trees for collection).

In smaller church forests with missing regeneration, many of the trees are aged and dying without any replacement. This may result in complete losses of valuable indigenous tree species and lead to the decline of species composition of the church forests (Wassie 2002). Hence, long-term conservation of biodiversity of individual patches and evolutionary potential of species may be threatened by isolation, small sizes of tree species populations, and disturbance, especially when considering climate change (Aerts et al. 2016). In largely degraded landscapes, small fragments and tree islands are at risk of elevated inbreeding when populations drop below species-specific threshold numbers (minimum viable population sizes). Absence of seedlings and saplings in such situations is a strong indication for the beginning of local extermination of species.

Because inbreeding depression may reduce the function and adaptation of populations, it is generally best to collect propagules from larger and higher-density populations. This means that in smaller church forests, populations are less numerous and more isolated, the collection of propagules from wider distances and multiple sources, but from similar ecozones, may be necessary for restoring these populations at risk (Gann et al. 2019). Not only for this reason but also standards and capacities must be developed and disseminated for properly collecting and storing seeds of native species.

At the moment, in church forests and in newly established gene conservation stands in situ and ex situ dynamic gene conservation seems to be the best option available. In the near future, genetic management of these fragmented tree populations in church forests might offer a significant solution approach in biodiversity conservation and could open up an important new research field (see Frankham et al. 2017).

For each native species, whether endangered or vulnerable, installment of seed orchards, clonal archives, and seed production areas is an imperative, which allows sufficient genetic diversity and adaptive adjustment in response to environmental change and evolution. This would be instrumental to support sustainable and sufficient provision of seeds and propagules to rebuild functional resilient communities (Gann et al. 2019). In case of shortages, seed mixtures within species from several genetically adequate source populations (proportionately weighted according to previously estimated provenance-specific recruitment successes) can be a tolerable way out (Fedriani et al. 2019).

In each of these areas, minimum viable population size has to be realized that the trees are able to mate and produce fruits and seeds, which can be used for seedling propagation in (church) nurseries, and for regeneration and recruitment in restoration. Processes linked with tree reproduction, such as pollination and seed dispersal, are of tremendous importance for many ecological functions in forests. Disruption in these processes is known to delay forest regeneration, diversity, and structural dynamics (Rocha Santos et al. 2020).

Establishment of (church) nurseries on local and regional scale in the vicinity of church forests, including the development of quality standards for nurseries, is highly recommended (see also Vidal et al. 2020).

Establishment of demonstration areas, farm and agroforestry practices with stewardship, and also in the vicinity of church forests might be a very promising option.

Institution building and knowledge transfer is a key issue in sustainable management of natural resources, which leads to capacity building and training of local experts, farmers, and the public.

12.7 Tree-Based Landscape Restoration and Its Socio-economic Relevance in Ethiopia

In our introduction, we mentioned an amount of 15 million ha commitment under AFR100 and Bonn Challenge Scheme. In the meanwhile, after preparing this text, we found a reference indicating that the commitments are much higher and encompass an area of more than 20 million ha (until 2035) (Adefa et al. 2019). Is the success of these commitments relevant?

Yes, it is vitally important for the country and its people (see also Weber et al. 2013).

If there is a demand of about 1.3 billion seedlings year^{-1} (1000 seedlings ha^{-1}), 13,000 nurseries with a productivity of 100,000 seedlings year^{-1} are needed. The area requirement for each of such nurseries is about 2000 m^2, which is manageable. Under these premises, 104,000 permanent and temporary jobs (8 workers 100,000^{-1} seedlings) can be created and the annual volume of sales could be 650 million USD (0.5$ seedling^{-1}, which is equivalent to 15 Ehiopian Birr seedling^{-1}).

12.8 Perspective on Church Forests

Church forests are an integrated part of a conceptual framework for landscape restoration in Ethiopia. According to Stanturf et al. (2019), the landscape approach will help to address the interactions, competition, and trade-offs between different land uses in Ethiopia, thereby, avoiding further degradation of land, ecosystems, and forests. Increasing competition for land for agriculture and other uses threatens the sustainability of FLR unless accompanied by attention to meeting local needs for food security (see also Fagan et al. 2020). Both, deep and broad knowledge, practical expertise, innovation, and experience exist within the scientific and professional community and countless communities living on the front lines of landscape degradation, climate change, and other challenges. Enhanced communication and collaboration are needed across forest science disciplines and between the scientific community and local communities, government agencies, NGOs, the private sector, and other organizations and movements, including the church forests (see Stanturf et al. 2019) (see Fig. 12.5).

Churches are designated sometimes as *Guardian Angels*. They fulfill a paramount duty as leaders in stewardship of creation, authorities of ethics, and education. Its forests and the biodiversity they contain resemble a *Noah's Ark*.

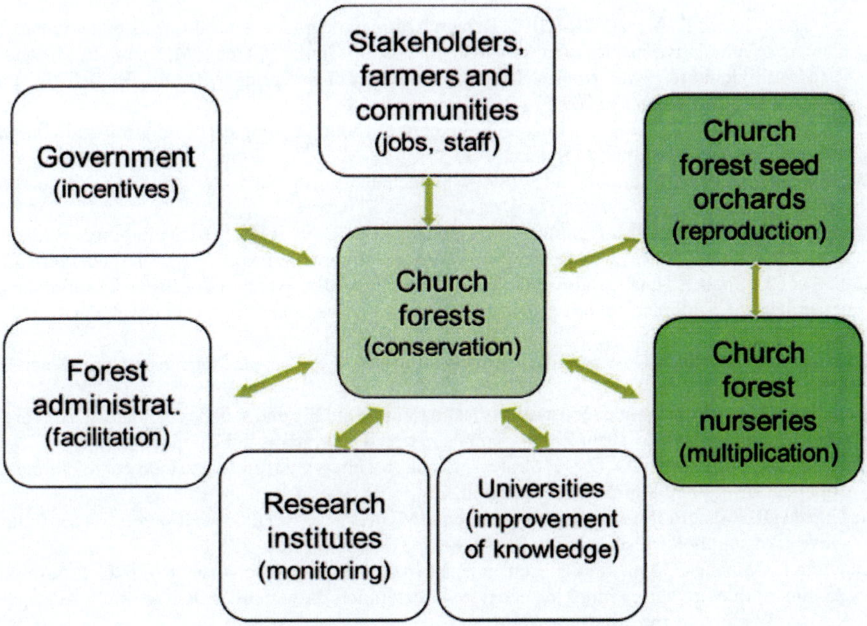

Fig. 12.5 Uniting Church, science, and policy for conservation and restoration (see also Cardelús et al. 2012)

References

Adefa M, Mulatu Y, Tadesse S (2019) Tree seed catalog. Tree seed technology coordination Unit (TSTCU), Ethiopian Environment and Forest Research Institute (EEFRI)

Aerts R, Van Overtveld K, November E., Wassie A, Abiyu A, Demissew S, Daye DD, Giday K, Haile M, TewoldeBerhan S, Teketay D, Teklehaimanot Z, Binggeli P, Deckers J, Friis I, Gratzer G, Hermy M, Heyn M, Honnay O, Paris M, Sterck FJ, Muys B, Bongers F, Healey JR (2016) Conservation of the Ethiopian church forests: threats, opportunities and implications for their management. Science of the Total Environment 551: 404-414

Aronson J, Goodwin N, Orlando L, Eisenberg C, Cross ATA (2020) World of possibilities: six restoration strategies to support the United Nation's decade on ecosystem restoration. Restor Ecol 28(4): 730-736

AFR100 (2020) Ethiopia. https://afr100.org/content/ethiopia. Accessed on 04 Apr 2020

Bannister JR, Vargas Gaete R, Ovalle JF, Acevedo M, Fuentes Ramirez A, Donoso PJ, Promis A, Smith Ramírez C (2018) Major bottlenecks for the restoration of natural forests in Chile. Restor Ecol 26(6):1039–1044

B. C. Min. For (2008) Ministry of Forests and Range—Glossary of Forestry Terms in British Columbia, March 2008: Restoration. Cited after: Dunster J, Dunster K (1996) Dictionary of natural resource management. UBC Press, Vancouver, B.C, p 266

Bekele Tesemma A (2007) Useful trees of Ethiopia: identification, propagation and management in 17 agroecological zones. RELMA in ICRAF Project, Nairobi, 552 p

Beltrán LC, Howe HF (2020) The frailty of tropical restoration plantings. Restor Ecol 28(1):16–21

Birhane E, Mengistu T, Seyoum Y, Hagazi N, Putzel L, Rannestad MM, Kassa H (2017) Exclosures as forest and landscape restoration tools: lessons from Tigray Region Ethiopia. Int Forestry Rev 19(4):37–50

Birru GA, El Kateb H, Mosandl R (2011) Rehabilitation of degraded natural forests by enrichment planting of four native species in Ethiopian highlands. In: Günter S, Weber M, Stimm B, Mosandl R (eds) Silviculture in the tropics. Tropical forestry, vol 8. Springer, Berlin, Heidelberg, pp 377-385, https://doi.org/10.1007/978-3-642-19986-8_25

Bongers F, Wassie A, Sterck FJ, Bekele T, Teketay D (2006) Ecological restoration and church forests in northern Ethiopia. J Drylands 1(1):35–44

Bonn Challenge (2020) Ethiopia. https://www.bonnchallenge.org/about-the-goal. Accessed on 06 Apr 2020

Bortoleto LA, Figueira CJM, Dunning JB Jr, Rodgers J, da Silva AM (2016) Suitability index for restoration in landscapes: An alternative proposal for restoration projects. Ecol Ind 60:724–735

Cardelús CL, Scull P, Hair J, Baimas-George M, Lowman MD, Wassie EA (2013) A preliminary assessment of Ethiopian sacred grove status at the landscape and ecosystem scales. Diversity 5(2):320–334

Cardelús CL, Lowman MD, Wassie EA (2012) Uniting church and science for conservation. Science 335(6071):915–917

Derero A (2011) Toward a tree seed system that guarantees quality and satisfies demand in Ethiopia. In: Derero A, Fantu W, Eshetu Z (eds) Trends in tree seed systems in Ethiopia, pp 43–56

Derero A, Mamo N, Kelemu K (2011) Strategic actions for integrated forest development in Ethiopia. Ethiopian Institute of Agricultural Research, 25pp

de Urzedo DI, Fisher R, Piña-Rodrigues FC, Freire JM, Junqueira RG (2019) How policies constrain native seed supply for restoration in Brazil. Restor Ecol 27(4):768–774

Duguma L, Minang P, Aynekulu E, Carsan S, Nzyoka J, Bah A, Jamnadass R (2020) From tree planting to tree growing: rethinking ecosystem restoration through trees. ICRAF working paper No 304. World Agroforestry

Dumroese RK, Landis TD, Pinto JR, Haase DL, Wilkinson KW, Davis AS (2016) Meeting forest restoration challenges: using the target plant concept. Reforesta 1:37–52

EC (1999) Council Directive 1999/105/EC of 22 December 1999 on the marketing of forest reproductive material. https://eur-lex.europa.eu/legal-content/EN/ALL/?uri=CELEX%3A31999L0105. Accessed on 02 Dec 2020

Evans J (1992) Plantation forestry in the tropics: tree planting for industrial, social, environmental, and agroforestry purposes. Oxford University Press, Oxford

FAO (2020) Forest Reproductive Material. http://www.fao.org/sustainable-forest-management/toolbox/modules/forest-reproductive-material/basic-knowledge/en/. Accessed on 02 Dec 2020

Fagan ME, Reid JL, Holland MB, Drew JG, Zahawi RA (2020) How feasible are global forest restoration commitments? Conserv Lett 13(3): e12700

Fedriani JM, Garrote PJ, Calvo G, Delibes M, Castilla AR, Żywiec M (2019) Combined effects of seed provenance, plant facilitation and restoration site on revegetation success. J Appl Ecol 56(4):996–1006

Frankham R, Ballou JD, Ralls K, Eldridge MDB, Dudash MR, Fenster CB, Lacy RC, Sunnnucks P (2017) Genetic management of fragmented animal and plant populations. Oxford Univerity Press

Gann GD, McDonald T, Walder B, Aronson J, Nelson CR, Jonson J, Hallett JG, Eisenberg C, Guariguata MR, Liu J, Hua F, Echeverría C, Gonzales E, Shaw N, DeCleer K, Dixon KW (2019) International principles and standards for the practice of ecological restoration. Second Edition. Restor Ecol 27:S1–S46

Grossnickle SC, MacDonald JE (2018) Seedling quality: history, application, and plant attributes. Forests 9(5):283

IBC (2012) FAO—The State of the World's Forest Genetic Resources—Country Report Ethiopia

Ivetic V, Novikov A (2019) The role of forest reproductive material quality in forest restoration. Forestry Eng J 9(2):58–67. https://doi.org/10.34220/issn.2222-7962/2019.2/7

Kassa H, Birhane E, Bekele M, Lemenih M, Tadesse W, Cronkleton P, Putzel L, Baral H (2017) Shared strengths and limitations of participatory forest management and area exclosure: two major state led landscape rehabilitation mechanisms in Ethiopia. Int Forestry Rev 19(4):51–61

Keiding H, Graudal L (1989 (rev. 1996)) Introduction to conservation of forest genetic resources. Lecture Note A-4. Danida Forest Seed Centre, Humlebaek, Denmark

Kildisheva OA, Dixon KW, Silveira FA, Chapman T, Di Sacco A, Mondoni A, Turner SA. Cross, AT (2020) Dormancy and germination: making every seed count in restoration. Restor Ecol 28: S256-S265

Klepeis P, Orlowska IA, Kent EF, Cardelús CL, Scull P, Wassie EA, Woods C (2016) Ethiopian church forests: a hybrid model of protection. Hum Ecol 44(6):715–730

Leischner B (2005) Phenology, seed production, germination and propagation of native tree species of the tropical mountainous rain forest in southern Ecuador—a contribution to reforestation. PhD thesis, Chair of Silviculture, Weihenstephan Center of Life Sciences, Technical University of Munich. https://mediatum.ub.tum.de/doc/603692/603692.pdf

Lemineh M, Bongers F (2011) Dry forests of Ethiopia and their silviculture In: Günter S, Weber M, Stimm B, Mosandl R (eds) Silviculture in the tropics. Tropical forestry, vol 8. Springer, Berlin, Heidelberg, pp 261–272. https://doi.org/10.1007/978-3-642-19986-8_17

León Lobos P, Bustamante Sánchez MA, Nelson CR, Alarcón D, Hasbún R, Way M, Pritchard HW, Armesto JJ (2020) Lack of adequate seed supply is a major bottleneck for effective ecosystem restoration in Chile: friendly amendment to Bannister et al. (2018). Restor Ecol 28(2):277–281

Mekonen AB, Gebreegziabher BG, Wassie WA, Tsegay BA (2019) Church forests—the green spots of Ethiopian highlands. Asian J Forestry 3(2):45–53

Ministry of Environment, Forest and Climate Change (2018) National Potential and Priority Maps for Tree-Based Landscape Restoration in Ethiopia (version 0.0): Technical Report. Addis Ababa: Ministry of Environment, Forest and Climate Change.

Mongabay (2018) Deforestation statistics for Ethiopia. rainforests.mongabay.com, Accessed on 04-04-2020

Pistorius T, Carodenuto S, Wathum G (2017) Implementing forest landscape restoration in Ethiopia. Forests 8(3): 61

Reynolds T, Sisay TS, Lowman M, Wassie EA (2015) Sacred natural sites provide ecological libraries for landscape restoration and institutional models for biodiversity conservation. https://sustainabledevelopment.un.org/

Reynolds TW, Collins CD, Wassie A, Liang J, Briggs W, Lowman M, Sisay TS, Adamu E (2017) Sacred natural sites as mensurative fragmentation experiments in long-inhabited multifunctional landscapes. Ecography 40(1):144–157

Rocha Santos L, Mayfield MM, Lopes AV, Pessoa MS, Talora DC, Faria D, Cazetta E (2020) The loss of functional diversity: a detrimental influence of landscape-scale deforestation on tree reproductive traits. J Ecol 108(1):212–223

Schmidt L (2000) Guide to handling of tropical and subtropical forest seed. http://dfsc.dk/Extensionstudy/EXT-823.htm. Accessed on 06 Apr 2020

Stanturf JA, Kleine M, Mansourian S, Parrotta J, Madsen P, Kant P, Burns J, Bolte A (2019) Implementing forest landscape restoration under the Bonn challenge: a systematic approach. Ann for Sci 76(2):50

Stimm B, Beck E, Günter S, Aguirre N, Cueva E, Mosandl R, Weber M (2008) Reforestation of abandoned pastures: seed ecology of native species and production of indigenous plant material. In: Beck E, Bendix J, Kottke I, Makeschin F. Mosandl R (eds) Gradients in a tropical mountain ecosystem of ecuador. Springer, Berlin-Heidelberg, Ecological Studies 198, pp 417–429

Teketay D (2011) Natural regeneration and management of *Podocarpus falcatus* (Thunb.) Mirb. in the Afromontane forests of Ethiopia. In: Günter S, Weber M, Stimm B, Mosandl R (eds) Silviculture in the tropics. Tropical forestry, vol 8. Springer, Berlin, Heidelberg, pp 325–337. https://doi.org/10.1007/978-3-642-19986-8_21

Teketay D, Bekele T, Haase G (2000) Forests and forest genetic resources of Ethiopia. Programme and abstracts of an international conference on "Ethiopia: a biodiversity challenge", organized by the Biological society of Ethiopia and the Linnean Society of London, 2–4 February 2000, pp 29–30

TSTCU-EEFRI (2020) Tree seed technology coordination http://www.eefri.org/researches/tree-seed-technology-coordination/. Accessed 05 Apr 2020

UNEP (2019) Spotlight on Ethiopia's tree-planting programme (https://www.unenvironment.org/news-and-stories/story/spotlight-ethiopias-tree-planting-programme. Accessed 04 Apr 2020

Urgiles N, Lojan P, Aguirre N, Blaschke H, Günter S, Stimm B, Kottke I (2009) Application of mycorrhizal roots improves growth of tropical tree seedlings in the nursery: a step towards reforestation with native species in the Andes of Ecuador. New Forests 38(3):229–239

Vidal CY, Naves RP, Viani RA, Rodrigues RR (2020) Assessment of the nursery species pool for restoring landscapes in southeastern Brazil. Restor Ecol 28(2):427–434

Vivero JL, Kelbessa E, Demissew S (2005) The red list of endemic trees & shrubs of Ethiopia and eritrea. Fauna & Flora International, Cambridge, UK

Wassie A (2002) Opportunities, constraints and prospects of EOTC in conserving forest resources: the case of churches in south Gonder, Northern Ethiopia. [MSc Thesis]. Swedish University of Agricultural Sciences, Skinnskatterberg, Sweden

Wassie AE (2007) Ethiopian church forests: opportunities and challenges for restoration. PhD thesis, Wageningen University and Research, Wageningen

Wassie A, Teketay D (2006) Soil seed banks in church forests of northern Ethiopia: implications for the conservation of woody plants. Flora Morphol Distrib Funct Ecol Plants 201(1):32–43

Wassie A, Sterck FJ, Teketay D, Bongers F (2009) Tree regeneration in church forests of Ethiopia: effects of microsites and management. Biotropica 41(1):110–119

Wassie A, Sterck FJ, Bongers F (2010) Species and structural diversity of church forests in a fragmented Ethiopian Highland landscape. J Veg Sci 21:938–948

Weber M, Stimm B, López, MF, Gerique A, Pohle P, Hildebrandt P, Knoke T, Palomeque X, Calvas B, Günter S, Aguirre N, Kübler D (2013) Conservation, management of natural forests and reforestation of pastures to retain and restore current provisioning services. In: Bendix J, Beck E, Bräuning A, Makeschin F, Mosandl R, Scheu S, Wilcke W (eds) Ecosystem services, biodiversity and environmental change in a tropical mountain ecosystem of south Ecuador, Ecological Studies 221, ISBN-13: 978-3-642-38137-9. Springer, Berlin, Heidelberg, pp 171–185

Willemen L, Barger NN, Bt B, Cantele M, Erasmus BFN, Fisher JL, Gardner T, Holland TG, Kohler F, Kotiaho JS, von Maltitz GP, Nangendo G, Pandit R, Parrotta JA, Potts MD, Prince SD, Sankaran M, Brainich A, Montanarella L, Scholes R (2020) How to halt the global decline of lands. Nat Sustain 3:164–166

Wubet T, Kottke I, Teketay D, Oberwinkler F (2003) Mycorrhizal status of indigenous trees in dry Afromontane forests of Ethiopia. For Ecol Manage 179(1–3):387–399

Chapter 13
Ecological Status and Plan for Connectivity of Fragmented Forests as a Means of Degraded Land Restoration in South Gonder, Ethiopia

Fikirte Demissie, Kumelachew Yeshitela, and Mengistie Kindu

Abstract Lack of networks of connected habitat patches affects the biodiversity of the area. Hence, this study assesses the status of fragmentation of the land use of Libokememek District and propose a plan for the ecological connectivity/corridor. Data from 2015 Land Use and Land Cover (LULC) results, Digital Elevation Model (DEM), all area closures, and church forests were used for the study. To analyze the fragmentations status of the area different metrics was calculated using fragstat4 software in ArcGIS10.3. The result reveals agricultural lands and grasslands occupied more than 80% of the land use compared to the forest land (844 hectare). This result shows that forest land use is the most fragmented land use. The planned ecological connectivity has a total of 2305 hectare of land. To reduce fragmentation focal patches such as area closure and church forest; sub patches like slope greater than 30%, degraded lands, and river stream with a buffer of 30 m are used. This will result in 29% of land is covered with vegetation/forest. Finally, soil and water conservation work that promotes the participation of the community and other stakeholders has been recommended. Likewise, the churches have to be encouraged and incentivized to keep their trees/forest land.

Keywords Fragmentation · Ecological connectivity · Focal patches · South Gonder · Riparian buffer

F. Demissie (✉) · K. Yeshitela
Ethiopian Institute of Architecture, Building Construction, and City Development (EiABC), Addis Ababa University, P.O. Box 518, Addis Ababa, Ethiopia
e-mail: fikirte.demissie@astu.edu.et

K. Yeshitela
e-mail: kumalachew.yeshitela@eiabc.edu.et

F. Demissie
Department of Architecture, Adama Science and Technology University, P.O.Box 1888, Adama, Ethiopia

M. Kindu
Institute of Forest Management, TUM School of Life Sciences Weihenstphan, Technische Universität München, Hans-Carl-von-Carlowitz-Platz 2, 85354 Freising, Germany
e-mail: mengistie@tum.de

M. Kindu et al. (eds.), *State of the Art in Ethiopian Church Forests and Restoration Options*, https://doi.org/10.1007/978-3-030-86626-6_13

13.1 Introduction

The implementation of area closures to rehabilitate degraded forestlands is grounded in classical biological conservation principles (Walpole and Goodwin 2001). That means area closures simply restrict clearing activities within a pre-specified range to make it possible for forests to recover natural vegetation without the need for human or animal intervention (Rahmato 2001). Hence, a growing body of empirical evidence further supports this claim and indicates that conservation efforts that lack such community support often produce unintended costs that are much greater than their intended forest conservation gains (Rashid et al. 2013).

There is increasing evidence that habitat regeneration on degraded land is affected by distance to the nearest intact edge (Young 2000). Hence, a basic feature of habitat fragmentation is a sharp increase in the amount of induced habitat edges (Laurance and Yensen 1991, Fahrig 2017). Similarly, Farhig (2020) explained that habitat fragmentation increases the number of habitat patches, through the removal of habitat. Consequently, plant and animal populations in fragmented habitats are not only reduced and subdivided but also are increasingly exposed to ecological changes associated with induced edges (Wilcove et al. 1986). According to Opdam and Wascher (2004) mentioned in a fragmented habitat network the individuals are not able to locate all the best quality habitat sites and the mortality loss during dispersal is much higher due to longer searching of favorable land through unfavorable landscape. Similarly, smaller patches may have a lower habitat quality, with less reproduction (Collinge 1996). This resulted in habitat degradation that is an important stressor of biodiversity on top of habitat loss (Thorn et al. 2020). Habitat loss has large, consistently negative effects on biodiversity; researchers who conceptualize and measure fragmentation as equivalent to habitat loss typically conclude that fragmentation has large negative effects (Fahrig 2003; Conceicao and Oliveira 2010).

Biodiversity is of value because ecosystems provide important services for human beings that enable our survival. Many of these are indirect, such as weather moderation, coastline stabilization, or water and air quality maintenance (Kingsland 2002). It is widely agreed that connectivity is important for biological conservation (Taylor et al. 1993; Hodgson et al. 2011). Fischer and Lindenmayer (2007) defined ecological connectivity as it is the connectedness of ecological processes across multiple scales, including trophic relationships, disturbance processes, and hydro-ecological flows (Tischendorf and Fahrig 2000). Connected remnants would be predicted to maintain the attributes of continuous habitat, and support a greater biological diversity than completely isolated remnants (Noss 1987; Bennett 1990; Saunders et al. 1991). Lack of continuous semi-natural habitat or networks of connected habitat patches can restrict the capacity of species to adjust to changing conditions (Rands et al 2010; Fagan et al 1999).

Similarly, Olson and Burnett (2009) identified the need for connectivity in managed forest landscapes, and provide examples where habitat connectivity issues are combined with and complement riparian management (Franklin 1993). Intensive forest management can also play a role in biodiversity management.

For instance, the maintenance of riparian vegetation along streams and rivers is critically important to prevent soil erosion, maintain high water quality and provide habitat for riparian specialists. Corridors of native vegetation which link remnants of similar vegetation may provide habitat and facilitate movement of plants and animals (Bennett 1999; Fischer and Lindenmayer 2007). Therefore, corridors should be viewed as one of a suite of strategies in planning projects aimed at habitat conservation or restoration (Collinge 1996; Cowling et al. 2003).

Over 80% of the rural population of Ethiopia relies on either rain-fed agriculture or pastoral grazing to sustain its livelihood. The spread of these practices combined with rapid population growth has led to devastating forest cover losses throughout the country (Tegene 2002; Garedew et al. 2012; Kindu et al. 2013, 2015, 2018; Temesgen et al. 2013; Desalegn et al. 2014; Demissie et al. 2019). This has sparked concern among government officials about the potential for political disruption due to the irreversible economic, ecological, and socio-cultural losses that tend to accompany such problems when left unaddressed (Taddese 2001). Lake Tana watershed exhibits alarming degradation of biota (riparian vegetation, grassland, and wetlands) crucial for maintaining the resilience of the ecosystem and improving livelihoods (Saha and Setegn 2015; Ligdi et al. 2010). Increasing sediment load to the Lake Tana may be due to physiographic, inappropriate intensive agricultural land use, absence/degradation of riparian vegetation, ecotones and buffer strips, and stony nature of the river banks (Ligdi et al. 2010). Setegn (2008) stated the land and water resources of the Lake Tana basin and the Lake Tana ecosystem are endangered among others by soil erosion, sediment transport, and land degradation.

This study seeks to contribute to the growing body of scholarship on area closures and/ protected forests and ecological connectivity (network) in the developing world. Fragmentation status of the study area is studied from the result of Land use and Land Cover (LULC) images classification. Land cover refers to the biophysical cover on the earth's surface including vegetation, water bodies, bare soil, and hard surfaces whereas land use is the utilization of land cover type by human activities for the purpose of agriculture, grazing, forestry, and settlement by altering land surface processes including biodiversity and hydrology (Di Gregorio and Jansen 2000). On the other hand, information on diversity, population dynamics, and regeneration status of woody species that are important to generate information in order to plan measures on conservation management and sustainable use of forest resources in the area were collected from plots in the field. The ecological connectivity/corridors of the study area are also proposed from the result of fragmentation and other results. Therefore, the study investigated the sustainable land use planning and management of area closure in order to establish ecological corridors to connect fragmented area closures and other forest patches with the participation of community in Libokemkem District, South Gonder, Ethiopia.

13.2 Materials and Methods

13.2.1 *Study Area*

The study was conducted in Libokemkem District, which is part of the Lake Tana watershed, located in the South Gondar Administrative Zone of the Amhara National Regional State at about 645 km Northwest of Addis Ababa, the capital of Ethiopia. The district lies within 11° 58′ 1.5″ and 12° 22′ 6.7′ latitude, and 37° 33′ 25.4′ and 37° 58′ 16.5′ longitude (Fig. 13.1). The District comprises a total of 32 Kebeles (the smallest administrative unit of Ethiopia) and an area of 1082km^2 (CSA 2007). According to the 2007 census, the population of the District was 198,435 (CSA 2007).

The study area is generally characterized by sub-humid highlands. The mean annual temperature of the region ranges from 18 to 25 °C. This area receives a unimodal rainfall of approximately 900–1400 mm per year, the majority of which falls between June and August. Volcanic rocks mainly basalt characterize the geology of the study area. Light dark, grey, whitish, reddish, or brown are the common rocks. The soils are mostly shallow and sandy and characterized by low organic matter due

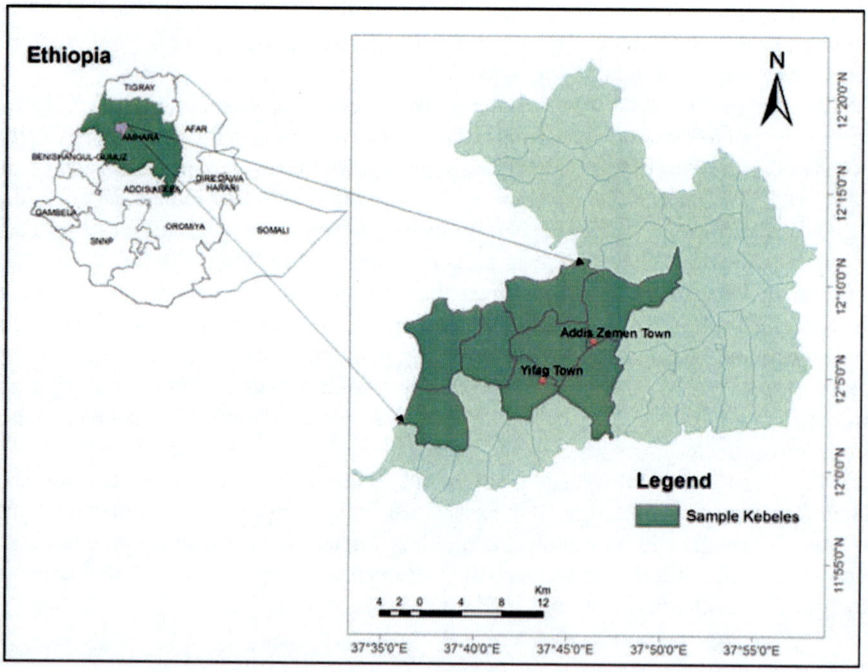

Fig. 13.1 Map of the Libokemkem District within Amhara Region of Ethiopia (*Source* CSA 2007)

to land degradation as a result of erosion and continuous cultivation (Zegeye et al. 2011).

Agriculture, including crop farming and animal husbandry, is the basis of livelihoods of the people of the study area and it is characterized by rain fed, oxen driven, small scale subsistence oriented, and labor intensive activities. Average land holding in the District is about one hectare/head. In the area various types of crops such as teff (*Eragrotis teff*), beans (*Phaseolus vulgaris L.*), wheat (*Triticum aestivum L.*), and barley (*Hordeum vulgare L*) are grown. Regarding the vegetation cover of the Lake Tana watershed, where the study area is found, it is indicated in Alelign et al. (2007), "the vegetation at Lake Tana has been considered as a transition type to humid evergreen forests of southwestern Ethiopia and also classified as upland dry evergreen forest (Friis 1986) and a special subtype of undifferentiated Afromontane forest (Friis 1992)". Furthermore, Friis et al. (2011) classification of vegetation based on altitude and rainfall, the Lake Tana region has a similar vegetation cover of Dry Evergreen Afromontane Forest (DAF) which is found within the altitudinal range of 1800–3000 m above sea level with rainfall <1700 mm.

13.2.2 Methodology

The result of LULC of 2015 was used as input data to calculate fragmentation of the forest cover of the study area using fragstat 4 to calculate different metrics that show the status of fragmentation of the area. In addition, data for potential ecological connectivity were extracted from the DEM (digital elevation model) of Ethio-GIS, all area closures, and the result of LULC of this study was also used.

To analyze the fragmentations status of the area under study different metrics was calculated using fragstat 4 software in ArcGIS10.3 (McGarigal and Marks 1994). The classified image of 2015 of the study area was converted from raster to a vector format before it was used to calculate the indices. Five indices of measures were considered: (1) class area (CA), (2) number of patches (NumP), (3) mean patch size (MPS), (4) area-weighted mean shape index (AWMSI), and (5) edge density (ED). CA, NumP, and MPS are non-spatial indices while AWMSI and ED are spatial indices.

On the other hand, to plan for ecological connectivity, the forest and bush and shrub land use were further classified into two groups, with an area greater than 0.5 hectare and less than 0.5 hectare. Based on this 746 ha was covered with fragmented forest area greater than 0.5 ha with 89 patches and 1256 ha of land with bush and shrub land with 661 patches. The rest 99 ha and 638 ha of land are under forest and bush and shrub land under 0.5 hectare, respectively. In addition, in the study area, there were 16 area closures that had total land coverage of 906.1 ha and church forests that account 93.05 ha of land (Table 13.2). Hence, the study area has a total area 2305.39 hectare of forest, bush, and shrub, area closure, and church forest land use. They are considered as focal patches and used to plan for potential ecological connectivity.

On the other hand, to reduce fragmentation of the focal patches and sub patches forest and bush and shrub land use less than 0.5 ha and degraded land are used. In addition, slope of the land greater than 30% was also extracted from DEM.

To calculate the change on the ecological connectivity status of the study area after the proposed plan, Fragstat 4 was used to see the fragmentation status. Similarly, area metrics patch density, size, and variation; edge metrics; shape metrics; nearest neighbor metrics, and interspersion (IJI) were calculated metrics. The result for the parameters of the metrics that show fragmentation status of the land use of 2015 after the proposed plan to increase connectivity and reduce fragmentation between and among patches, subpatches, and river stream with 30 m buffer zone of the study area.

To analyze and suggest the potential ecological connectivity for the study area the result of the LULC of 2015 of the study area, first the land use area classified into forest and non-forest land use and slope of the study area greater than 30% that was extracted from DEM using ARCGIS10.3. In the forest land cover type bush and shrub land use is included whereas the rest are classified under non-forest land uses. Forest and bush and shrub land use of the area greater than 0.5hectare were extracted and used as focal patches and the "forest" and bush and shrub land less than 0.5 hectare and slope are extracted from the LULC of 2015, degraded land, slope greater than 30%, and river streams were also identified and used for the analysis.

13.3 Results

13.3.1 Status of Ecological Connectivity

As Table 13.1 shows Percent of landscape (PLAND) of agricultural lands and grasslands occupied 56.17% and 26.75% of land use, respectively. However, forest lands occupied 3.18% of the total land use of the study area. Largest Patch index (LPI) of land use shows agriculture land has 24.46 followed by grass land (3.85). The remaining land uses have LPI value less than one. Regarding patch density, size, and variability metrics, the number of patch (NP) for Bush and shrub land, settlement, grasslands, agricultural land, and degraded land were 3504, 3070, 2618, 2058, and

Table 13.1 Focal patches to plan for ecological connectivity

Focal patches	Total land use in Ha	Number of patches
Forest land > 0.5Ha	746.16	89
Shrub and Bush land use > 0.5 Ha	1256.21	661
Area closure	906.1	16
Church forest	93.05	3

1389, respectively. On the other hand, forest land use and wetland had 526 and 19 patches respectively. The Mean patch size (MPS) of wetland is 5.16 ha followed by agricultural land (2.38 hectare), grassland (1.22 hectare), and forestland (0.96 hectare). The remaining land uses have less than 1 hectareMPS. Patch size standard deviation (PSSD) of agricultural land use is 161.10 and wetland and grass land have the same PSSD value of 28 followed by forest land (13.33). The Patch size coefficient of variation (PSCoV) values for agricultural land and grass land were 2220.47 and 1056.88, respectively. Forest land and settlement had a value of 830.17 and 756.73, respectively.

Regarding the edge metrics, ED and TE were calculated for the respective land uses of 2015 (Table 13.1). ED of agricultural land and grassland were 108.85 and 85.61, respectively, whereas forest land and wetlands had edge density of 9.26 and 0.90 respectively. MSI, AWMSI, and LSI of the shape metrics were also calculated, the result of MSI for all land use in the study area is greater than one. AWMSI results showed 19.22, 8.51, 4.06, 3.57 agricultural land, grassland, wetlands, and forest land respectively (Table 25). LSI value for agricultural land, grassland, settlements, bush, and shrub land is greater than 60 whereas, for degraded land, forest land and wetland accounts 40.19, 21.25, and 6.69 respectively.

Nearest neighbor metrics were also computed for the land use of 2015 the study area, MNN result showed wetland has 229.4 followed by forest land (152.10), degraded land (138.79), and bush and shrubland (102.42). NNSD result revealed wetland has 510.09 followed by forest land (174.67) and degraded land (154.92) and the same is true for NNCV. On the other hand, MPI was calculated in the radius of 500 m. Hence, agricultural land has 3417.74 followed by grassland (249.65), wetlands (60.73), and forest land (37.64). The rest land uses have a value of less than 10.

Finally, interspersion and juxtaposition (IJI) were calculated. Table 13.1 showed bush and shrub land has IJI of 69.89, agricultural land has 67.61 followed by settlement (61.46) and forest land (61.21). Degraded land, grassland, and wet land have IJI values of 57.72, 55.21, and 27.64 respectively.

13.3.2 Plan for Potential Ecological Connectivity

LULC result of 2015 revealed that the study area had a total of 2740.10 hectare or (10%) of forest and bush and shrub land use. The remaining are non-forest land use, such as agriculture, grassland, settlement, degraded land, and wetlands (Fig. 13.2).

The forest and bush and shrub land use were further classified into two groups, with an area greater than 0.5 hectare and less than 0.5 hectare. Based on this 746 hectare was covered with fragmented forest area greater than 0.5 hectare with 89 patches and 1256 hectare of land with bush and shrub land with 661 patches. The rest 99 hectare and 638 hectare of land are under forest and bush and shrub land under 0.5 hectare, respectively. In addition, in the study area, there were 16 area closures that had total land coverage of 906.1 hectare and church forests that account 93.05

Table 13.2 Calculation result for landscape metrics by the year 2015 land cover map at class level

Index	Class name						
	Agricultural lands	Wetlands	Degraded lands	Settlements	Bush and shrub lands	Grassland	Forest lands
Area metrics							
CA	14,930.82	173.34	581.13	1105.83	1834.20	7111.08	844.38
LPI	24.46	0.48	0.07	0.41	0.35	3.85	0.95
PLAND	56.17	0.65	2.19	4.16	6.90	26.75	3.18
Patch density, size, and variability metrics							
NP	2058	19	1389	3504	3070	2618	526
PD	7.74	0.07	5.23	13.18	11.55	9.85	1.98
MPS	7.26	9.12	0.42	0.32	0.60	2.72	1.61
PSSD	161.10	28.19	1.14	2.39	2.39	28.71	13.33
PSCoV	2220.47	308.96	273.41	756.73	400.37	1056.88	830.17
Edge metrics							
ED	108.85	0.90	14.57	32.49	41.92	85.61	9.26
TE	2,893,230	24,000	387,180	863,640	1,114,260	2,275,710	246,060
Shape metrics							
MSI	1.30	1.46	1.14	1.13	1.21	1.28	1.18
AWMSI	19.22	4.06	1.63	2.60	2.26	8.51	3.57
LSI	61.03	6.69	40.19	65.31	65.40	68.17	21.25
Nearest neighbor metrics							
MNN	72.94	229.40	138.79	102.42	98.92	86.55	152.10
NNSD	24.08	510.09	154.92	69.73	67.12	49.35	174.67
NNCV	33.01	222.36	111.62	68.08	67.86	57.02	114.84
MPI (500 m)	3417.74	60.73	2.57	5.30	6.29	249.65	37.64
Interspersion metrics							
IJI	67.61	27.64	57.72	61.46	69.89	55.21	61.21

Class name: Land use type; CA: Class area; LPI: Largest Patch index; PLAND: Percent of landscape; NP: Number of patches; PD: Patch Density; MPS: Mean patch size; PSSD: Patch size standard deviation; PSCoV: Patch size coefficient of variation; ED: Edge Density; TE: Total Edge; MSI:AWMSI: Area-weighted mean shape index; LSI: Landscape Shape Index; MNN: Mean nearest neighbor distance; NNSD: Nearest neighbor standard deviation; NNCV: Nearest neighbor coefficient of variation; MPI: Mean proximity index (500 M); IJI: Interspersion and Juxtaposition index

hectare of land (Table 13.2). Hence, the study area has a total area 2305.39 hectare of forest, bush, and shrub, area closure, and church forest land use. They are considered as focal patches and used to plan for potential ecological connectivity (Fig. 13.3, Table 13.3)

Fig. 13.2 Forest and bush and shrub land; and non-forest land use of 2015 of the study area

On the other hand, to reduce fragmentation of the focal patches and sub patches forest and bush and shrub land use less than 0.5 hectare and degraded land are used. In addition, slope of the land greater than 30% was also extracted from DEM. As Table 13.3 revealed, total of 5025 hectare or (19%) of land use of the study area lies above 30% slope; of these 40.06% and 30.19% of the area were agricultural land and grassland, respectively. Bush and shrub land and forest land account 14.53% and 11.13%, respectively. About 2.46%, 1.57%, and 0.07% of the land in the respective slope were settlements, degraded land, and wetland, respectively.

Table 13.3 Focal patches to plan for ecological connectivity

Focal patches	Total land use in Hectare	Number of patches
Forest land > 0.5 hectare	746.16	89
Shrub and Bush land use > 0.5 hectare	1256.21	661
Area closure	906.1	16
Church forest	93.05	3

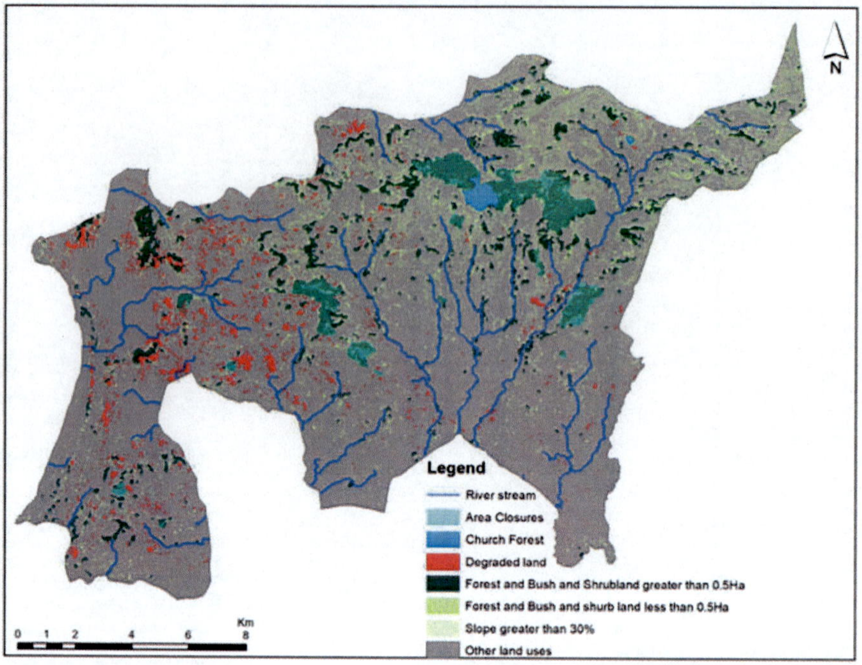

Fig. 13.3 Potential land use for ecological connectivity in Libokemekem District 2015

As it is also shown in Fig. 13.3, forest and bush and shrub land use greater than 0.5 hectare, area closures, and church forest are all the focal patches while land use of slope of the area greater than 30%, degraded land, forest and shrub land use less than 0.5 ha are sub patches. Moreover, river streams that are extracted from DEM are used to link all the patches.

Figure 13.3 shows that the land use that is categorized into major focal patches, sub patches, other land uses, and river streams. Then connectivity plan that used river stream as a link is proposed. In addition, to reduce fragmentation and increase the potential for ecological connectivity, area with slope greater than 30%, patches of forest and bush and shrub land less than 0.5 hectare, and degraded lands of the study site are used as sub patches. Therefore, river stream with buffer of 30 m is used to link among and between focal patches and sub patches.

According to Table 13.4, 29.58% of the land use of the study area will be covered with vegetation/forest. Of these 8.7% are focal patches, 3.23% are river streams with 30 m buffer and 17.64% are sub patches.

Table 13.4 Size of LULC of the study area (in Hectare) greater than 30% slope

Land use	Area (hectare)	Percentage
Agricultural lands	2013	40.06
Wetlands	4	0.07
Degraded lands	79	1.57
Settlements	123	2.46
Bush and Shrublands	730	14.53
Grasslands	1517	30.19
Forest land	559	11.13
Total	5025	100.00

13.3.3 Ecological Status of the Study Area After the Proposed Ecological Connectivity Plans

To calculate the change on the ecological connectivity status of the study area after the proposed plan, Fragstat 4 was used to see the fragmentation status. Similarly, area metrics patch density, size, and variation; edge metrics; shape metrics; nearest neighbor metrics, and interspersion (IJI) were calculated metrics. The result for the parameters of the metrics that show fragmentation status of the land use of 2015 after the proposed plan to increase connectivity and reduce fragmentation between and among patches, sub patches, and river stream with 30 m buffer zone of the study area as it is explained here.

As Table 13.5 stated, PLAND of agricultural lands and grasslands occupies 46.19% and 20.16% of land use respectively. Vegetation (forest and bush and shrub land) occupies 29.75% of the total land use of the study area while settlements and wetland occupy 3.45% and 0.63% respectively. LPI of land use shows vegetation land has 19.13 followed by agricultural land (4.28) and grassland (2.67). The rest land uses have LPI value less than one. Regarding patch density, size, and variability metrics, the number of patch (NP) for vegetation, grasslands, agricultural land, and settlement, were 3837, 3704, 3626, and 3023 respectively. On the other hand, wetland had 22 numbers of patches respectively. The MPS of wetland is 7.57% followed by agricultural land (3.38%), vegetation (2.05%), and grassland (1.45%). Settlement has a value less than 1 MPS. PSSD of vegetation land use is 82.38 followed by agricultural land use that is 41.96. Wetland and grassland have 15.70 and 14.84 respectively. The

Table 13.5 Proposed land use after the ecological connectivity plan and conservation work for degraded land use of the study area

Class name	Area in Ha	Percentage
Focal patches	2313.72	8.70
River stream	859.14	3.23
Sub patches	4689.54	17.64
Other land use	18,727.11	70.43
	26,589.51	100.00

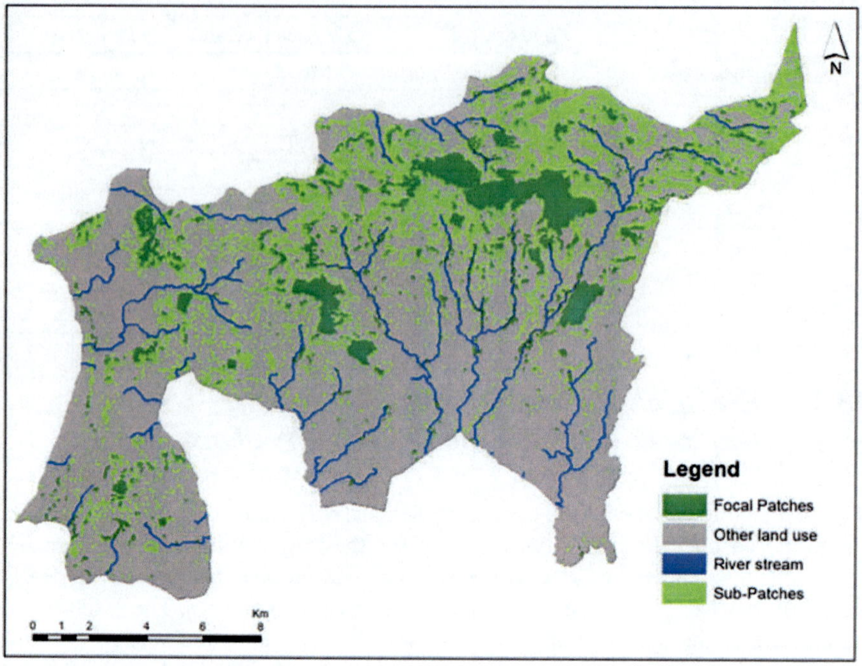

Fig. 13.4 Plan for ecological connectives

PSCoV value for vegetation is 4020.32 while agricultural land and grass land were 1239.90 and 1025.18 respectively. Settlement and wetland had a value of 544.36 and 207.36 respectively.

Regarding the edge metrics, ED and TE were calculated for the respective land uses of 2015 (Table 13.5; Fig. 13.4). ED of agricultural land and vegetation is 116.68 and 107.65 respectively; whereas, grassland, settlement, and wetlands will have edge density of 85.63, 27.08, and 1.01 respectively. MSI, AWMSI, and LSI of the shape metrics were also calculated, the result of MSI for all land use in the study area is greater than one. AWMSI results showed 81.32, 78.38, 71.81 for vegetation, grassland, and agricultural land respectively (Table 13.5). LSI value for vegetation land, agricultural land, and grassland are greater than 60, whereas for settlement and wetland accounts 59.88 and 7.22 respectively.

Nearest neighbor metrics were also computed for the land use of 2015 of the study area after the planning recommendation, MNN result showed wetland has 210.39 followed by settlement (108.91), grassland (81.83), vegetation (80.42), and agriculture land (70.91). NNSD result revealed wetland has 476.45 followed by settlement (76.62) and grassland (45.27) and the same is true for NNCV. On the other hand, MPI was calculated in the radius of 500 m. Hence, vegetation land has 2855.60 followed by agricultural land (230.03) while grasslands, wetland, and settlements have values of 84.66, 32.65, and 3.33 respectively.

Finally, interspersion and juxtaposition (IJI) were also calculated. Table 13.6 showed settlement has IJI of 72.14, agricultural land has 71.99 followed by vegetation land (64.14) grassland (61.21), and wetlands (61.76).

Table 13.6 Calculation result for Landscape metrics by the year 2015 land cover map at class level after the plan for ecological connectivity

Index	Class name				
	Agricultural lands	Wetlands	Settlements	Grassland	Vegetation land
Area Metrics					
CA	12,280.95	166.59	917.91	5361.66	7862.4
LPI	4.28	0.27	0.24	2.67	19.13
PLAND	46.19	0.63	3.45	20.16	29.57
Patch density, size, and variability metrics					
NP	3629	22	3023	3704	3837
PD	13.65	0.08	11.37	13.93	14.43
MPS	3.38	7.57	0.30	1.45	2.05
PSSD	41.96	15.70	1.65	14.84	82.38
PSCoV	1239.90	207.36	544.36	1025.18	4020.32
Edge metrics					
ED	116.68	1.01	27.08	85.63	107.65
TE	3,102,360	26,730	720,120	2,276,970	2,862,480
Shape metrics					
MSI	1.30	1.52	1.12	1.26	1.19
AWMSI	7.64	2.91	2.23	6.28	32.26
LSI	71.81	7.22	59.88	78.38	81.32
Nearest neighbor metrics					
MNN	70.91	210.39	108.91	81.83	80.42
NNSD	21.67	476.45	76.62	45.27	37.07
NNCV	30.56	226.46	70.35	55.32	46.10
MPI (500 m)	230.03	32.65	3.33	84.66	2855.60
Interspersion metrics					
IJI	71.99	61.76	72.14	63.89	64.14

Class name: Land use type; CA: Class area; LPI: Largest Patch index; PLAND: Percent of landscape; NP: Number of patches; PD: Patch Density; MPS: Mean patch size; PSSD: Patch size standard deviation; PSCoV: Patch size coefficient of variation; ED: Edge Density; TE: Total Edge; MSI:AWMSI: Area-weighted mean shape index; LSI: Landscape Shape Index; MNN: Mean nearest neighbor distance; NNSD: Nearest neighbor standard deviation; NNCV: Nearest neighbor coefficient of variation; MPI: Mean proximity index (500M); IJI: Interspersion and Juxtaposition index

13.4 Discussion

13.4.1 The 2015 Status of Land Use/Land Cover in the Study Area

Based on the result of the area metrics indicates, Class area (CA) and Percent of landscape (PLAND) of agricultural land have high value followed by grassland. This implies that the dynamics of some ecological processess of the forest land use are likely to be quite different than the two land uses and have an implication of extinction of certain species of forest and wildlife. According Seibold et al. (2017) patch size was the most important attribute that can be influencing different measures of species composition and anthropogenic disturbances of an area. Hence, Largest Patch index (LPI) of agricultural land and grassland is 20 times that of forest land use. Kim and Pauleit (2005) stated larger patches are considered to be important for ecological integrity of landscapes but the smallest patches can have some advantages like acting as ‘stepping stones’ for species movement. The largest patch index (LPI) indicates that forest land is fragmented into smaller patches compared to agricultural and grassland use. Hence, keeping large patches, supplemented with small patches, is an important element of biodiversity conservation.

The Number of Patch (NP) and patch density depicted that settlement, bush and shrubland, grassland and agricultural land can be considered more fragmented compared to wetland and forest land. According to Gokyer (2013) when the NP values increase it is understood fragmentation increases in the field and vice versa. In contrast to these indices, Mean patch size (MPS) does a good job of ranking that settlement, degraded land, bush and shrubland, and forestland are most fragmented. However, wetland, agricultural land, and grassland are least fragmented. In line with Kirstein and Netzband (2001) implied smaller values points to a higher fragmentation of the area. However, MPS is more meaningful when it is interpreted with Number of Patch (NP), Patch Density (PD), and patch size variability. According to Fan and Myint (2014) PD measures the number of patches per hectares. It is one of the best indicators of landscape fragmentation because it implies how a particular class is fragmented within the area. PSSD of degraded land, settlement, bush and shrubland, and forest land have fragmented land use compared to agricultural land, grassland, and wetlands. Generally, these results implied that wetland, bush and shrubland, and forestland have been subjected to human disturbance undergone considerable fragmentation due to expansion of agricultural land and grazing land. Herzog et al. (2001) stated that mean patch size (MPS) decrease as natural landscape units became smaller due to the increased proportion of patches that are cut arbitrarily by natural landscape unit boundaries. Moreover, land uses became more heterogonous mainly due to fragmentation (Shi et al. 2008).

Edge metrics are usually best considered as representing landscape configuration and are important to many ecological phenomena (McGarigal and Marks 1994). In the study area, edge density and total edge for agricultural land and grassland have higher values. This implies this land use is highly fragmented compared to forest land and wetland use, both have the least Edge Density (ED) and Total Edge (TE). However, according to McGarigal and Marks (1994) TE and ED used to model habitat suitability and in this case, forest land and wetlands are the least suitable land for habitats uses compared to the others.

According to McGarigal and Marks (1994) the shape metrics of the study area land use can influence a number of important ecological processes. Mean shape index (MSI) values for all the land use types are greater than one; this implies the average patch shape in all land use types is non-circular. This result also showed that forestland, degraded land, and settlements are highly fragmented compared to the rest land use types. In addition, Landscape Shape Index (LSI) is another indicator for landscape complexity, according to Herzog and Lausch (2001) with increasing size, the patch shapes simplify. LSI is also very effective in characterizing landscape fragmentation (Fan and Myint 2014).

Hence, except wetlands and forest land use types, the rest land use types have the largest LSI. The smaller the size for wetland and forest land use types are mainly due to an increase in agricultural and grassland use of the study area at the expense of forest and wetlands (Demissie et al. 2017). The area-weight mean shape index (AWMSI) supports this conclusion.

Nearest neighbor metrics is defined as the distance from a patch to the nearest neighboring patches of the same types and can influence a number of important ecological processes (McGarigal and Marks 1994). Hence, the Mean nearest neighbor distance (MNN) is greatest for wetland followed by forest landscape; this implies that forestland and wetland patches are most isolated in this area compared to the patches of the other land use types. Similarly, the Nearest neighbor standard deviation (NNSD) and Nearest neighbor coefficient of variation (NNCV) result showed, wetland and forest land have the greater values of these variability indices. This implies there is irregular or uneven distribution of patches. That is, the distribution of patches may reflect underlying processes or human caused disturbance patterns. The Mean proximity index (MPI) is inversely related MNN based on a 500 m search radius and indicates that degraded land, settlement, and bush and shrubland are most fragmented and insular. These nearest neighbor indices indicated that agricultural land, grassland, and wetland are less fragmented than forest land.

The arrangement of patches and land use types in the study area are assessed via interspersion and juxtaposition (IJI). It approaches 0 when adjacencies are unevenly distributed; IJI = 100 if all patch types are equally adjacent to all other patch types (McGarigal and Marks 1994). The interspersion and juxtaposition (IJI) indicate that bush and shrubland, agricultural land, settlement, and forest land have high value respectively. This implies these land use types are relatively more equitable distributed among patches than wetland (lowest IJI values), grasslands, and degraded lands. In line with this Herzog et al. (2001) stated that arable lands, permanent grasslands, and forest land adjacent to most other land use types had higher values of IJI whereas linear elements such as riparian wood which were almost always adjacent to grassland and water had lower IJI values. This mostly occurs only in particular locations and share borders with a reduced number of land use (Herzog and Lausch (2001).

13.4.2 Plan for Potential Ecological Connectivity

To drive possible ecological connectivity and potential ecological connection, river streams of the study area are used as connectors. In addition, forest land and bush and shrub lands less than 0.5 hectare, degraded land, and area of slope greater than 30% are used to reduce fragmentation between and among focal patches and themselves. Connectivity of habitat provided by corridors can play a critical role in the maintenance of local diversity (Metzger and Decamps 1997; Damschen and Brudvig 2012). For the study area, as shown in Fig. 13.3, focal patches are area closures, forest area and bush and shrubland greater than 0.5 hectares, and church forests. In line with this Huber et al. (2010) stated in the design of ecological connectivity and core area identified by conservation planning processes (Walz 2011) are linked by corridors meant to permit the movement of animals and plants between the cores.

Therefore, Fig. 13.4 shows the possible connection of the area mainly using the river streams. According to FDRE (2012) river side protection is suggested to prevent further erosion of the upland areas and to reduce gully widening and increase tree cover. Because physical and ecological function of the landscape can be provided when the riparian vegetation are conserved and restored (Peak and Thompson 2006). Once riparian vegetation is conserved and restored, they can easily facilitate wildlife movement between patches by serving as ecological corridors (Hilty et al. 2012; Peak and Thompson 2006). Hence, the study also suggested buffer zone for river streams not only to increase movement among focal patches but also to reduce sediment load that comes from modern agricultural practices to the Lake Tana water body. Similarly, FDRE (2012) explained well established and managed river catchments reduce sediment load, nutrients, and other pollutants loads to the water body (Polyakov et al. 2005). Then the slope of the area greater than 30%, patches of forest area and bush and shrub land less than 0.5 hectare of 2015, and degraded land use of 2015 used to reduce fragmentation. Slope, an important element of landform, plays an important role such as to avoid soil erosion and other problems derived from the use of the land (Kibret et al. 2016). According to Jahn et al. (2006) suggested slopes greater than 30% described as steep slope and not capable of supporting cultivation of crop. They are good for growing grasses, forestry, and supporting wild life (AbdelRahman et al. 2016).

Corridors such as river buffer will ease movement of resources but there are no clear rules and regulation on the size of buffer for the river stream and other corridors but the wider the buffer, the more effective it will be (Hilty et al. 2012). According to Polyakov et al. (2005) there are two approaches for designing buffer zones. The first one is the fixed width approach when a minimum width is defined according to regional conditions and government bodies' recommendations (Lee et al. 2004). This requires minimum planning. The second approach is spatially variable riparian buffer designed to achieve specific water conservation goal of reduction of non-point source pollutant via optimizing its characteristics with respect to runoff contributing area, slope, soil type and use, and climate in that particular location (Polyakov et al. 2005). Therefore, based on this argument the study suggests the second approach, which is spatially variable riparian buffer and buffer zone greater than or equal to 30 m is suggested buffer zone.

Therefore, based on the suggested river buffer zone and after soil and water conservation measures on degraded land and slope of the area greater than 30%, the vegetation cover of the area will increase from 10.08% to 29.57%. In addition, the LPI of vegetation will increase by more than 14 folds which imply on ecological integrity of landscape and reduced fragmentation (Kim and Pauleit 2005). However, NP is high for vegetation followed by agricultural land use after the plan that implies for high fragmentation compared to other land use. In contrast, MPS shows that agricultural land, wetlands, and vegetation land is least fragmented compared to settlement and grasslands. PD is one of the best indicators, PSSD of vegetation is 82.38 implies that it is least fragmented followed by agricultural land compared to other land use. On the other hand, edge density and total edge of agricultural land and vegetation are higher compared to other land use that implies high fragmentation and according to McGarigal and Marks (1994) stated the highest ED and TE has the highest suitable land use compared to others that have low ED and TE. In line with this ED metrics also identifies the level of provision of continuity of existence of a variety of species and their natural habitat (Sowińska-Świerkosz and Soszyński 2014). In addition, these metrics have also been proven to be vital in the watershed for water quality and affecting nutrient and organic matter runoff.

In conclusion, the six basic metrics such as area metrics, patch density, edge metrics, shape metrics, nearest neighbor metrics, and interspersion metrics were considered for landscape planning that are useful tools of conservation efficiency analysis at class level. The result shows the fragmentation status of vegetation cover is improved. This improvement is seen after planning with possible patches to increase the vegetation cover of the study area. Jha (2006) reports that biosphere reserves are important in rehabilitating and restoring the ecology of modified or degraded landscape and maintaining of long-term health of representative ecosystems. In 2014 part of the watershed with an area of 5000 square kilometers including the study area has been delineated for biosphere reserves by regulation number 125/2014 (Zikre Hig 2014). This can be used as an opportunity to implement the suggested connectivity approach in this study.

13.5 Conclusions

The study also analyzed the current ecological connectivity status of the study area and the result of the calculated metrics revealed forest land and bush and shrub land use are the most threatened land use that occupies 3.18% and 6.90% of the total land of the area respectively. The MPS, PSSD and MPS result shows land use such as forest, bush and shrub land and wetlands are most fragmented due to human disturbance for the expansion of agricultural land and grazing land use. This implies forest land and bush and shrub land can be accessed by anyone as "free good" and this encourages the land to be easily fragmented, degraded, and encroached by other land use.

However, to reverse the fragmented status of forest and bush and shrub land use, the study uses forest and bush and shrub land use less than 0.5 hectare, river with 30 m buffer, slope of the area greater than 30%, and degraded land use. For the river side, slope greater than 30% and degraded land need soil and conservation works. Then after the application of this scenario, the calculated result shows the vegetation cover of the study area increased from 10 to 29%. The LPI also increased by 14 folds. The ED and TE result is higher for vegetation, which has a positive implication on suitability of the land use for the existence of a variety of species and their natural habitat. This all implies ecological integrity of landscape for the flow of resources will be improved by reducing fragmentation among the vegetation land use after the proposed development strategies.

13.6 Recommendation

- Soil and water conservation work have to be recognized and promoted with the participation of the community and other stakeholders on the degraded land, slope greater than 30%, and river side. Consequently, community participation is very essential for the sustainable use and conservation of the resources of the area specially the forest and other common resources.
- In the study area, there is a beginning in promoting zero grazing by encouraging community to feed their livestock on their yards from the communal grazing land and area closure by cutting and carrying, and feeding straw. This should be well organized and recommended to continue for the betterment of the degraded grazing area and improvement of the milk and meat product from the livestock.
- The government needs to give unambiguous tree/forest use right for the household that plants trees in their backyard and farm land by considering the existing land use right. The use right has to take into account the need of the community that can encourage to plant more trees on the available spaces they have. In addition, it is also recommended to plant these trees and other indigenous trees that are important to protect the river side (buffer zone) and can reduce further degradation and the widening of the gulley.

References

AbdelRahman MA, Natarajan A, Hegde R (2016) Assessment of land suitability and capability by integrating remote sensing and GIS for agriculture in Chamarajanagar district, Karnataka, India. Egypt J Remote Sens Space Sci 19(1):125–141

Alelign A, Teketay D, Yemshaw Y, Edwards S (2007) Diversity and status of regeneration of woody plants on the peninsula of Zegie, northwestern Ethiopia. Trop Ecol 48:37–49

Bennett AF (1990) Habitat corridors and the conservation of small mammals in a fragmented forest environment. Landscape Ecol 4(2–3):109–122
Bennett AF (1999). Linkages in the landscape: the role of corridors and connectivity in wildlife conservation (No. 1). Iucn
Central Statistical Agency of Ethiopia (CSA) (2007) Population and housing census of Ethiopia: statistical report for Amhara region. Addis Ababa, Ethiopia: Central Statistical Agency of Ethiopia
Collinge SK (1996) Ecological consequences of habitat fragmentation: implications for landscape architecture and planning. Landsc Urban Plan 36(1):59–77
Conceicao KS, De Oliveira VM (2010) Habitat fragmentation effects on biodiversity patterns. Physica A 389(17):3496–3502
Cowling RM, Pressey RL, Rouget M, Lombard AT (2003) A conservation plan for a global biodiversity hotspot—the Cape Floristic Region South Africa. Biol Conserv 112(1–2):191–216
Damschen EI, Brudvig LA (2012) Landscape connectivity strengthens local–regional richness relationships in successional plant communities. Ecology 93(4):704–710
Demissie F, Yeshitila K, Kindu M, Schneider T (2017) Land use/land cover changes and their causes in Libokemkem District of South Gonder, Ethiopia. Remote Sens Appl Soc Environ 8:224–230
Demissie F, Yeshitela K, Rouleau M, Flaspohler D, Kindu M (2019) Socio-economic importance of forest resources and their conservation measures in Ethiopia: the case of area closure in South Gonder of Ethiopia. Environ Monit Assess 191:437
Desalegn T, Cruz F, Kindu M, Turrión MB, Gonzalo J (2014) Land-use/land-cover (LULC) change and socioeconomic conditions of local community in the central highlands of Ethiopia. Int J Sust Dev World 21:406–413
Di Gregorio A, Jansen LJ (2000) Land cover classification system: LCCS: classification concepts and user manual. Food and Agriculture Organization of the United Nations Rome
Fagan WF, Cantrell RS, Cosner C (1999) How habitat edges change species interactions. Am Nat 153(2):165–182
Fahrig L (2020) Why do several small patches hold more species than few large patches? Glob Ecol Biogeogr 29(4):615–628
Fahrig L (2017) Ecological responses to habitat fragmentation per se. Annu Rev Ecol Evol Syst 48:1–23
Fahrig L (2003) Effects of habitat fragmentation on biodiversity. Annu Rev Ecol Evol Syst 34(1):487–515
Fan C, Myint S (2014) A comparison of spatial autocorrelation indices and landscape metrics in measuring urban landscape fragmentation. Landsc Urban Plan 121:117–128
Federal Democratic Republic of Ethiopia (FDRE) (2012) Urban water supply and sanitation project: consultancy service for master plan review, catchment rehabilitation and awareness creation for Geffersa, Legedadi, and Dire catchment areas: Catchment rehabilitation: Volume 1: report
Fischer J, Lindenmayer DB (2007) Landscape modification and habitat fragmentation: a synthesis. Glob Ecol Biogeogr 16(3):265–280
Franklin JF (1993) Preserving biodiversity: species, ecosystems, or landscapes? Ecol Appl 3(2):202–205
Friis I, Demissew S, Van BP (2011) Atlas of the potential vegetation of Ethiopia. Addis Ababa University Press and Shama Book, Addis Ababa
Friis IB (1986) The forest vegetation of Ethiopia. Symbolae Botanicae Upsalienses 26(2):31–47
Friis IB (1992) Forests and forest trees of northeast tropical Africa: their natural habitats and distribution patterns in Ethiopia. HMSO, Djibouti and Somalia
Garedew E, Sandewall M, Soderberg U (2012) A dynamic simulation model of land-use, population, and rural livelihoods in the Central Rift Valley of Ethiopia. Environ Manage 49:151–162
Gokyer E (2013) Understanding landscape structure using landscape metrics. In: Advances in landscape architecture. IntechOpen
Herzog F, Lausch A (2001) Supplementing land-use statistics with landscape metrics: some methodological considerations. Environ Monit Assess 72(1):37–50

Herzog F, Lausch A, MÜller E, Thulke HH, Steinhardt UTA, Lehmann S (2001) Landscape metrics for assessment of landscape destruction and rehabilitation. Environ Manage 27(1), 91–107
Hilty JA, Lidicker WZ Jr, Merenlender AM (2012) Corridor ecology: the science and practice of linking landscapes for biodiversity conservation. Island Press
Hodgson JA, Moilanen A, Wintle BA, Thomas CD (2011) Habitat area, quality and connectivity: striking the balance for efficient conservation. J Appl Ecol 48(1):148–152
Huber PR, Greco SE, Thorne JH (2010) Spatial scale effects on conservation network design: trade-offs and omissions in regional versus local scale planning. Landscape Ecol 25(5):683–695
Jahn R, Blume HP, Asio VB, Spaargaren O, Schad P (2006) Guidelines for soil description. FAO
Jha SG (2006) An overview of the biosphere reserve concept and its application to Nepal. Banko Janakari 16(1):21–24
Kibret KS, Marohn C, Cadisch G (2016) Assessment of land use and land cover change in South Central Ethiopia during four decades based on integrated analysis of multi-temporal images and geospatial vector data. Remote Sens Appl Soc Environ 3:1–19
Kim KH, Pauleit S (2005) Landscape metrics to assess the ecological conditions of city regions: application to Kwangju City, South Korea. Int J Sustain Dev World Ecol 12(3):227–244
Kindu M, Schneider T, Teketay D, Knoke T (2013) Land use/land cover change analysis using object-based classification approach in Munessa-Shashemene landscape of the Ethiopian highlands. Remote Sens 5:2411–2435
Kindu M, Schneider T, Teketay D, Knoke T (2015) Drivers of land use/land cover changes in Munessa-Shashemene landscape of the south-central highlands of Ethiopia. Environ Monit Assess 187(7):60
Kindu M, Schneider T, Döllerer M, Teketay D, Knoke T (2018) Scenario modelling of land use/land cover changes in Munessa-Shashemene landscape of the Ethiopian highlands. Sci Total Environ 622–623:534–546
Kingsland S (2002) Designing nature reserves: adapting ecology to real-world problems. Endeavour 26(1)
Kirstein W, Netzband M (2001) Landscape metrics as a tool for the comparison of different Urban Regions. Remote Sensing of Urban Areas/Fernerkundung in urbanen Räumen. Regensburger Geographische Schriften (35)
Laurance WF, Yensen E (1991) Predicting the impacts of edge effects in fragmented habitats. Biol Cons 55(1):77–92
Lee P, Smyth C, Boutin S (2004) Quantitative review of riparian buffer width guidelines from Canada and the United States. J Environ Manage 70(2):165–180
Ligdi E, Kahloun ME, Meire P (2010) Ecohydrological status of Lake Tana: a shallow highland lake in the Blue Nile (Abbay) basin in Ethiopia: Review. Ecohydrol Hydrobiol 10:109–122
McGarigal K, Marks BJ (1994) FRAGSTATS. Spatial pattern analysis program for quantifying landscape structure. Version 2.0. Forest Science Department, Oregon State University, Corvallis, 67
Metzger JP, Décamps H (1997) The structural connectivity threshold: an hypothesis in conservation biology at the landscape scale. Acta Oecologica 18(1):1–12
Noss RF (1987) Corridors in real landscapes: a reply to Simberloff and Cox. Conserv Biol 1(2):159–164
Olson DH, Burnett KM (2009) Design and management of linkage areas across headwater drainages to conserve biodiversity in forest ecosystems. For Ecol Manage 258:S117–S126
Opdam P, Wascher D (2004) Climate change meets habitat fragmentation: linking landscape and biogeographical scale levels in research and conservation. Biol Cons 117(3):285–297
Peak RG, Thompson FR III (2006) Factors affecting avian species richness and density in riparian areas. J Wildl Manag 70(1):173–179
Polyakov V, Fares A, Ryder MH (2005) Precision riparian buffers for the control of nonpoint source pollutant loading into surface water: a review. Environ Rev 13(3):129–144

Rahmato D (ed.) (2001) Environmental change and state policy in Ethiopia: lessons from past experience. Forum for social studies. Addis Ababa, Addis Ababa University, p 121 (Series No. 2)

Rands MRW, Adams WM, Bennun L, Butchart SHM, Clements A, Coomes D, Entwistle A, Hodge I, Kapos V, Scharlemann JPW (2010) Biodiversity conservation: challenges beyond 2010. Science 329:1298–1303

Rashid M, Craig D, Mukul SA, Khan NA (2013) A journey towards shared governance: Status and prospects for collaborative management in the protected areas of Bangladesh. J Forest Res 24:599–605

Saha AK, Setegn SG (2015) Ecohydrology for sustainability of iwrm: a tropical/subtropical perspective. In: Sustainability of integrated water resources management. Springer, Cham, pp 163–178

Saunders DA, Hobbs RJ, Margules CR (1991) Biological consequences of ecosystem fragmentation: a review. Conserv Biol 5(1):18–32

Setegn SG (2008). Hydrological and sediment yield modelling in Lake Tana basin, Blue Nile Ethiopia (Doctoral dissertation, KTH)

Seibold S, Bässler C, Brandl R, Fahrig L, Förster B, Heurich M, Hothorn T, Scheipl F, Thorn S, Müller J (2017) An experimental test of the habitat-amount hypothesis for saproxylic beetles in a forested region. Ecology 98(6):1613–1622

Shi Y, Xiao J, Shen Y (2008) Landscape pattern change and associated environmental implications in the Haihe River Basin China. ISPRS J Photogramm Remote Sens 37:569–574

Sowińska-Świerkosz BN, Soszyński D (2014) Landscape structure versus the effectiveness of nature conservation: Roztocze region case study (Poland). Ecol Ind 43:143–153

Taddese G (2001) Land degradation: a challenge to Ethiopia. Environ Manage 27:815–824

Taylor PD, Fahrig L, Henein K, Merriam G (1993) Connectivity is a vital element of landscape structure. Oikos, 571–573

Tegene B (2002) Land-cover/land-use changes in the derekolli Catchment of the South Welo Zone of Amhara Region, Ethiopia. East Afr Soc Sci Res Rev 18:1–20

Temesgen H, Nyssen J, Zenebe A, Haregeweyn N, Kindu M, Lemenih M, Haile M (2013) Ecological succession and land use changes in a lake retreat area (Main Ethiopian Rift Valley). J Arid Environ 91:53–60

Tischendorf L, Fahrig L (2000) On the usage and measurement of landscape connectivity. Oikos 90(1):7–19

Thorn S, Seibold S, Leverkus AB, Michler T, Müller J, Noss RF, Stork N, Vogel S, Lindenmayer DB (2020) The living dead: acknowledging life after tree death to stop forest degradation. Front Ecol Environ 18(9):505–512

Walpole MJ, Goodwin HJ (2001) Local attitudes towards conservation and tourism around Komodo National Park Indonesia. Environ Conserv 28(2):160–166

Walz U (2011) Landscape structure, landscape metrics and biodiversity. Living Rev Landscape Res 5(3):1–35

Wilcove DS, McLellan CH, Dobson AP (1986) Habitat fragmentation in the temperate zone. Conserv Biol 6:237–256

Young TP (2000) Restoration ecology and conservation biology. Biol Cons 92(1):73–83

Zegeye H, Teketay D, Kelbessa E (2011) Diversity and regeneration status of woody species in Tara Gedam and Abebaye forests, northwestern Ethiopia. J Forest Res 22:315–328

Zikre Hig (2014) A regulation provided for administration of biosphere reserves of lake tana regulation number 125/2014 issued at 'Zikre Hig' publication, the Amhara National regional state, Amhara region, Bahir dar, Ethiopia

Chapter 14
Identifying Priority Areas for Conservation in Mojo River Watershed of Ethiopia Using GIS-Based Erosion Risk Evaluation

Kiflu Gudeta, Mekuria Argaw, and Mengistie Kindu

Abstract Identification of priority areas for the establishment of conservation measures is the first step in conservation planning. Resources may constrain launching of watershed management activities all over a watershed at the same time, hence methods to prioritize intervention are essential. Intensity of land degradation may be one of the key factors to consider in the process of prioritization. This study investigated prioritization of Micro-watersheds (MWs) using soil erosion risk and tested using Mojo River watershed as a case study area. The Revised Universal Soil Loss Equation (RUSLE) and Multi-Criteria Evaluation (MCE) approaches were integrated in GIS environment using remotely sensed and other ancillary data. The analysis showed that RUSLE and MCE help to categorize landscape units into different levels of erosion risk and identify areas that require priority for conservation measures. Based on the RUSLE, MW-level average annual soil loss could be estimated, severity level assessed, and the area covered under various severity levels estimated to support planning. Based on the approach, MW-wise Composite Erosion Index (CEI) could be estimated. As a result, the critical MWs under very high and severe categories were recommended for immediate conservation intervention to reduce on-site and off-site soil loss effects.

Keywords MCE · Conservation priority area · Mojo River Watershed RUSLE · Soil erosion

K. Gudeta (✉)
Ministry of Agriculture, Addis Ababa, Ethiopia

M. Argaw
Center for Environmental Science, College of Natural and Computational Sciences, Addis Ababa University, Addis Ababa, Ethiopia

M. Kindu
TUM School of Life Sciences Weihenstephan, Institute of Forest Management, Technical University of Munich, Hans-Carl-von-Carlowitz-Platz 2, 85354 Freising, Germany
e-mail: mengistie@tum.de

M. Kindu et al. (eds.), *State of the Art in Ethiopian Church Forests and Restoration Options*, https://doi.org/10.1007/978-3-030-86626-6_14

14.1 Introduction

Soil erosion is the most widespread and multifaceted global land degradation process which leads to a decline in ecosystem services and functions (Adimassu et al. 2014; Haregeweyn et al. 2015). Its on-site and off-site effects threaten food security and economic growth (Hurni 1993; Sutcliff 1993; Tamene 2005; FAO 2019).

In Ethiopia, several studies investigated historical land use land cover (LULC) changes (Bewket 2002; Kindu et al. 2013; Temesgen et al. 2013; Demissie et al. 2017; Gashaw et al. 2018). These studies revealed a worrying trend of LULC changes with consequent soil erosion and land degradation (Tegene 2002; Desalegn et al. 2014; Kindu et al. 2018). The recorded annual soil erosion in Ethiopia ranges from 16 to 300 t/ha/year (Hawando 1995) depending mainly on slope, land cover, and rainfall intensity.

As a consequence of soil degradation, the productive capacity of soils in the Ethiopian highlands has been declining at a rate of 2–3% annually (Hurni 1993). Apart from the adverse effects on land productivity, soil erosion also causes off-site effect and adversely affects irrigation and hydropower generation capacity through sedimentation (Haregeweyn et al. 2015). For instance, sedimentation in Koka hydropower dam of Ethiopia caused potential storage capacity loss. According to EEPC (2002), the rate of siltation in the reservoir had grown from 857 tons/km^2 in 1970 to 2115 tons/km^2/yr. This situation has lowered water volume from the designed live storage capacity of 1667 M m^3 in 1959 to 1186 M m^3 in 1998, which is a loss of 30% of the total storage volume of the reservoir. (EEPC 2002; Elias 2003).

In order to prevent further degradation of upper catchments and to address its off-site effects in lower catchment, information on the extent and spatial distribution of erosion source areas is of paramount importance (Deore 2005; Shi et al. 2003; Tripathi et al. 2003). Because of differences in environmental attributes across landscapes, often few areas of the watershed are responsible to instigate erosion and soil losses. Moreover, limited financial resources often exclude the application of conservation measures all over affected areas at the same time (Tamene 2007; Tripathi et al. 2003). Identification of hot-spot areas of erosion and prioritizing intervention is important to effectively deal with erosion related problems (Khan et al. 2001; Tamene 2005; Kindu et al. 2015).

Various erosion models and/or multi-criteria evaluation approaches integrated with Remote Sensing and GIS have been successfully used in various studies (Tripathi et al. 2003; Deore 2005; Tamene 2007; Asis et al. 2008; Shi et al. 2003; Gashaw et al. 2018; Kindu et al. 2016, 2018; Yahya et al. 2013). This study investigates application of Revised Universal Soil loss Equation (RUSLE) model and multi-criteria evaluation (MCE) approaches integrated with GIS in identifying priority areas for soil erosion conservation using Mojo River Watershed as a case study area. To achieve the research objective, climatological, pedological, topographic, anthropogenic (ground cover) parameters and potential location for gully formation (Tamene 2007) were utilized in the analysis.

14.2 Materials and Methods

14.2.1 Study Area

Mojo River Watershed is located in the main Ethiopian Rift Valley extending up to the eastern escarpment. It is part of the upper Awash River basin in the Eastern Showa Zone of Oromia regional state. Geographically the study area lies between 38°54′ 10″ to 39° 17′ 30″ E and 8° 24′ 57″ to 9° 5′ 47″ N. The watershed covers an area of 1680.41 km^2. Elevations of the watershed vary from 1591 to 3069 m asl. Topography of the study area is generally characterized by gently undulating terrain. Of the total area, 61% (1024 km^2) lies within the slope range of 2–10% which is gently flat to undulating topography. Thirty three years (1986–2018) rainfall records for 6 stations within and adjacent to the study area show an average annual rainfall between 772.73 mm and 990.49 mm. The annual temperature of the study area is 19.5 °C with average winter high temperature of 24.1 °C in May and an average low temperature of 14.8 °C in December. Lack of vegetation cover associated with other erosion factors are exposing the area to high rates of soil erosion and loss of soil fertility which inducing heavy silt loads in rivers (Elias 2003).

Secondary data used in location map (Fig. 14.1) gathered from various organizations: towns and administrative boundary from Central Statistics Agency (CSA) (2007); road network from Ethiopian Road Authority (2007); river basin, lakes and soil types (Table 14.1) from Ministry of Water Resources (MoWR) (1999). River network and Mojo River Watershed boundary generated from Digital Elevation Model (DEM) using ArcSWAT.

14.2.2 Data and Preprocessing

In this study, cloud free Landsat OLI imagery acquired on February 1, 2019 (Path/Row is 168/54) was used for mapping land cover and estimating RUSLE *C*-factor. The digital numbers (DN) of the imagery were first converted to at-sensor radiance by using the radiometric calibration parameters in ENVI software. The Fast Line-of-sight Atmospheric Analysis of Spectral Hypercubes (FLAASH) algorithm was then used to convert radiance to reflectance and perform atmospheric correction. Climatic data such as annual rainfall (1986–2018) were gathered from National Meteorological Agency. In addition, a digital elevation model (DEM) with a spatial resolution of 30 m was downloaded from the United States Geological Survey's (USGS) Earth Explorer website (http://earthexplorer.usgs.gov/) for deriving various topographic indices. Field work was carried out during dry season in February 2019 to collect information about vegetation and erosion features in the watershed.

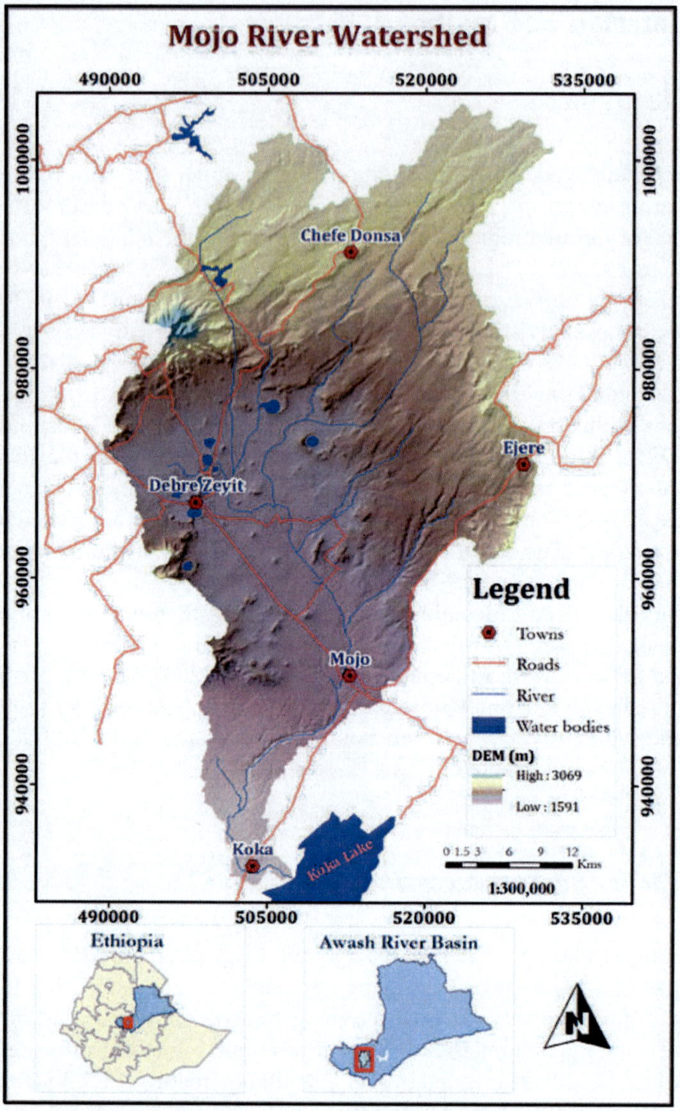

Fig. 14.1 Location map of the study area

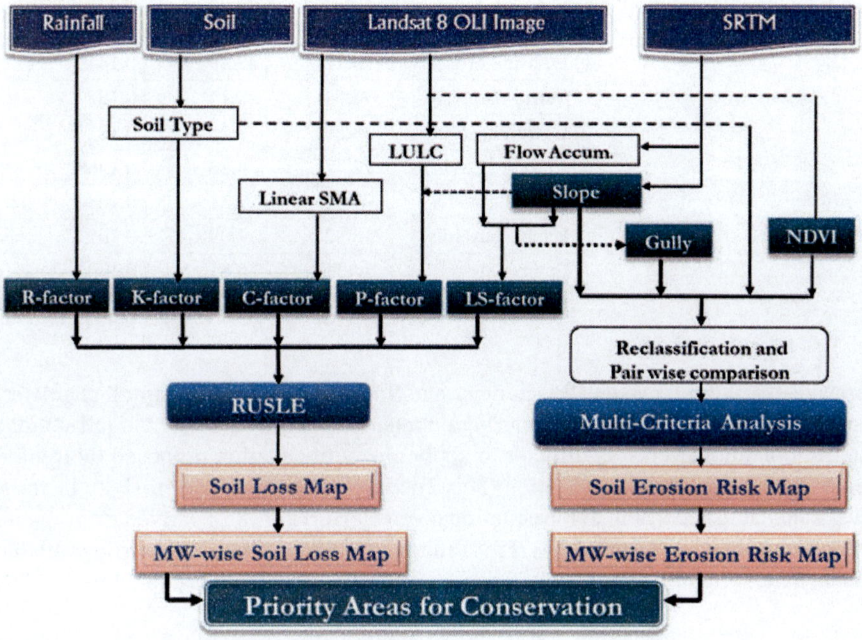

Fig. 14.2 Flow chart of research methodology

14.2.3 *Methods of Data Integration and Analysis*

14.2.3.1 Micro-watershed Delineation for Identifying Priority Areas

From the 30 m resolution DEM data, Mojo River Watershed and 22 micro-watersheds were delineated using ArcSWAT software. In this study, two approaches were adopted for identifying priority areas in the watershed based on soil erosion-proneness of micro-watersheds. RUSLE was used for estimating soil loss (Renard et al. 1997; FAO 2019; Yahya et al. 2013; Shi et al. 2003) and MCE approach was utilized for mapping erosion risk (Deore 2005; Tamene 2007). The overall methodology is illustrated in Fig. 14.2.

14.2.3.2 Erosion Factors Generation

Rainfall Erosivity (*R*) Factor

Potential ability of rain to cause erosion is known as erosivity (*R*-factor). It is defined as the product of kinetic energy and the maximum 30 min intensity and shows the

Table 14.1 Soil types and calculated erodibility factor value

Soil type	Area (km^2)	Percentage	*K* factor value
Euthric Vertisols	729.12	43.39	0.1440
Vertic Cambisols	674.54	40.14	0.1522
Luvic Phaeozems	138.59	8.25	0.1629
Chromic Luvisols	72.77	4.33	0.1568
Lithic Leptisols	20.41	1.21	0.1632
Haplic Luvisols	18.32	1.09	0.1659
Euthric Fluvisols	14.31	0.85	0.1702
Mollic Andosols	10.57	0.63	0.1660

erosivity of rainfall events (Wischmeier and Smith 1978). Due to rainfall characteristics and absence of automatic hourly rain intensity records in many rainfall stations in Ethiopia, however, it is difficult to apply erosivity equation proposed by Renard et al. (1997) for Ethiopia condition (Nyssen 2001). Therefore, the erosivity factor *R* was calculated according to the equation given by Hurni (1985), derived from spatial regression analysis (Hellden 1987) for Ethiopian conditions based on easily available mean annual rainfall (*P*). The *R*-factor is given by a regression Eq. (14.1):

$$\boldsymbol{R} = -8.12 + 0.562\boldsymbol{P} \tag{14.1}$$

To determine the value of the R-factor, the average of 33-years annual historic rainfall event (1986–2018) was collected from six meteorological stations located within and near the study area. Spline interpolation was done to generate an estimated surface from these scattered sets of point data.

Soil Erodibility (*K*) Factor

Vulnerability of the soils to get eroded is referred to as erodibility of soils. The *K*-factor is defined as the rate of soil loss per unit of R-factor on a unit plot (Renard et al. 1997). For Ethiopian condition, Hellden (1987) proposed the *K* values of the soil based on their color by adapting different sources. Using Eqs. 14.2–14.6 below, the *K* factor value was estimated for eight soil types identified in the study area (Table 14.1) based on a formula adapted from Yahya et al. (2013) using the FAO harmonized digital soil data.

$$K_{\mathrm{USLE}} = f_{\mathrm{csand}} \cdot f_{\mathrm{cl-si}} \cdot f_{\mathrm{orgC}} \cdot f_{\mathrm{hisand}} \tag{14.2}$$

$$f_{\mathrm{csand}} = \left(0.2 + 0.3\,.\,\mathrm{Exp}\left[-0.256\,.\,\mathrm{m_s}\,.\left(1 + \frac{m_{\mathrm{silt}}}{100}\right)\right]\right) \tag{14.3}$$

$$\text{cl - si} = \left(\frac{m_{\text{silt}}}{m_c + m_{\text{silt}}}\right)^{0.3} \tag{14.4}$$

$$f_{\text{orgC}} = \left(1 - \frac{0.256.\text{orgC}}{\text{orgC} + \text{Exp}\left[3.72 - 2.95.\ \text{orgC}\right]}\right) \tag{14.5}$$

$$f_{\text{hisand}} = 1 - \frac{0.7.\left(1 - \frac{m_s}{100}\right)}{\left(1 - \frac{m_s}{100}\right) + \text{Exp}\left[5.51 + 22.9\left(1 - \frac{m_s}{100}\right)\right]} \tag{14.6}$$

where f_{csand} is a factor that lowers the K indicator in soils with a high proportion of coarse-sand content and higher for soils with little sand; $f_{\text{cl-si}}$ gives low soil erodibility factors for soils with a high clay-to-silt ratio; f_{orgC} reduces the K values in soils with a high organic carbon content while f_{hisand} reduce the K value of soil classes with high sand contents.

Topographic (LS) Factors

The combined slope length and slope angle (LS-factor) describes the effect of topography on soil erosion. The steeper and the longer the slope, the higher is the rate of erosion due to the greater accumulation of runoff (Wischmeier and Smith 1978; Alexakis et al. 2013). In RUSLE, Slope length is defined as the horizontal distance from the origin of overland flow to the point where deposition begins or where runoff flows into a defined channel (Renard et al. 1997; Yahya et al. 2013). However, in a real two-dimensional landscape, overland flow and the resulting soil loss do not depend on the distance, rather on the area per unit of contour length contributing runoff to that point. For this reason, the slope length unit replaced by the unit-contributing area (Desmet and Govers 1996) from digital elevation models (DEMs). Slope length factor (L) is given by the following expression:

$$L = (\lambda/22.13)^m;\ m = \beta/(1+\beta);\ \beta = \frac{\left(\frac{\text{Sin}\theta}{0.0896}\right)}{\left(3\left(\text{Sin}\theta^{0.8}\right) + 0.56\right.} \tag{14.7}$$

where λ is the slope length (m), m is the slope length exponent, β is a factor that varies with slope gradient and θ is slope angle. Replacing slope length with unit-contributing areas, the slope length factor $L_{i,j}$ (Desmet and Govers 1996) written as:

$$L_{i,j} = \frac{\left(A_{i,j} + D^2\right)^{m+1} - A_{i,j}^{m+1}}{D^{m+2}.\ X_{i,j}^{m}.\ 2.13^m} \tag{14.8}$$

where $A_{i,j}$ is unit-contributing area at the inlet of grid cell, D is grid cell size which is 30 m in this case and $X_{i,j} = \sin\alpha_{i,j} + \cos\alpha_{i,j}$, the $\alpha_{i,j}$ is the aspect direction of the grid cell (i, j).

To obtain a better representation of the Slope Steepness factor, calculation of the S-factor proposed by Wischmeier and Smith (1978)was modified in RUSLE considering ratio of the rill and inter-rill erosion.

$$\begin{cases} S = 10.8\sin + 0.03, \tan < 0.09 \\ 16.8\sin\ -\ 0.50,\ \tan\ \geq 0.09 \end{cases} \tag{14.9}$$

For calculating the LS factor, LS-TOOL proposed by Zhang et al. (2017) was used in this study.

Cover (C) Factor

The cover and management factor (C) reflects the effect of cropping and management practices on soil erosion rates (Renard et al. 1997). The C-factor is defined as the ratio of soil loss from land with specific vegetation to the corresponding soil loss from continuous fallow (Wischmeier and Smith 1978).

In this study, the C factor was derived from Landsat OLI imagery using Linear Spectral Mixture Analysis (LSMA) approach (De Asis et al. 2008). LSMA has been frequently used to derive subpixel information from remotely sensed imagery (Adams et al. 1995; Lu et al. 2003; He et al. 2010). The basic premise of mixture modeling is that within a given scene, the surface is dominated by a small number of distinct materials called endmembers. The fractions in which the endmembers appear in a mixed pixel are called fractional abundances (Adams et al. 1995; Asis et al. 2008). The selection of endmembers is a critical component for successful application of mixture modeling. The minimum noise fraction (MNF) transformation was applied to the reflectance image to improve quality of fraction images through decorrelation. The result of MNF transformation is then used to calculate the pixel purity index (PPI) to determine the most spectrally pure pixels (endmembers). N-Dimensional Visualizer tool in ENVI software then applied to select the endmembers (He et al. 2010). In reality, three or four endmembers (e.g., green vegetation (GV), shade, soil, and non-photosynthetic vegetation (NPV)) can be used to characterize the variance in the image for LSMA (Asis et al. 2008). Mathematically LSMA model can be expressed as:

$$R_i = \sum f_j r_{ij} + \varepsilon_i; \quad \sum f_j = 1; \quad 0 < f_j < 1 \tag{14.10}$$

where R_i is the spectral reflectance of the mixed pixel in band i,f_j is the fraction of the pixel area covered by the endmember j, r_{ij} denotes the reflectance of the endmember j in band i, and ε_i is the root mean square (RMS) in band i.

Table 14.2 Conservation practice (*P*) factor values

Land cover type	Slope (%)	*P* factor
Agriculture land	0–5	0.1
	5–10	0.12
	10–20	0.14
	20–30	0.19
	30–50	0.25
	50–100	0.33
Other land	All	1.00

In this study, the fractional abundance of bare soil and vegetation to define a bare soil to vegetation cover ratio was used as an indicator of susceptibility to soil erosion as follows:

$$C = \frac{F_{bs}}{1 + F_{veg}} \tag{14.11}$$

where, C is RUSLE C-factor, F_{bs} and F_{veg} are the fractions of bare soil and vegetation respectively. The equation assumes that soil erosion occurs only when there are exposed soils that are subject to soil detachment by raindrop impact and surface runoff. The addition of 1 in the denominator limits the C value between 0 and 1.

Conservation Practice (*P*) Factor

Specific cultivation practices affect erosion by modifying the flow pattern and direction of runoff and by reducing the amount of runoff (Renard et al. 1997). The Conservation practice (*P*) factor is the ratio of soil loss with a specific support practice to the corresponding loss with up slope and down slope cultivation (Wischmeier and Smith 1978). Since there is no complete data on the conservation structures and most of the structures in the study area are not functional due to lack of regular maintenance, the *P*-factor for this study was determined using slope and land cover (Wischmeier and Smith 1978) as shown in Table 14.2.

14.2.3.3 Potential Locations for Ephemeral Gully Formation

To predict the susceptibility of a particular field to ephemeral gully formation, a threshold concept has been adopted (Tamene 2005; Daba et al. 2003). Generally, the gully incision is expected to appear when contributing area together with local slope exceeds a given threshold (Poesen et al. 2003). To predict the potential location and spatial patterns of gullies in the study area, the method proposed by Moore et al. (1988) were used (Fig. 14.3). Upslope contributing area or flow accumulation (As)

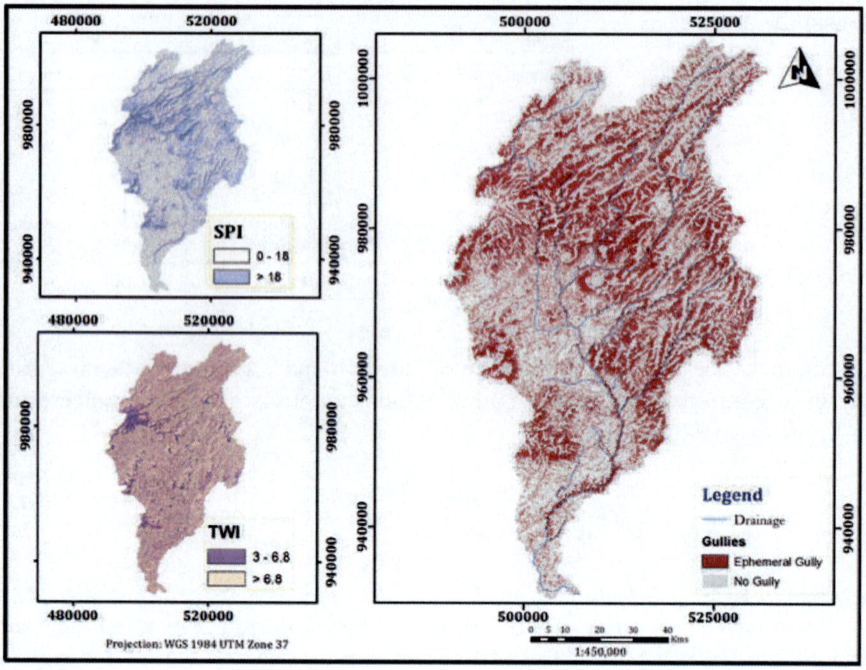

Fig. 14.3 Spatial location of ephemeral gullies

and local slope (β) were generated from DEM of the study area to generate stream power index (SPI) and wetness index (WI) maps using the following equations. Previous studies (Moore et al. 1988; Daba et al. 2003; Tamene 2005) used these indices to predict potential areas of initiation of ephemeral gullies when SPI > 18 and WI > 6.8 (Daba et al. 2003; Tamene 2005).

$$\text{SPI} = \text{As}(\tan\beta) \tag{14.12}$$

$$\text{WI} = \ln(\text{As}/\tan\beta) \tag{14.13}$$

where As = the unit-contributing area (m^2/m), β = the local slope (m/m), SPI = stream power index and WI = wetness index.

14.3 Results and Discussion

14.3.1 Potential Soil Loss Based on RUSLE

The various erosion factors (*R*, *K*, LS, *C* and *P*) for input into the RUSLE and estimate potential annual soil loss for the Mojo River Watershed are presented in Fig. 14.4.

Based on the analysis, the entire watershed loses a total of about 44,992,460.42 tons of soil annually from1680.41 km^2 of land. Poor vegetation cover is exposing the area to high rates of erosion. As shown in Fig. 14.5, estimated annual soil loss was classified into six erosion severity classes (Singh et al. 1992; Gara et al 2011). MW-wise average soil loss ranges from 2.36 to 47.99 t/ha/year with a mean annual soil loss of 25.83 t/ha/year and standard deviation 12.04.

Table 14.3 shows estimated area of the study area based on annual soil loss in relation to the prevailing slope steepness. A large portion of the study area is classified as slight erosion and it is more than 62% (1045.6 Km^2). This is due to the low average surface slope with plain topography. From the total area, 128.8 Km^2 (7.7%) and 137.4 km^2 (8.2%) reported severe (40–80 t/ha/year) and very severe (>80 t/ha/year) annual erosion rates respectively in all slope classes. Of which, 58.4 Km^2 (3.47%)

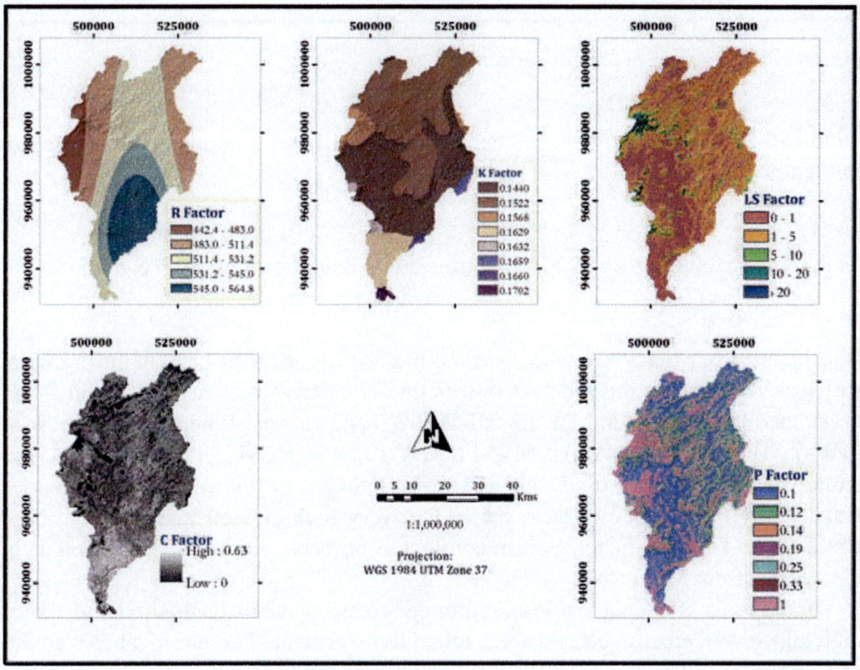

Fig. 14.4 RUSLE factors (*R*, *K*, *LS*, *C* and *P*) Map

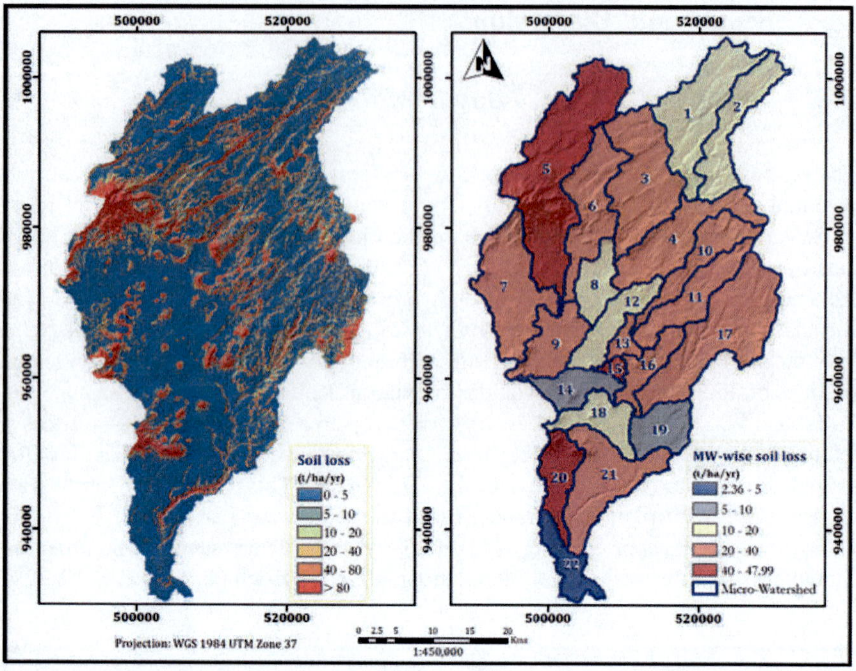

Fig. 14.5 Soil loss map of the study area

within 15–30% slope gradient (moderately steep surface) is under a very severe soil loss class.

14.3.1.1 Prioritization of Micro-watersheds Based on Potential Soil Loss

Prioritization of Micro-watersheds (MWs) has been done on the basis of mean annual soil loss. The result showed that out of the 22 micro-watersheds in Mojo River Watershed, three (MW-5, MW-15 and MW-20) and eleven of them (MW-3, MW-4, MW-6, MW-7, MW-9, MW-10, MW-11, MW-13, MW-16, MW-17, and MW-21) fell under severe (40–47.99 t/ha/year) and very high (20–40 t/ha/year) erosion classes respectively (Table 14.2). These severe and very high erosion rate is due to high contribution from *LS* factor. Severe soil loss is observed in the valley as well as in the ridges of the Mojo River Watershed.

On the basis of annual soil losses, fourteen micro-watersheds that fell under very high and severe erosion classes were found to be critical. These critical MWs with annual mean soil loss greater than 20 t/ha/year were given top and first priority for conservation. As a result, the critical MWs were recommended to adopt the management measures to reduce the soil and nutrient losses.

Table 14.3 Tabulated area (Km^2) of soil loss within existing slope gradients

Soil loss class									
Slope class description*	Slope (%)	Slight	Moderate	High	Very high	Severe	Very severe	Total area (Km^2)	Percent
Flat to very gently sloping	0–2	386.5	0.0	0.0	0.0	0.0	0.0	386.5	23.0
Gently sloping	2–5	552.8	2.7	0.8	45.3	7.5	0.0	609.1	36.2
Sloping	5–10	106.0	167.7	13.6	43.7	78.0	6.7	415.7	24.7
Strongly sloping	10–15	0.2	24.7	31.3	11.7	23.9	36.0	127.7	7.6
Moderately steep	15–30	0.2	0.5	9.2	16.9	17.9	**58.4**	103.1	6.1
Steep	30–60	0.0	0.0	0.0	0.5	1.5	33.1	35.1	2.1
Very steep	>60	0.0	0.0	0.0	0.0	0.0	3.2	3.2	0.2
Area (Km^2)		1045.6	195.6	54.8	118.1	**128.8**	**137.4**	1680.4	100.0
Percent		62.2	11.6	3.3	7.0	**7.7**	**8.2**	100.0	

* Modified from FAO (2006)

Previous studies conducted on soil erosion assessment in Ethiopia shows different rate of soil erosion. For example, studies conducted by FAO (1986), in the Ethiopian high lands shows 100 t/ha/year soil loss from cropped lands taking into consideration the redeposit. Another survey conducted on soil and water resources of Ethiopia revealed that the annual soil loss rate ranges from 16 to 300 t/ha/year (Hawando 1995) and 18 to 214.8 t/ha/year (SCRP 1996). According to the Ministry of Water Resource (MoWR 2008), estimates for soil erosion based on the sediment curve at Mojo gauging station and on an average sedimentation rating in Koka reservoir are 15–25 t/ha/year. This has resulted in a loss of 30% of the total storage volume of the reservoir due to siltation (EEPC 2002; Elias 2003).

Removal of vegetation cover due to LULC change and lack of adequate soil and water conservation has largely contributed to increased rates of soil erosion and loss of soil nutrients and top soil (Tegene 2002; Desalegn et al. 2014; Kindu et al. 2018). The implications of soil erosion extend beyond the removal of valuable topsoil. It causes a decline in crop yields. National level estimates reveal a 2% average annual reduction of the agricultural GDP due to erosion (FAO 1986).

14.3.2 Erosion Risk in the Watershed Based on MCE

14.3.2.1 Composite Erosion Index (CEI)

The weight and rating system used for soil erosion intensity map is based on the relative importance of various causative factors (Deore 2005). Four selected criteria layers (slope, NDVI, soil type and gully) were used in multi-criteria analysis to generate Composite Erosion Index (CEI) by Weighted Linear Combination (WLC) as follows:

$$\mathrm{CEI} = (\mathrm{W}_1 \times \mathrm{Slope}) + (\mathrm{W}_2 \times \mathrm{NDVI}) + (\mathrm{W}_3 \times \mathrm{Soil}) + (\mathrm{W}_{14} \times \mathrm{Gully}) \quad (14.14)$$

where CEI is Composite Erosion Index; W_1, W_2, W_3 and W_4 are pairwise weights for reclassified layers of slope, NDVI, soil type, and gully respectively.

Values of CEI in the study area ranges between 1 and 4.41 (dimensionless). Minimal and low erosion potential was present under dense vegetation when the slope gradient was also low but increased with higher slope values. There were few cells with extreme erosion potential, and these were usually restricted along stream channels and ridges with very high slope values. Table 14.4 shows area and proportion of the study area categorized in a classified CEI map.

Assessments of gully erosion volumes in Ethiopia are rare. Using photogrammetric techniques, Daba et al. (2003) estimate that between 1966 and 1996, 700,000 tons of soils were lost by gully erosion from a 9-km^2 watershed in the eastern highlands (26 t/ha/year). Using monitoring and interview techniques to establish average long-term soil loss rates by gully erosion obtained 5 t/ha/year in Central Tigray

Table 14.4 Classification of micro-watersheds based on soil loss

Soil erosion class*	Mean soil loss (t/ha/year)	No. of MWs	Area (Km^2)	Percentage
Slight	0–5	1	36.35	2.16
Moderate	5–10	2	89.56	5.33
High	10–20	5	339.17	20.18
Very high	20–40	11	940.89	55.99
Severe	40–47.99	3	274.44	16.33

* Adapted from Singh et al. (1992) and Gara et al. (2011)

(Nyssen 2001). To reduce further expansion of gullies, buffer plantation along gully sides in the study area is required.

14.3.2.2 Prioritization of Micro-watersheds Based on CEI

The intensity of soil erosion indicated by Composite Erosion Index in the MWs is considered for their prioritization for selection and implementation of conservation measures and plan appropriate land use to minimize the soil losses in them (Tripathi et al. 2003; Deore 2005; Khan et al. 2001). All 22 micro-watershed were grouped into five CEI classes (Fig. 14.6) based on the data natural breaks. Micro-watershed wise mean CEI indicates that MW-3, MW-4, MW-6, and MW-12 are under very high CEI category with a value above 2.16. The area covered under this category is 389.77 Km^2 (23.2%). High CEI category is represented by six MWs covering 622.85 Km^2 area (37.07%). Most parts of these micro-watersheds are under low vegetation cover and high proportion of ephemeral gullies which causing high erosion intensity. CEI class in Table 14.5 was just based on Natural breaks (Jenks) classification mehtod used in ArcGIS symbology.

14.4 Conclusions

Application of RUSLE model and MCE in GIS environment can help watershed managers in assessing and identifying erosion prone areas for undertaking required conservation measures. The landscape positions where steep slope, poor surface cover, erodible soil, and gully erosion coincided show high erosion risk compared to others. Based on the RUSLE model, MW-wise average annual soil loss ranges from 2.36 to 47.99 t/ha/year with a mean annual soil loss of 25.83 t/ha/year and standard deviation 12.04. Based on MCE approach, micro-watershed wise mean CEI ranges from 1.71 to 2.31 with a mean value of 2.06. Four micro-watersheds (MW-3, MW-4, MW-6, and MW-12) are under very high CEI category covering 389.77 Km^2 (23.2%) with a value above 2.16. Most parts of these micro-watersheds are under cultivated

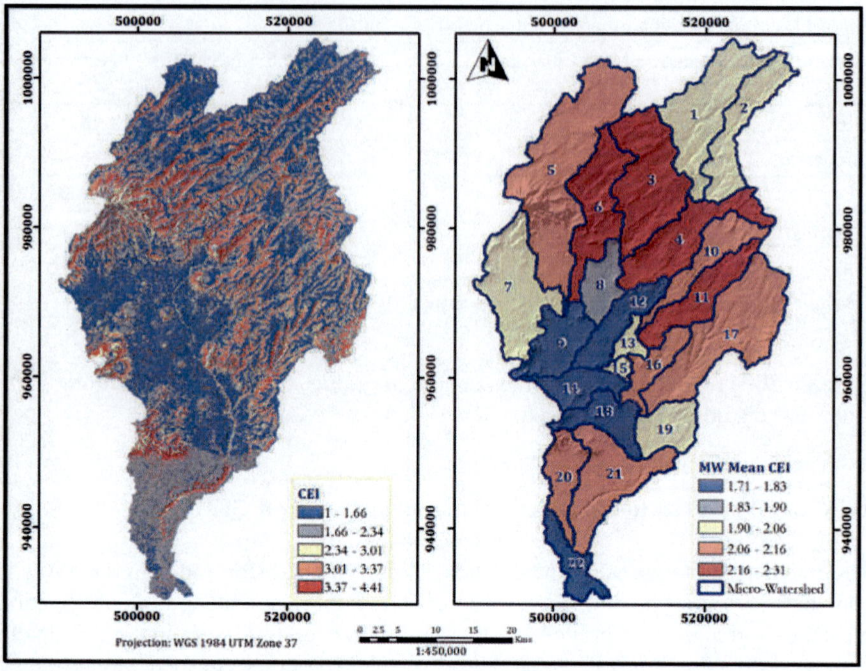

Fig. 14.6 Micro-watershed wise Composite Erosion Index map

Table 14.5 Categorization and prioritization of Micro-watersheds based on CEI

CEI class	Mean CEI	No. of MWs	Area (Km^2)	Percentage
Very low	1.71–1.83	5	239.89	14.28
Low	1.83–1.90	1	42.49	2.53
Moderate	1.90–2.06	6	385.41	22.94
High	2.06–2.16	6	622.85	37.07
Very high	2.16–2.31	4	389.77	23.20

and bare land and high proportion of ephemeral gullies which causing high erosion intensity.

Based on the result, the micro-watersheds which fall under very high and high categories need immediate attention in their order of soil erosion potential. Therefore, for reducing further degradation, reclaiming the degraded areas and improving the land productivity of the watershed, and reducing their sediment delivery to low lands and reservoirs, hot-spot areas having a large rate of erosion should be given first priority for well-designed conservation interventions in the study area.

Acknowledgments This work was financially supported by Horn of Africa regional environment program. The authors gratefully indebted to Dr. Lulseged Tamene and those organizations that support all kinds of resources in one way or another.

References

Adams JB, Sabol DE, Kapos V, Filho RA, Roberts DA, Smith MO, Gillespie A (1995) Classification of multispectral images based on fractions of endmembers: application to land-cover change in the the Brazilian Amazon. Remote Sens Environ 52(2):137–154

Adimassu Z, Mekonnen K, Yirga C, Kessler A (2014) Effect of soil bunds on runoff, soil and nutrient losses, and crop yield in the central highlands of Ethiopia. Land Degrad Dev 25:554–564

Alexakis D, Hadjimitsis D, Agapiou A (2013) Integrated use of remote sensing, GIS and precipitation data for the assessment of soil erosion rate in the catchment area of "Yialias" in Cyprus. Atmos Res 131:108–124

Bewket W (2002) Land cover dynamics since the 1950s in Chemoga Watershed, Blue Nile Basin, Ethiopia. Mt Res Dev 22:263–269

Central Statistical Agency (CSA) (2007). Population and housing census of Ethiopia. Addis Ababa, Ethiopia

Daba S, Rieger W, Strauss P (2003) Assessment of gully erosion in eastern Ethiopia using photogrammetric techniques. Catena 50:273–291

De Asis AM, Omasa K, Oki K, Shimizu Y (2008) Accuracy and applicability of linear spectral unmixing in delineating potential erosion areas in tropical watersheds. Int J Remote Sens 29(14):4151–4171

Daba S, Rieger W, & Strauss P (2003) Assessment of gully erosion in eastern Ethiopia using photogrammetric techniques. CATENA 50(2-4):273–291

Demissie F, Yeshitila K, Kindu M, Schneider T (2017) Land use/land cover changes and their causes in Libokemkem District of South Gonder, Ethiopia. Remote Sens Appl Soc Environ 8:224–230

Deore SJ (2005) Prioritization of Micro-watersheds of Upper Bhama Basin on the basis of soil erosion risk using remote sensing and GIS technology. PhD thesis, University of Pune, Pune

Desalegn T, Cruz F, Kindu M, Turrión M, Gonzalo J (2014) Land-use/land-cover (LULC) change and socioeconomic conditions of local community in the central highlands of Ethiopia. Int J Sust Dev World 21(5):406–413

Desmet PJJ, Govers, G. (1996) A GIS procedure for automatically calculating the USLE LS factor on topographically complex landscape units. J Soil Water Conserv 51(5):427–433

EEPC (2002) Koka Dam sedimentation study: recommendations report. Ethiopian electric power corporation. Addis Ababa, June 2002

Elias E (2003) Case studies: genetic diversity, coffee and soil erosion in Ethiopia. Roles of agriculture project international conference, Rome, Italy, 20–22 Oct 2003

Ethiopian Roads Authority (2007) Ethiopian road map. Non-authoritative border. Data from 1984-2005. Addis Ababa, Ethiopia

FAO (1986) Ethiopia Highlands Reclamation Study. Final Report, vol 1. FAO, Rome

FAO (2019) Soil erosion: the greatest challenge to sustainable soil management. Licence: CC BY-NC-SA 3.0 IGO, Rome, p 100

Gara A, Suzuki S, Watanabe F, Himadas S, Toyoda H (2011) Soil erosion assessment using USLE in East Shewa, Ethiopia. J. Agric. Sci. Tokyo Univ. Agric. 56(1), 1–8

Gashaw T, Taffa T, Argaw M, Worqlul AW, Tolessa T, Kindu M (2018) Estimating the impacts of land use/land cover changes on ecosystem service values: the case of the Andassa watershed in the Upper Blue Nile basin of Ethiopia. Ecosyst Serv 31:219–228

Haregeweyn N, Tsunekawa A, Nyssen J, Tsubo M, Meshesha DT, Schutt B, Adgo E, Tegegne F (2015) Soil erosion and conservation in Ethiopia: a review. Prog Phys Geogr 39(6):750–774

Hawando T (1995) The survey of the soil and water resources of Ethiopia. NU/Toko
He M, Zhao B, Ouyang Z, Yan Y, Li B (2010) Linear spectral mixture analysis of Landsat TM data for monitoring invasive exotic plants in estuarine wetlands. Int J Remote Sens 31(16):4319–4333
Hellden U (1987) An assessment of woody biomass, community forests, land use and soil erosion in Ethiopia. Lund University Press
Hurni H (1985) Soil conservation manual for Ethiopia. First draft. Ministry of Agriculture, Natural Resources Conservation and Development Department, Community Forests and Soil Conservation Development Department, Addis Ababa
Hurni H (1993) Land Degradation, famine and resource scenarios in Ethiopia. In: Pimentel D (ed) World soil erosion and conservation, Cambridge University press, Cambridge
Khan MA, Gupta VP, Moharana PC (2001) Watershed prioritization using remote sensing and geographical information system: a case study from Guhiya, India. J Arid Environ 49:465–475
Kindu M, Schneider T, Teketay D, Knoke T (2013) Land use/land cover change analysis using object-based classification approach in Munessa-Shashemene landscape of the Ethiopian highlands. Remote Sens 5(5):2411–2435
Kindu M, Schneider T, Teketay D, Knoke T (2015) Drivers of land use/land cover changes in munessa-shashemene landscape of the south-central highlands of Ethiopia. Environmental Monitoring and Assessment 187(7):452
Kindu M, Schneider T, Teketay D, Knoke T (2016) Changes of ecosystem service values in response to land use/land cover dynamics in Munessa-Shashemene Landscape of the Ethiopian highlands. Sci Total Environ 547:137–147
Kindu M, Schneider T, Döllerer M, Teketay D, Knoke T (2018) Scenario modelling of land use/land cover changes in Munessa-Shashemene landscape of the Ethiopian highlands. Sci Total Environ 622–623:534–546
Lu D, Moran E, Batistella M (2003) Linear mixture model applied to Amazonian vegetation classification. Remote Sens Environ 87:456–469
Ministry of Water Resources (MoWR) (1999) Survey of the awash river basin. Vol II. Soils and agronomy. Integrated Development Master Plan Study. Addis Ababa, Ethiopia
Moore ID, Burch GJ, Mackenzie DH (1988) Topographic effects on the distribution of surface soil water and the location of ephemeral gullies. Trans ASAE 31(4):1098–1107
MoWR (2008) Awash River basin flood control and watershed management project. Feasibility report for Mojo River badlands rehabilitation. Main report
Nyssen J (2001) Erosion processes and soil conservation in a tropical mountain catchment under threat of anthropogenic desertification – a case study from Northern Ethiopia. PhD thesis, Department of Geography, K.U. Leuven, Belgium
Poesen J, Nachtergaele J, Verstraeten G, Valentin C (2003) Gully erosion and environmental change: importance and research needs. Catena 50:91–133
Renard KG, Foster GR, Weesies GA, McCool DK, Yoder DC (1997) Predicting soil erosion by water: a guide to conservation planning with the revised universal soil loss equation (RUSLE). US Department of Agriculture, Agricultural Handbook Number, vol. 703. Government Printing Office, Washington
SCRP (1996) Soil erosion hazard assessment for land evaluation. University of Bern, Switzerland, in association with Ministry of Agriculture, Government of Ethiopia
Shi ZH, Cai CF, Ding SW, Wang TW, Chow TL (2003) Soil conservation planning at the small watershed level using RUSLE with GIS: a case study in the Three Gorge Area of China. Catena 55:33–48
Singh G, Babu R, Narayan P (1992) Soil erosion rates in India. J Soil Water Conserv 47:97–99
Sutcliff, J.P. (1993). Economic assessment of land degradation in the Ethiopian highlands: a case study. National conservation strategy secretariat, Ministry of planning and economic development, Addis Ababa, Ethiopia
Tamene, L. (2005). Reservoir siltation in Ethiopia: causes, source areas, and management options. Ecology and Development Series No. 30. Center for Development Research, Bonn, Germany

Tamene L, Vlek PLG (2007) Assessing the potential of changing land use for reducing soil erosion and sediment yield of catchments: a case study in the highlands of northern Ethiopia. Soil Use Manag 23:82–91

Tegene B (2002) Land-cover/land-use changes in the derekolli catchment of the South Welo Zone of Amhara Region, Ethiopia. East Afr Soc Sci Res Rev 18:1–20

Temesgen H, Nyssen J, Zenebe A, Haregewey N, Kindu M, Lemenih M, Haile M (2013) Ecological succession and land use changes in a lake retreat area (Main Ethiopian Rift Valley). J Arid Environ 91, 53–60

Tripathi MP, Panda RK, Raghuwanshi NS (2003) Identification and prioritization of critical sub-watersheds for soil conservation management using the SWAT model. Biosys Eng 85(3):365–379

Wischmeier WH, Smith DD (1978) Predicting rainfall erosion losses: a guide to conservation planning. In: Agriculture handbook No. 537. US Department of Agriculture, Washington DC

Yahya F, Zregat D, Farhan I (2013) Spatial estimation of soil erosion risk using RUSLE approach, RS, and GIS techniques: a case study of Kufranja Watershed Northern Jordan. J Water Res Prot 5:1247–1261

Zhang H, Wei J, Yang Q, Baartman JEM, Gai L, Xiaomei Yang X, Li S, Yu J, Ritsema CJ, Geissen V (2017) An improved method for calculating slope length (λ) and the LS parameters of the revised universal soil loss equation for large watersheds. Geoderma 308(2017):36–45

Chapter 15
Ethiopian Church Forests as Monitoring Towers in Reconstructing Climate Change and Its Impacts and to Make Evidence-Based Climate-Smart Restoration Efforts

Mulugeta Mokira, Aster Gebrekirstos, Abrham Abiyu, Kiros Hadgu, Niguse Hagazi, and Achim Bräuning

Abstract Climate change has become one of the biggest threats to nature and society, due to extreme weather events including floods, droughts, and increased frequency and intensity of dry spells. This has been causing devastating impacts on agriculture and forestry, which are the main livelihood sources in Africa. However, there is insufficient scientific information on climate change, especially as it concerns related changes in growth patterns, water use efficiency, and stress-induced mortality of tree species. Furthermore, predictions of climate change in Africa are characterized by a high degree of uncertainty due to the limited availability of long-term and high-quality climate data. Dendrochronological data can be useful to better understand forest growth dynamics, tree-to-forest level responses to changing climate, and to reconstruct multi-century climate information and characterize extreme climate events beyond instrumental periods. Although Ethiopia has lost most of its natural forest cover, church forests provide a unique source of old-growth trees that can serve as monitoring towers in multiple ways. We addressed opportunities and challenges of dendroecological application and summarized results from different studies on *Juniperus procera* growing in church forests. We describe dendrochronological applications for studying current climate change effects, hydroclimatic reconstructions, forest ecological research, forest rehabilitation, and carbon management. Reviewed hydroclimate and dendroecological studies confirmed that church forests are valuable archives of regional hydroclimate information and important to understand the growth dynamic of old-growth Afromontane forests. This information is required

M. Mokira (✉) · A. Abiyu · K. Hadgu · N. Hagazi
World Agroforestry (ICRAF), C/O ILRI Campus, Gurd Shola, P.O. Box 5689, Addis Ababa, Ethiopia
e-mail: m.mokria@cgiar.org

A. Gebrekirstos
World Agroforestry (ICRAF), United Nations Avenue, P.O. Box 30677-00100, Nairobi, Kenya

A. Bräuning
Institute of Geography, Friedrich-Alexander-University Erlangen-Nuremberg, Wetterkreuz 15, 91058 Erlangen, Germany

M. Kindu et al. (eds.), *State of the Art in Ethiopian Church Forests and Restoration Options*, https://doi.org/10.1007/978-3-030-86626-6_15

to assist evidence-based climate-smart restoration efforts. Multiple tree-ring formation per year was reported, particularly in areas with ambiguous rainfall seasonality. However, such challenges can be resolved through proper site selection, the advancement of technologies, and using newly emerging methodologies.

Keywords Dendroecology · Church forests · Climate reconstruction · River flow · Remote climate-teleconnection · Growth dynamics · Forest restoration · Carbon dynamics · Ethiopia

15.1 Introduction

Climate change is a global phenomenon, but its impacts are heterogeneous across regions of the world (IPCC 2014). The projected trend in global climate change is expected to exacerbate drought and flood frequency in Africa (IPCC 2014; Funk et al. 2015) and is likely to reverse developmental gains, which may set countries to 10–20 years back (World Bank 2017). On the other hand, the continent's climate is one of the world's most understudied (Future Climate for Africa 2016), with less than 300 weather stations that meet the World Meteorological Organization (WMO) observation standards, corresponding to only 13% of the required density (World Bank 2017). More importantly, climate prediction models tend to disagree in their representation of the historical climate in Sub-Saharan Africa (SSA), leading to high uncertainty on the current and future climate conditions (Future Climate for Africa 2016). These differences are mainly due to a lack of long-term instrumental climate information, along with limited understandings of the region's climate system (Nash et al. 2016). Despite regional climate data shortage, several studies on the climate of SSA, including Ethiopia have been published in the last two decades (Degefu et al. 2017), but only a few studies tried to extend the rather short instrumental climate and hydrological data series by using proxy data. Effective hydroclimate services can also offer a solution for challenges in accessing quality climate information in the future. However, it is unlikely to resolve the problems emanating from the lack of long-term and past climate information, which is required to make reliable climate predictions across the region. Therefore, to fill the existing gaps in climate information, proxy data that combine precise dating with a high temporal resolution have special relevance for the SSA region. In this context, tree-rings are among the most useful and readily available proxy data sources of past hydroclimate information (Singh and Yadav 2013; Mokria et al. 2017, 2018). More importantly, multi-century tree-ring records can be developed from vast geographic areas of Africa at low cost, and thereby provide high-resolution hydroclimate information which is indispensable to accurately forecast future changes in the climate and hydrological system (Gebrekirstos et al. 2014; Schofield et al. 2016; Briffa et al. 2009).

Regarding forests' responses to climate change, studies have shown that climate-induced drought has become apparent across SSA countries, suggesting that eastern African dry forests are increasingly vulnerable to climate-induced forest degradation and biodiversity loss (Allen et al. 2015, 2017; Mokria et al. 2015). Earlier studies conducted in dry tropical forests, including sacred forest areas of Ethiopia, have generated important information on population structure, biodiversity, and natural regeneration (Aerts et al. 2016; Wassie et al. 2010). Nevertheless, the potential responses of forests to the predicted changes in climate, past disturbance events, their age structure, growth performance under current and past environmental conditions, as well as their potential contribution to adaptation and mitigation of the impacts of climate change are still vaguely understood. However, such information is critical to design forest conservation strategies that help to reduce the adverse impacts of climate change (Allen et al. 2017; Mokria et al. 2015; Abiyu et al. 2018). In this regard, the region could benefit from dendrochronology (Gebrekirstos et al. 2014). Tree-ring analysis can be used to investigate both environmental and climatic factors influencing tree growth, as well as to document stand age, population dynamics, carbon dynamics, wood production, and rotation ages while providing time estimates required to restore degraded landscapes (Abiyu et al. 2018; Gebrekirstos et al. 2008; Tolera et al. 2013; Mokria et al. 2015). Besides, adjustments of wood structures in response to extreme climate events are permanently stored in tree-rings (Bräuning et al. 2016). These characteristics enable the use of dated tree rings to reconstruct how trees have been growing and functioning in the past (Sass-Klaassen et al. 2016), and consequently reflect the resilience and acclimation strategies of trees in a changing climate (Gebrekirstos et al. 2011; Abiyu et al. 2018; Mokria et al. 2016). However, due to intense deforestation and land-use changes during the past centuries, Ethiopia has lost most of its natural forest cover. Hence, old trees that provide the potential to derive long-term climate reconstructions from tree-rings, are often restricted to relict forest patches under protection, as represented by sacred church forest groves in Ethiopia. Therefore, this chapter aims to illustrate the importance of church forests to reconstruct long-term hydro-climate records reflecting past climate variations, and to understand the growth performance of old-growth remnant Afromontane forest, through (1) developing high-resolution proxy climate records, and (2) using dendroecological applications to quantify tree age, growth performance as well as environmental costs of forest degradation as well as to balance carbon gains and losses.

15.2 Ethiopian Church Forests as Climate and Environmental Monitoring Towers

15.2.1 Dendrochronological Opportunities and Challenges of Church Forest

African pencil cedar (*Juniperus procera*) is the dominant tree species Afromontane church forest of Ethiopia (Eshete 2007). The species is able to produce anatomically clear annual growth rings in response to a distinct annual cycle of wet and dry seasons (Wils et al. 2009), which provides ample opportunities for dendrochronological studies (Gebrekirstos et al. 2014; Worbes 2002). Tree growth periodicity is modulated by climatic factors, hence the signal in the tree-ring can be used as a proxy to reconstruct multiple moisture-related information, such as precipitation or river flow (Bräuning et al. 2008). In the tropics, trees that are sensitive to temporal soil moisture fluctuations are typically grown on sloppy and mountainous areas, or in areas with distinct climatic seasonality and drier regions. In the dry Afromontane forests of Ethiopia, tree-ring studies from church forests have shown that African juniper is a prime candidate for dendrohydroclimate applications because it is highly sensitive to moisture availability. The trees are growing on mountain slopes, where the only water source is seasonal rainfall. Furthermore, juniper is one of the foundation tree species, which can determine fundamental ecosystem processes and services of the church and Afromontane forest areas (Eshete 2007; Gebregeorgis et al. 2020; Mokria et al. 2017, 2018; Wils et al. 2010a). The annual growth ring formation of *J. procera,* growing in many parts of Ethiopia, has been confirmed by radiocarbon dating and significant inter-series correlation between different tree-ring records (Gebregeorgis et al. 2020; Mokria et al. 2017; Wils et al. 2010a). The formation of a distinct wood anatomical growth ring boundary (Fig. 15.1) is ascribed to the distinct rainfall seasonality in the study region (Schöngart et al. 2017; Gebrekirstos et al. 2008). Generally, in water-limited environments, a long dry season is the main driver for cambial dormancy, leading to the formation of a tree-ring boundary (Mokria et al. 2016; Gebrekirstos et al. 2006).

Recent studies reported precisely dated tree-ring series extracted from different Ethiopian church forests (Gebregeorgis et al. 2020; Mokria et al. 2017; Wils et al.

Fig. 15.1 Annual growth rings of *Juniperus procera* from a church forest in northern Ethiopia. **a** Stem disc collected from an old log found in the church compound, **b** microscopic thin section, **c** increment core collected from a standing tree. The black arrows and dots indicate growth direction and tree-ring boundaries, as indicated by a darker latewood band, respectively (*Source* Mokria et al. unpublished)

2010a). The cross-dated tree-ring series from different church forests showed statistical characteristics indicating consistent growth patterns across sites, such as high series intercorrelation, mean sensitivity, signal-to-noise ratio, and high mean tree-to-tree relationships, implying that each church forest has captured common regional climate signals (Mokria et al. 2017; Wils et al. 2010b, c). Hence, a regional chronology developed by synthesizing the tree-ring series from different church forests can be used as a reliable proxy record of regional climate variability. Recent studies developed tree-ring chronologies from different church forests covering the periods 1665–2014 (Mokria et al. 2018, 2017), 1758–2013 (Gebregeorgis et al. 2020), and 1836–2010 (Wils et al. 2010a). These chronologies were found to be climate-sensitive and recorded long-term inter-annual climate variability, which is typical for chronologies from moisture-sensitive regions.

The major challenges in applying dendrochronology in Ethiopia using samples from Afromontane church forest are the complex nature of ring formation, particularly in areas with ambiguous rainfall seasonality. A detailed review report has identified four main types of growth ring formation, such as anatomically indistinct rings, multiple rings per year, annual rings, and multiple missing rings (Wils et al. 2010a; 2009). This complex nature of growth ring formation might be explained by inter-and intra-annual rainfall variation across sites, and differential sensitivity of trees to temporal changes in soil water availability, such as rainfall in one or two days followed by dry periods of 2 to 3 weeks (Mokria et al. 2016; Wils et al. 2010a, 2009; Worbes 1995). However, dendrochronological challenges emanating from the complex periodicity of growth ring formation can be resolved by selecting more mesic sites with an unambiguously unimodal rainfall regime. Besides, the advancement of technology and emergence of new methods including cambium wounding, dendrometer measurements, regular microcoring of the cambium, and quantitative microscopic analysis of the cells formed between sampling intervals, radiocarbon dating, and intra-annual stable isotope analysis could play a substantial role in minimizing most of the challenges encountered in dendrochronological studies in the tropics, including Ethiopia (Worbes 1995; Mokria et al. 2016; Krepkowski et al. 2011; Verheyden et al. 2004; Gebrekirstos et al. 2014).

15.2.2 Response of Church Forest to Regional Climate Variability

Concerns on forest degradation are increasing worldwide because the frequency of drought events and extreme temperatures are projected to increase in the future (Allen et al. 2017; IPCC 2014). However, tropical forest responses to changing climate vary from region to region, mainly due to complex biosphere-climate interactions (Worbes 2002; Worbes et al. 2013). On the other hand, the potential responses of tropical dry forests to the predicted changes in climate are still coarsely understood (Corlett 2011). Thus, analysis of tree-rings from church forests provides a new opportunity to study the responses of the remnant Afromontane forests to climate variability across northern Ethiopia (Gebregeorgis et al. 2020; Mokria et al. 2017, 2018; Wils et al. 2010a). Mokria et al. (2017) reported a significant positive correlation between juniper growth and total rainy season rainfall (JJAS) ($r = 0.64$, $P < 0.01$), confirming that wet season total rainfall is the dominant factor influencing Afromontane forest tree growth across northern Ethiopia. Other studies also reported that precipitation is the main growth driving factor in the drier parts of Africa (Gebrekirstos et al. 2008). The significant response to total wet season rainfall may indicate that the soil-plant system is buffered at seasonal timescales, which might be ascribed to the phenological characteristics of the evergreen *J. procera,* always allowing radial growth when soil moisture is sufficient to support cambial activity. Hence, its ring width variations rather reflect the total amount of water in the soil reservoir during the entire growing season, than water availability during a specific period (Krepkowski et al. 2011; Gebrekirstos et al. 2011). Mokria et al. (2017) also found a negative relationship between temperature and juniper growth, pointing to a negative impact of high temperatures on tree growth performance probably by increasing drought stress due to enhanced evapotranspiration. Moreover, the strong spatial correlation between the regional juniper chronology and precipitation (Fig. 15.2) further confirms that the chronology has captured climate signals, which represent nearly the complete northern Ethiopian highlands as well as parts of the Sahel belt and eastern equatorial Africa (Mokria et al. 2017). This also indicates that the atmospheric flow of moisture affects a forest area at a wider geographic scale and is not limited by

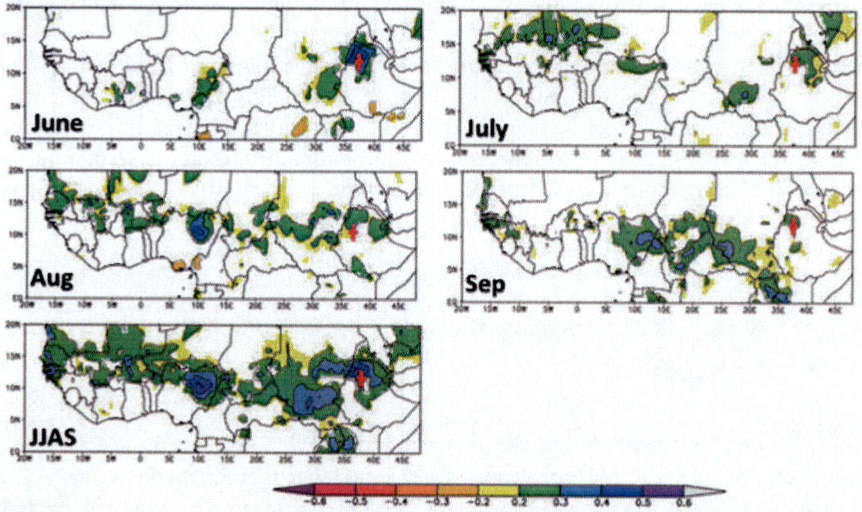

Fig. 15.2 Spatial correlations between a regional juniper chronology from northern Ethiopia and June, July, August, September, and total wet season (JJAS) precipitation for the period 1901–2014. In Fig. 15.2, the red cross indicates the location of the church forest in Ethiopia (*Source* Mokria et al. 2017)

watershed boundaries (Mokria et al. 2017). This in turn can be explained by the recycling of atmospheric moisture from evapotranspiration, which precipitates as rainfall across atmospherically teleconnected locations (Ellison et al. 2017; Mokria et al. 2017). With further study, such research findings could play a considerable role to quantify the proportion of rainfall across northern parts of Ethiopia that is directly derived from the equatorial Pacific and the Indian Ocean and of terrestrially recycled moisture from a distant location, such as from the White Nile swamps of Southern Sudan and source areas around Lake Victoria (Ellison et al. 2017; Mokria et al. 2017). More importantly, understanding and quantifying the contribution of forest and wetlands on rainfall amount and seasonality beyond their geographical location would stimulate regional collaboration to develop cross-regional planning for sustainable forest management and strategies to guide restoration interventions. Generally, research findings included in this document highlighted that church forests in northern Ethiopia can serve as a monitoring tower for current climate change effects and are of extreme importance as witnesses of past climate variability since they represent invaluable archives for the reconstruction of regional climate information (Mokria et al. 2017).

15.2.3 The Reconstruction of Multi-century Hydroclimate Data and Extreme Climate Events

Across the SSA region, the high uncertainty in projected climate change is caused by the lack of climate information and poses difficulties in evaluating the implications of future climate changes on regional water resources and vegetation over space and time (Future Climate for Africa 2016; Nash et al. 2016). Therefore, reconstructed long-term hydroclimate information is relevant for the development of plans and strategies to mitigate and to adapt to climate-related risks, and for sustainable water and forest resource utilization in the region. Dendrochronology is a well-established method for reconstructing rainfall and river flow records, especially in water-limited growing environments, where both tree growth and river discharge respond positively to precipitation. The regional juniper tree-ring chronology from church forests does not only show a significant positive relationship with seasonal precipitation ($r = 0.64$, $P < 0.01$) (Mokria et al. 2017), but also with Blue Nile river flow ($r = 0.75$, $p < 0.01$) and Tekeze-Atbara river flow ($r = 0.66$, $p < 0.01$) (Mokria et al. 2018; Wils et al. 2010a). The significant correlations between tree-ring width, river discharge, and precipitation suggest that precipitation controls both the growth of trees and the amount of water that drains into the river. Other studies reported that trees growing on sloppy areas, high above the river channel and headwater source areas are typically very sensitive to climate variability (Axelson et al. 2009; Woodhouse and Lukas 2006). Despite a long distance between the gauge station and the tree-ring sites (church forests), they showed a significant positive relationship, indicating that the atmospheric flow of moisture affecting both tree growth and river flow is the same and not limited by watershed boundaries. Thus, forests that are located even at a great distance from a particular meteorological station can be used to reconstruct climate variability across northern Ethiopia and beyond.

Tree-ring chronologies from church forests have been used to reconstruct regional rainfall until 1811 (Fig. 15.3) (Mokria et al. 2017), as well as Blue Nile River flow until 1836 (Wils et al. 2010a). These data have been extended to 1784 for Blue Nile and Tekeze-Atbara river flow (Mokria et al. 2018). These precipitation and river flow reconstructions were the first for the Great Horn of Africa. Among other benefits, the reconstructed rainfall and river flow series offers new opportunities to understand the long-term regional river flow and rainfall variability. For instance, Mokria et al. (2017) reported that since 1784, high-frequency variations of rainfall and river flow at 2- to 4-year periodicity are the dominant mode of moisture variability in the region. These findings are confirmed by other studies, indicating that high inter-annual variability is a prominent characteristic of hydroclimate in SSA region (Dutra et al. 2013; Conway et al. 2008).

The presented rainfall and river flow reconstructions can considerably contribute to improving the understanding of long-term patterns of climate change in the data-poor SSA region. Although not yet common in Africa, tree-ring-based water-related decision-making and water management have attracted much attention in different parts of the world (Rice et al. 2009). For instance, tree-ring reconstructed river flow data has been used to develop worst-case scenarios for water supply (Axelson et al. 2009), and in forecasting, if current river flow will satisfy future freshwater demand (Chen et al. 2016). More importantly, the data can be of great benefit for regional hydrological flux studies, which are important to improve operational activities,

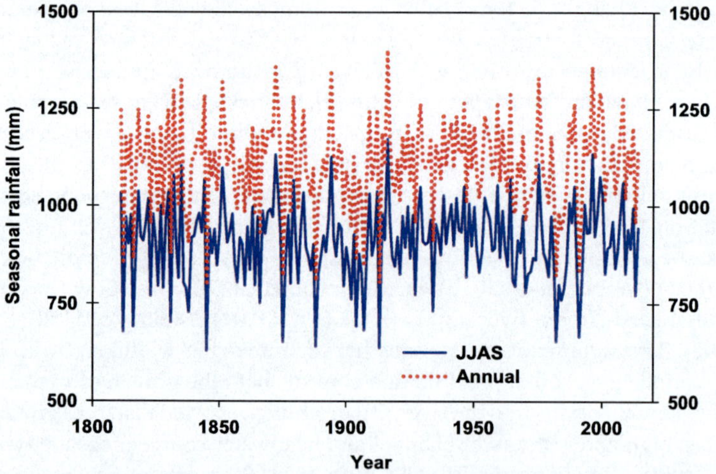

Fig. 15.3 Reconstructed JJAS (smooth line) and annual (broken line) rainfall using tree-ring data from Ethiopian church forests (*Source* modified from Mokria et al. 2017)

including water supply for domestic and agricultural purposes as well as hydroelectric production, and reduce uncertainty in projecting impacts of climate change on regional water resources (Schofield et al. 2016).

In addition to longer climate data, the changes in the intensity and frequency of extreme climate events are particularly important for policymakers. Tree-ring reconstructed rainfall and river flow data provide opportunities to explore extreme climatic events and drought-related famine years beyond the instrumental period. Studies have consistently found that tree-ring reconstructed dry and wet events considerably match with the few available historical records of drought/drought-related famine and flooding years across eastern Africa (Gebregeorgis et al. 2020; Mokria et al. 2017; Wils et al. 2010a), confirming the reliability of reconstructed extreme climate events. Mokria et al. (2018) presented a list of reconstructed extreme climate events, as well as a compilation of historic climate events in eastern Africa (Table 15.1). The reconstructed dry and wet periods also contained dry/wet years that have never been reported, especially in the nineteenth century, indicating the value of proxy-based hydroclimate information to fill the existing data gap in Africa (Mokria et al. 2017). For independent verification of reconstructed extreme climate events, Mokria et al. (2017) compared the reconstructed rainfall with historical records of semi-quantitative African rainfall data spanning 1801–1900 (accessed from Nicholson et al. 2012). They found that 20–35% of matching for rainfall regions located between 10 °N and 20 °N (i.e., regions: 13–22, 27, 85, 88) and 30–65% for East African rainfall regions (i.e., regions: 31–33, 37–39, 8) (see Nicholson et al. 2012, for the exact location of the rainfall regions). However, matching considerably increased after considering data from the substitutable rainfall regions (see Nicholson et al. 2012, for the list of substitutable rainfall regions). For the Sahel region (east Africa), the report showed matching of 85% (50%), 59% (68%), and 33% (50%) for normal, wet, and very wet years, and 73% (91%), and 67% (56%) for dry and very dry years, respectively. In general, reconstructed rainfall provided long-term regional-scale climate information, highlighting its potential to improve the quality of future paleoclimate studies in the region (Mokria et al. 2017; Wils et al. 2010a).

15.2.4 Indications of Regional Climate Teleconnections Inferred from Tree-Rings

Teleconnections refer to climate variability links between non-contiguous geographic regions. Teleconnection patterns are extracted from the analysis of sea-level/tropospheric pressure variations on monthly or/and weekly timescales (Yuan et al. 2018). Tree-ring reconstructed multi-century rainfall and river flow series provided new opportunities to assess the connection between north Ethiopian rainfall and the equatorial Pacific Ocean Sea Surface Temperatures (SST) over the past two centuries. Reconstructed rainfall by Mokria et al. (2017) showed a significant negative correlation with SST (Fig. 15.4), suggesting that the inter-annual variability

Table 15.1 Lists of tree-ring reconstructed extreme climate events from 1811 to 2014 and compiled historical records of drought/famine and wet/flooding in Ethiopia and the Great Horn of Africa.

Centuries	Reconstructed extreme dry/dry years	Historically recorded drought or drought-related famine years		
		Ethiopia	Sudan	Kenya
19th	*Extreme dry*: 1812, 1816, 1823, 1831, 1837–1838, 1845, 1866, 1875, 1888 *Dry*: 1826, 1828, 1834, 1836 1849, 1856–1857 1860, 1862 1864, 1876, 1878, 1881, 1885–1890, 1896, 1899	1812–1816, 1826–1829, 1831, 1833–1834, 1835–1838, 1864–1866, 1876–1878, 1880, 1888–1892, 1895–1896, 1899–1900	1833, 1835–1838, 1855–1861, 1864, 1868, 1873, 1877, 1885, 1886, 1888–1890	1814, 1823, 1830, 1833–1834, 1836, 1851, 1862, 1863, 1871, 1876–1877, 1883, 1889–1890, 1894–1895, 1896–1900
20th	*Extreme dry*: 1902, 1904, 1907, 1913, 1927, 1934, 1970, 1982, 1984, 1991 *Dry*: 1901, 1906, 1910, 1918–1919 1921, 1925, 1947, 1952, 1960, 1963, 1965–1966, 1971–1974, 1978, 1979, 1981, 1983, 1985, 1986, 1990, 1992	1913–1914, 1918–1922, 1932–1934, 1953, 1957–1958, 1962–1963, 1965–1966, 1969, 1971–1975, 1978–1979, 1982, 1984–1985, 1987–1988, 1989–1990, 1991–1992	1913–1914, 1927, 1932–1935, 1937–1938, 1942, 1949, 1951, 1957, 1966, 1968–1973, 1979–1983, 1984–1985, 1987–1988, 1991–1992, 1998, 2000	1907–1911, 1913–1919, 1921, 1925, 1933–1934, 1937, 1939, 1942–1944, 1947–1950, 1952–1955, 1960–1961, 1965, 1972, 1973–1976, 1980, 1981–1983, 1984, 1987, 1992–1994, 1999–2000
21st	*Extreme dry*: 2013 *Dry*: 2001, 2005, 2009	2001–2002, 2005–2006, 2008–2009, 2010–2011, 2012	2001–2002, 2008–2009, 2010–2011	2001–2002, 2005–2006, 2008–2009, 2010–2011, 2012

(continued)

Table 15.1 (continued)

Centuries	Reconstructed extreme wet/wet years	Historically recorded wet or rainfall-induced flooding years		
19th	*Extreme wet*: 1818, 1832, 1835, 1844 1851, 1872–1873, 1879, 1883, 1894 *Wet years*: 1811, 1813, 1815, 1822, 1824, 1827, 1829–1830, 1841–1842, 1852, 1855, 1863, 1865, 1867, 1869–1871, 1877, 1893		1840–1841, 1863, 1866, 1869, 1874, 1878, 1887, 1889, 1894	1894
20th	*Extreme wet*: 1912, 1916, 1929, 1946, 1959, 1964, 1975, 1989, 1996, 1999 *Wet*: 1909, 1914, 1917, 1922, 1924, 1926, 1928, 1933, 1938, 1940–1941, 1943, 1948, 1950, 1954–1956, 1961, 1969, 1976, 1980, 1987–1988, 1993, 1998	1916–1918, 1924, 1928–1930, 1946–1948, 1961, 1964, 1968–1970, 1988, 1997–1998	1916–1917, 1924, 1946, 1961–1964, 1988, 1997–1998	1961, 1963/1964, 1988, 1997–1998
21st	*Extreme wet*: 2008 *Wet years*: 2000, 2004, 2007, 2012	2007, 2010	2006–2007, 2009,	2002, 2003, 2007

(*Source* Mokria et al. 2017)

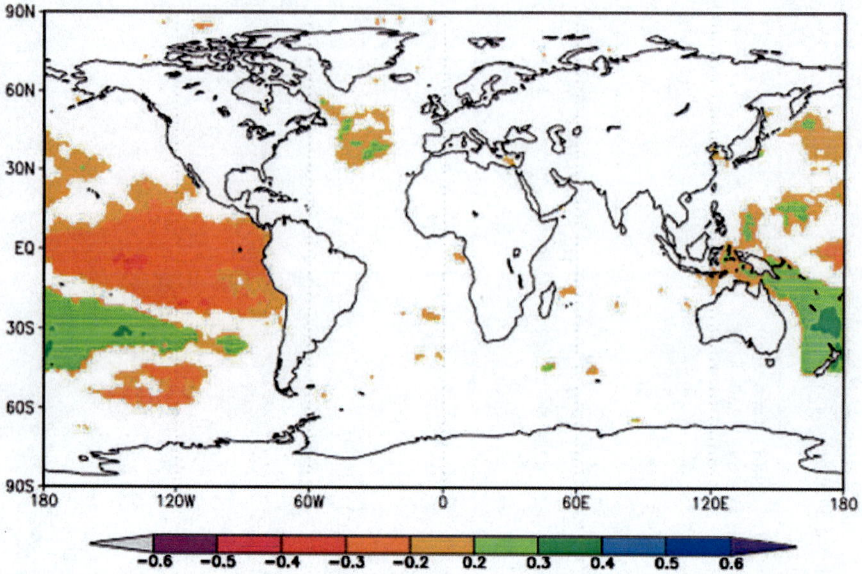

Fig. 15.4 Correlation patterns of reconstructed summer (JJAS) season rainfall in northern Ethiopia with gridded sea surface temperature (SST) data of HadISST1 over their overlapping period from 1870–2014. $P < 10\%$ (*Source* modified from Mokria et al. 2017)

of rainfall and the climate system of eastern Africa, specifically northern Ethiopia, is influenced by remote ocean-atmosphere interactions. The teleconnection between regional rainfall and SST was further supported by the strong coincidence between El Niño and La Niña events that coincided with 38–40 and 50–59% of reconstructed extreme dry/low flow and extreme wet/high flow events (Mokria et al. 2018, 2017). Hence, the reconstructed climate information from proxy data might play a significant role in understanding a remote ocean and atmospheric interaction, as well as to improve the understanding of mechanisms affecting regional weather conditions and the extent of global weather teleconnections. Other studies have also reported a strong influence of El Niño–Southern Oscillation (ENSO) during the June–September rainfall period in the eastern Africa region (Degefu et al. 2017). More importantly, ENSO influences northern Ethiopian rainfall, which is strongest after the onset of the rainy season (Mokria et al. 2017), suggests that El Niño events have serious consequences on the national food security of the region because local food production highly depends on sufficient June–September rainfall.

15.3 Tree-Ring Information for Landscape Restoration and Forest Conservation

Forest landscape restoration and conservation of remnant forests are complementary activities in sustainable forest management. Tree-ring analyses provide crucial information on growth performance for selecting the right tree species for the right place and purpose, including resilience to climate change during forest restoration. Information obtained from tree rings provides evidence to identify ecosystems and tree species that need conservation priority. The absence of growth rings in many tropical tree species has made the availability of tree-ring information scanty, and hence related benefits of this field of study. However, the presence of distinct growth rings in several species (Brienen et al. 2016; Schöngart et al. 2017), and methodological advancement in the dendrochronological fields will increase the understanding of tropical forest growth dynamic, that will eventually be used to guide restoration and conservation intervention in the tropics, including Ethiopia.

15.3.1 Forest Disturbance and Carbon Dynamics Inferred from Tree-Rings

A key aspect of sustainable forest management and conservation is the monitoring of the forest status, including the assessment of forest disturbances and forest degradation (Brienen and Zuidema 2006). Among other definitions, the Food and Agriculture Organization of the United Nations (FAO) defines forest degradation as "natural and human-induced changes within the forest which negatively affect the structure or function of the stand and thereby lower the capacity to supply products and/or services for longer temporal scale", while disturbance was defined as "naturally-induced environmental fluctuation and destructive event that affects forest health, structure, and/or changes resource or physical environment at any spatial or temporal scale" (Schoene et al. 2007). The scale of deforestation in SSA underscores the need for conserving old-growth forests and landscape restoration. However, basic information for evidence-based restoration options is scarce. Information obtained from tree rings provides evidence to identify ecosystems and tree species that need conservation priority. Thus, tree-ring analysis is a reliable and cost-effective approach to generate information on the stand age structure of the current populations, growth dynamics, and growth trajectories. Furthermore, information about past disturbance events and long-lasting impacts of forest degradation on regional carbon fluxes and forest ecosystem services can be inferred (Mokria et al. 2015; Schöngart et al. 2011; Abiyu et al. 2018). In line with this, church forests are the last refuge of old-growth Afromontane forest, hence they potentially provide a unique treasure in the field of dendroecology. Due to the destruction of natural vegetation, most of the church forests host by far the oldest trees. A tree-ring study on *J. procera* from church forests by Abiyu et al. (2018) showed that the age of the standing trees varied between 49 and

244 years. The mean annual radial increment of *J. procera* varied between 0.8 mm and 4.0 mm, with an overall mean of 1.6 (±0.7) mm year^{-1} (Mokria et al. 2015; Sass-Klaassen et al. 2008; Wils et al. 2010a). Abiyu et al. (2018) found a significant ($P < 0.05$) differences in growth rates between young and old generations of *J. procera* trees (Fig. 15.5). Across the studied church forests, generation-specific time (# years) required to reach 30 cm diameter was at least 105, 81, and 57 years, for old, middle, and young tree generations (Fig. 15.5), respectively (Abiyu et al. 2018). Another study reported that *J. procera*, grown in the northern Afromontane forest of Ethiopia reached a medium-sized stem diameter (i.e., 20–25 cm) after no less than 100 years (Mokria et al. 2015), indicating that the effect of forest degradation on carbon sequestration potential and ecosystem function is long-lasting. To this line, the slow-growth performance of Afromontane church forests implies the presence of resource limitations that are important for tree growth. This in turn urges the need to characterize specific site conditions, while planning and implementing forest management and restoration interventions. On the other hand, such findings clarify the importance of protecting church and remnant old-growth forest to maintain the quality of the environment and to reduce efforts and cost for forest restoration after major loss (Mokria et al. 2015).

Slow growth rate and homogeneity of the growth pattern in the old tree generation are indications of inter-tree competition under a dense canopy and undisturbed natural environment (Abiyu et al. 2018). The observed high growth rates in the younger tree generation suggest that trees are currently growing in less complex structured forests, more open stands, and under less competition for light conditions than in earlier centuries. In line with this, 78% of the old generation trees, 60% of middle generation trees, and 100% of young generation trees have experienced disturbances (Abiyu et al. 2018). This, further clarify that higher growth rates in the younger generation are the result of accelerated disturbance frequency during the last two centuries. This may harm the genetic diversity at the population level and may affect the ecosystem services, which could have been obtained from dense forests. The increased growth rate in the younger generation may also help trees reach the upper canopy faster on the cost of reduced investment in defense, wood density, and mechanical strength, hydraulic resistance, and vigor to grow under stress, which ultimately shortens the longevity and may subsequently affect future forest population dynamics (Abiyu et al. 2018; Bigler 2016). On the other hand, accelerated growth may create opportunities for fast accumulation of aboveground biomass, which possibly enhances the climate change adaptation and mitigation potential of dry Afromontane forests.

The estimation of the annual carbon sequestration potential of individual trees can be obtained by integrating tree-ring and biomass estimation models (Köhl et al. 2017; Mokria et al. 2015; Schöngart et al. 2011). Abiyu et al. (2018) reported that individual *J. procera* trees from church forests considerably varied in their carbon accumulation rates (Fig. 15.5). At the forest level, the carbon sequestration rate ranged from 2.4–17.2 kg C tree^{-1}. Year^{-1}, with the overall mean (±SE) of 8.1 ± 4.1 kg C tree^{-1}. Year^{-1}). Generation-wise, the calculated carbon sequestration potential of *J. procera* ranges (mean ± SD) were 5.1–17.1 (11.6 ± 4.9 kg C tree^{-1} Year^{-1}), 3.1–17.2 (8.3 ± 3.7 kg C tree^{-1} Year^{-1}), and 2.4–12.3 (6.4 ± 2.5 kg C tree^{-1} Year^{-1})

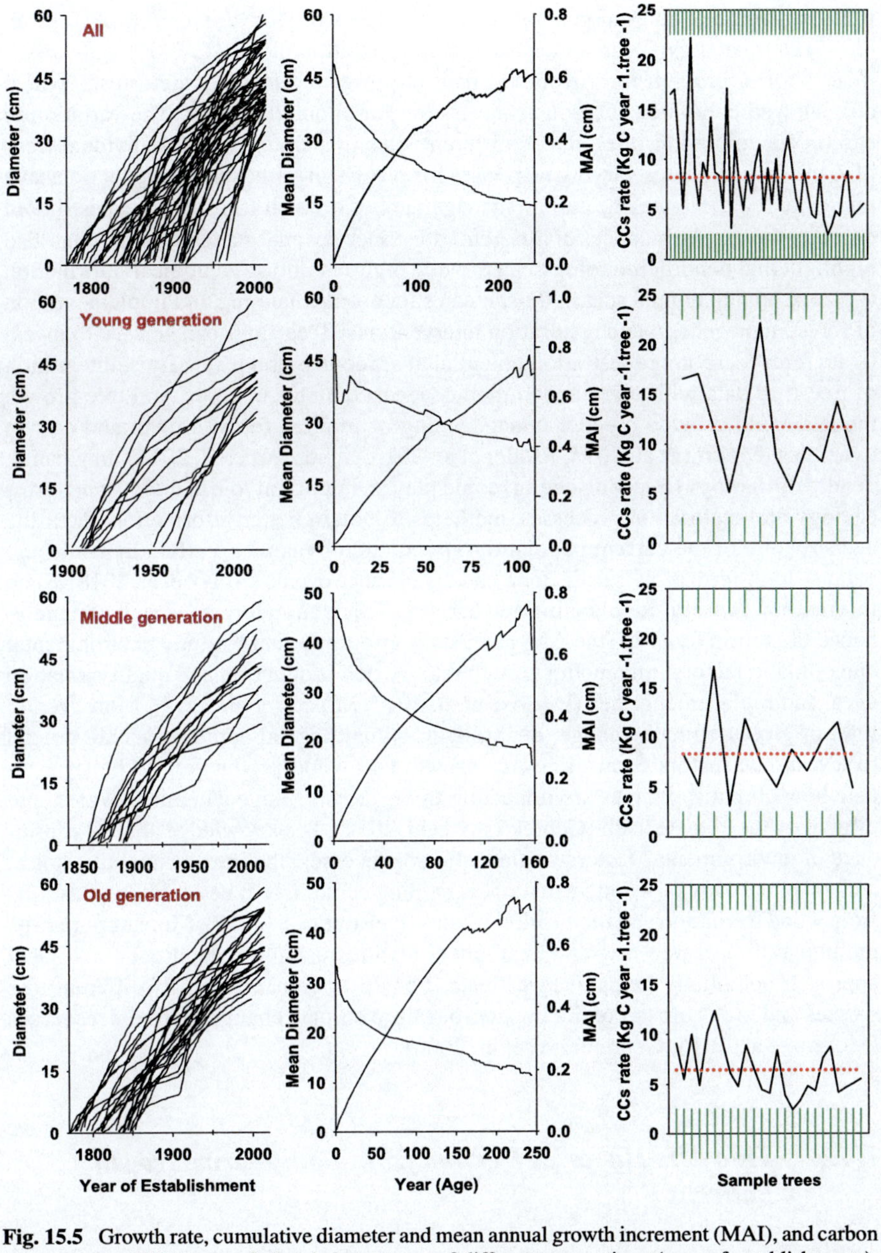

Fig. 15.5 Growth rate, cumulative diameter and mean annual growth increment (MAI), and carbon sequestration (CCs) rates of church forest trees of different generations (year of establishments). The broken horizontal line indicates the mean CCs rate. (*Source* modified from Abiyu et al. 2018)

in young, middle, and old generation trees, respectively. In this line, Mokria et al. (2015) estimated the mean annual carbon sequestration of 1.12 (±0.05) kg C tree^{-1} Year^{-1} for *J. procera trees* from remnant old-growth Afromontane forests, which also showed continuous carbon sequestration for about 200 years. The variation in carbon accumulation rates across different sites, generations, and individual trees of *J. procera* indicates varying responses for plant-soil–atmosphere interactions at a microsite scale. Hence, the carbon management of tropical forest landscapes should consider the carbon balances of different forest successional stages. Our findings also highlight that dendrochronology can provide high-resolution ecological information (i.e., seasonal to annual scale); thus, it can support the planning and implementation of forest management and restoration interventions. Tree-rings can be used to investigate forest carbon sequestration potential at a tree-to-stand level. Thus, integration of tree-ring data with other contemporary measurements of forest and tree growth models could provide a better understanding of tropical forest growth and carbon dynamics (Mokria et al. 2015; Ridder et al. 2013; Schöngart et al. 2011). In general, dendrochronology (tree-ring dating) could play an important role to understand forest ecology and restoration processes and helps to obtain better information about the age structure of the current populations, population dynamics, carbon dynamic and balance long-term growth trajectory, and disturbance events (Abiyu et al. 2018). Such information is useful for planning and harvest of plantation forests as well as maintenance of natural forests. Tree-ring analysis is also important to study environmental and climatic factors influencing growth rates, wood production and quality, rotation ages, and replacement rates (DeRose et al. 2017; Mokria et al. 2017). Moreover, in light of global climate change, dendrohydroclimatological applications are crucial to investigate factors controlling tree growth and wood production, to better evaluate how climatic changes are impacting trees, forests, biogeochemical cycles, and ultimately the climate itself (Gebrekirstos et al. 2014). The knowledge about the influence of environmental factors on the cellular processes, the intra-annual dynamics, and the phenology of wood formation, is helping ecologists to better interpret results from wood formation and monitoring studies (DeRose et al. 2017). More importantly, the integration of tree-ring and anatomical features using the dendrochronological approach potentially helps to investigate drought tolerance levels of different tree species and may help to predict impacts of future climate change on forest resources (Mokria et al. 2016; Gebrekirstos et al. 2006).

15.3.2 Tree Life Histories Derived from Long-Term Growth Patterns

The long-term growth pattern of a tree can help to understand the weather and environmental conditions of that growth period at annual to multidecadal time scales. We found different types of radial growth patterns in junipers from Ethiopian church forests (Fig. 15.6) (Abiyu et al. 2018), indicating temporal changes in microsite

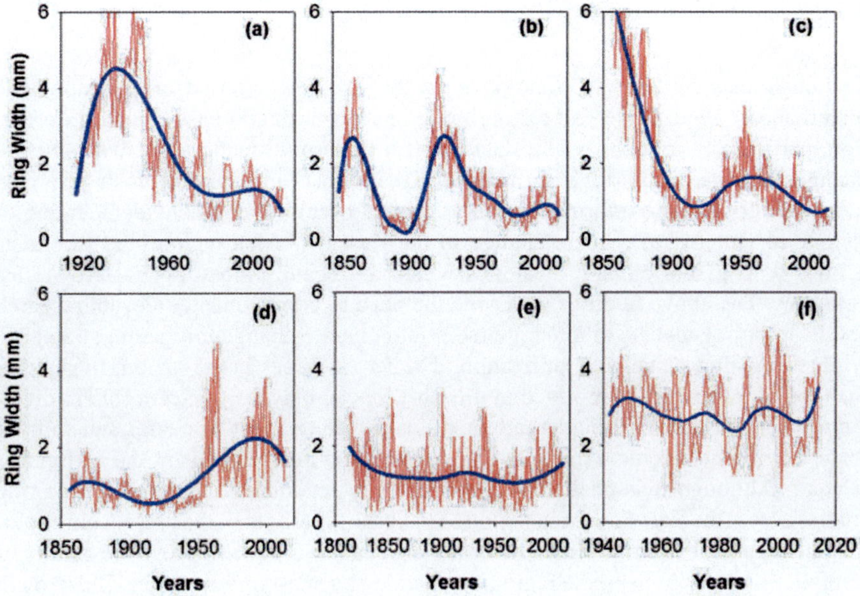

Fig. 15.6 Observed radial growth patterns of church forest inferred from tree-ring. Fig. **a**, **b**, **c**, **d**, **e**, and **f** represent unimodal, bimodal, descending, ascending, stable, and fuzzy radial growth patterns, respectively. (*Source* modified from Abiyu et al. 2018)

conditions, including competition for light and nutrients, and canopy gaps due to disturbances (Rozas 2001; Abiyu et al. 2018). Radial growth patterns of individual trees reflect growth history across the life span of the trees under changing micro-climate and environmental conditions. For instance, suppressed radial growth is an indication of competition under a dense canopy and undisturbed natural environment (Fig. 15.6d) at the early stage of tree growth. Release and reduced suppression reflect signs of local disturbance and canopy gap appearances (Rozas 2001; Abiyu et al. 2018). The growth pattern can be roughly used to investigate temporal environmental changes, such as periodic disturbance events, and the magnitude of competition and canopy cover through the life span of the tree which might be difficult to explain without the application of dendrochronological methods. Moreover, growth patterns of tree-rings are used as indicators of mortality risk, which is often adopted in tree mortality models as well as to relate growth patterns before death to biotic and abiotic stress factors (Bigler 2003). Hence, developing restoration and sustainable management options for dry Afromontane forests dominated by *J. procera* requires knowledge on growth dynamics, age structure of the population, and response to changing climate conditions (Abiyu et al. 2018; Mokria et al. 2015).

15.4 Conclusions and Recommendations

Church forests contain old standing trees that are less exposed to anthropogenic disturbances. Hence, they are a storehouse of long-term and past climate information, particularly in areas where data about long-term climatic and environmental changes are scarce and urgently needed. Tree-rings from church forests provided multi-century climate information that is critical to improving the understanding of long-term patterns of climate change in the Eastern Africa region, and the interactions with global climate dynamics which are heavily affecting Eastern Africa countries. The above findings underline the need to communicate relevant research results to policy makers to integrate them into climate change mitigation strategies. while enhancing sustainable utilization of water resources in the region. In general, the type of information presented in this chapter can play a significant role in diversifying comprehensive drought and pluvial risk management strategies and shifting from a crisis management mode to ex-ante strategic interventions in the region and beyond. Although it is challenging, we strongly recommend expanding tree-ring studies across the Eastern Africa region to increase the length of the present chronologies and to establish a tree-ring network, which can be used to improve the quality of rainfall reconstructions in the region conducted at a sub-rainfall region. Old-growth church forests may be important cornerstones of such a network in an otherwise deforested environment, where trees are harvested at young ages under various forest management strategies. Furthermore, historical perspectives on regional rainfall variability allow backcasting as part of climate model improvement and thus can help to improve the interpretation of recent climate trends and to validate future climate scenarios derived from global climate models.

Tree-ring based growth performance analysis showed that *Juniperus procera* from church and old-growth remnant Afromontane forests requires at least 801 years to reach 30 cm stem diameter, implying that the impact of forest degradation on the ecosystem services of church forests as well as Afromontane forests are long-lasting in the study region. The information substantiates the importance of protecting Afromontane forest areas to maintain the quality of the environment and to reduce efforts and costs for forest restoration after a major loss. Thus, we strongly recommend the conservation of remnant Afromontane forests which would avoid unnecessary costs to be incurred in restoring degraded landscapes. This chapter also demonstrated that tree-ring data from church forests are extremely valuable sources of information that would help to better understand forest environmental benefits with high temporal resolution (i.e., seasonal to annual time scale). Furthermore, the integration of tree-ring with other ecological information may provide an extraordinary opportunity for a better understanding of forest productivity and visualizing the consequences of forest degradation through the calculation of the required time to recover degraded landscapes. Church forest islands are, thus, towers for monitoring and reconstructing past climate change information. Moreover, church forests provide unique archives and important references on the environment which enhances informed decision-making to support designs and scaling-up of massive climate-smart restoration initiatives across the SSA region and beyond.

Acknowledgments We thank the Ethiopian Orthodox Tewahido Church and Congregants for allowing us to collect wood samples. This study was conducted with the financial support of the CUOMO FOUNDATION through an IPCC scholarship, ICRAF as part of the Forests, Trees, and Agroforestry research program of the CGIAR, and the German Research Council for supporting project BR1895/24-1. The contents of this document are solely the responsibility of the authors and do not represent the institutional position of the Cuomo Foundation and/or the IPCC.

References

Abiyu A et al (2018) Tree-ring record in Ethiopian church forests reveals successive generation differences in growth rates and disturbance events. For Ecol Manage 409:835–844. https://doi.org/10.1016/j.foreco.2017.12.015

Aerts R et al (2016) Conservation of the Ethiopian church forests: threats, opportunities and implications for their management. Sci Total Environ 551–552:404–414. https://doi.org/10.1016/j.scitotenv.2016.02.034

Allen CD, Breshers DD, McDowell NG (2015) On underestimation of global vulnerability to tree mortality and forest die-off from hotter drought in the Anthropocene. Ecosphere 6(April):1–55

Allen K et al (2017) Will seasonally dry tropical forests be sensitive or resistant to future changes in rainfall regimes? Environ Res Lett 12(2):0–19

Axelson JN, Sauchyn DJ, Barichivich J (2009) New reconstructions of streamflow variability in the South Saskatchewan River Basin from a network of tree ring chronologies, Alberta, Canada. Water Resour Res 45(9):1–10. https://doi.org/10.1029/2008WR007639

Bigler C (2016) Trade-offs between growth rate, tree size and lifespan of mountain pine (Pinus Montana) in the swiss national park. PLoS ONE 11(3):1–18. https://doi.org/10.1371/journal.pone.0150402

Bigler CJ (2003) Growth-dependent tree mortality: ecological processes and modeling approaches based on tree-ring data 15145

Bräuning A, Von Schnakenburg P, Peters T (2008) Seasonal growth dynamics and its climate forcing in a tropical mountain rain forest in Southern Ecuador. TRACE 6(March 2006):27–30

Bräuning A et al (2016) Tree-ring features: indicators of extreme event impacts. IAWA J 37(2):206–231. https://doi.org/10.1163/22941932-20160131

Brienen RJW, Schöngart J, Zuidema PA (2016) tree rings in the tropics: insights into the ecology and climate sensitivity of tropical trees. In: Tropical Tree Physiology, pp 439–461. https://doi.org/10.1007/978-3-319-27422-5_20

Brienen RJW, Zuidema PA (2006) The use of tree rings in tropical forest management: projecting timber yields of four Bolivian tree species. For Ecol Manage 226(1–3):256–267. https://doi.org/10.1016/j.foreco.2006.01.038

Briffa KR et al (2009) High-resolution palaeoclimatology of the last millennium: a review of current status and future prospects. Holocene 19(1):3–49. https://doi.org/10.1177/0959683608098952

Chen F et al (2016) 'Reconstruction of a long streamflow record using tree rings in the upper Kurshab river (Pamir-Alai mountains) and its application to water resources management. Int J Water Resour Dev 33(6):976–986. https://doi.org/10.1080/07900627.2016.1238347

Conway D et al (2008) Rainfall and water resources variability in Sub-Saharan Africa during the 20th century. Hydrometeorology (July):33. https://doi.org/10.1175/2008JHM1004.1

Corlett RT (2011) Impacts of warming on tropical lowland rainforests. Trends Ecol Evol 26(11):606–613. https://doi.org/10.1016/j.tree.2011.06.015

Degefu MA, Rowell DP, Bewket W (2017) Teleconnections between Ethiopian rainfall variability and global SSTs: observations and methods for model evaluation. Meteorol Atmos Phys 129(2):173–186. https://doi.org/10.1007/s00703-016-0466-9

De Ridder M et al (2013) Tree-ring analysis of an African long-lived pioneer species as a tool for sustainable forest management. For Ecol Manage 304:417–426. https://doi.org/10.1016/j.foreco.2013.05.007

DeRose RJ, Shaw JD, Long JN (2017) Building the forest inventory and analysis tree-ring data set. J Forest 115(4):283–291. https://doi.org/10.5849/jof.15-097

Dutra E et al (2013) The 2010–2011 drought in the Horn of Africa in ECMWF reanalysis and seasonal forecast products. Int J Climatol 33(7):1720–1729. https://doi.org/10.1002/joc.3545

Ellison D et al (2017) Trees, forests and water: Cool insights for a hot world. Glob Environ Chang 43:51–61. https://doi.org/10.1016/j.gloenvcha.2017.01.002

Eshete AW (2007) Ethiopian Church forests opportunities and challenges for restoration. Doctoral dissertation. Wageningen University, Wageningen, The Netherlands

Funk C et al (2015) The climate hazards infrared precipitation with stations—a new environmental record for monitoring extremes. Sci Data 2:150066. https://doi.org/10.1038/sdata.2015.66

Future Climate for Africa (2016) Africa's climate helping decision-makers make sense of climate information. www.futureclimateafrica.org

Gebrekirstos A et al (2006) Adaptation of five co-occurring tree and shrub species to water stress and its implication in restoration of degraded lands. For Ecol Manage 229(1–3):259–267. https://doi.org/10.1016/j.foreco.2006.04.029

Gebrekirstos A et al (2008) Climate-growth relationships of the dominant tree species from semi-arid savanna woodland in Ethiopia. Trees 22(5):631–641. https://doi.org/10.1007/s00468-008-0221-z

Gebrekirstos A et al (2011) Relationships of stable carbon isotopes, plant water potential and growth an approach to asses water use efficiency and growth strategies of dry land agroforestry species. Tree 25:95–102. https://doi.org/10.1007/s00468-010-0467-0

Gebrekirstos A et al (2014) Opportunities and applications of dendrochronology in Africa. Curr Opin Environ Sustain 6:48–53. https://doi.org/10.1016/j.cosust.2013.10.011

Gebregeorgis EG et al (2020) Historical droughts recorded in extended Juniperus procera ring-width chronologies from the Ethiopian Highlands. Int J Biometeorol (Degefu 1988). https://doi.org/10.1007/s00484-020-01863-7

IPCC (2014) Climate change 2014 synthesis report summary chapter for policymakers. https://doi.org/10.1017/CBO9781107415324

Köhl M, Neupane PR, Lotfiomran N (2017) The impact of tree age on biomass growth and carbon accumulation capacity: a retrospective analysis using tree ring data of three tropical tree species grown in natural forests of Suriname. PLoS ONE 12(8):e0181187. https://doi.org/10.1371/journal.pone.0181187

Krepkowski J et al (2011) Cambial growth dynamics and climatic control of different tree life forms in tropical mountain forest in Ethiopia. Trees 25(1):59–70. https://doi.org/10.1007/s00468-010-0460-7

Mokria M et al (2015) Tree dieback affects climate change mitigation potential of a dry afromontane forest in northern Ethiopia. For Ecol Manage 344:73–83. https://doi.org/10.1016/j.foreco.2015.02.008

Mokria M et al (2016) The frankincense tree Boswellia neglecta reveals high potential for restoration of woodlands in the Horn of Africa. For Ecol Manage 385:16–24. https://doi.org/10.1016/j.foreco.2016.11.020

Mokria M et al (2017) Multi-century tree-ring precipitation record reveals increasing frequency of extreme dry events in the upper Blue Nile river catchment. Glob Change Biol 23:2222–2249. https://doi.org/10.1111/gcb.13809

Mokria M et al (2018) Upper Nile river flow reconstructed to A.D. 1784 from tree-rings for a long-term perspective on hydrologic-extremes and effective water resource management. Quatern Sci Rev 199:126–143. https://doi.org/10.1016/j.quascirev.2018.09.011

Nash DJ et al (2016) African hydroclimatic variability during the last 2000 years. Quatern Sci Rev 154:1–22. https://doi.org/10.1016/j.quascirev.2016.10.012

Nicholson SE, Klotter D, Dezfuli AK (2012) Spatial reconstruction of semi-quantitative precipitation fields over Africa during the nineteenth century from documentary evidence and gauge data. Quatern Res 78(1):13–23. https://doi.org/10.1016/j.yqres.2012.03

Rice JL, Woodhouse CA, Lukas JJ (2009) Science and decision making: water management and tree-ring data in the Western United States. J Am Water Resour Assoc 45(5):1248–1259. https://doi.org/10.1111/j.1752-1688.2009.00358.x

Rozas V (2001) Detecting the impact of climate and disturbances on tree-rings of *Fagus sylvatica* L. and *Quercus robur* L. in a lowland forest in Cantabria, Northern Spain. Ann For Sci 58(3):237–251. https://doi.org/10.1051/forest:2001123

Sass-Klaassen U et al (2016) A tree-centered approach to assess impacts of extreme climatic events on forests. Front Plant Sci 7(July 2016):1–6. https://doi.org/10.3389/fpls.2016.01069

Sass-Klaassen U et al (2008) Juniper from Ethiopia contains a large-scale precipitation signal. Int J Plant Sci 169(8):1057–1065. https://doi.org/10.1086/590473

Schofield MR et al (2016) A model-based approach to climate reconstruction using tree-ring data. J Am Stat Assoc 111(513):93–106. https://doi.org/10.1080/01621459.2015.1110524

Schöngart J et al (2011) Age-related and stand-wise estimates of carbon stocks and sequestration in the aboveground coarse wood biomass of wetland forests in the northern Pantanal, Brazil. Biogeosciences 8(11):3407–3421. https://doi.org/10.5194/bg-8-3407-2011

Schoene D et al (2007) Forest and climate change working paper 5: definitional issues related to reducing emissions from deforestation in developing countries. http://www.fao.org/3/j9345e/j9345e00.htm

Schöngart J et al (2017) Dendroecological studies in the neotropics: history, status and future challenges. In: Dendroecology: tree-ring analyses applied to ecological studies, pp. 35–73. https://doi.org/10.1007/978-3-319-61669-8

Singh J, Yadav RR (2013) Tree-ring-based seven century long flow records of Satluj river, Western Himalaya, India. Quatern Int 304:156–162. https://doi.org/10.1016/j.quaint.2013.03.024

Tolera M et al (2013) Frankincense tree recruitment failed over the past half century. For Ecol Manage 304:65–72. https://doi.org/10.1016/j.foreco.2013.04.036

Verheyden A et al (2004) Growth rings, growth ring formation and age determination in the mangrove Rhizophora mucronata. Ann Bot 94(1):59–66. https://doi.org/10.1093/aob/mch115

Wassie A, Sterck FJ, Bongers F (2010) Species and structural diversity of church forests in a fragmented Ethiopian Highland landscape. J Veg Sci 21(5):938–948. https://doi.org/10.1111/j.1654-1103.2010.01202.x

Wils THG et al (2009) Periodicity of growth rings in Juniperus procera from Ethiopia inferred from crossdating and radiocarbon dating. Dendrochronologia (Elsevier) 1–14. https://doi.org/10.1016/j.dendro.2008.08.002

Wils THG, Robertson I, Eshetu Z, Touchan R et al (2010a) Crossdating Juniperus procera from North Gondar, Ethiopia. Trees—Struct Func 25(1):71–82. https://doi.org/10.1007/s00468-010-0475-0

Wils THG, Sass-Klaassen UGW et al (2010b) Dendrochronology in the dry tropics: the Ethiopian case. Tree 25(3):345–354. https://doi.org/10.1007/s00468-010-0521-y

Wils THG, Robertson I, Eshetu Z, Koprowski M et al (2010c) Towards a reconstruction of Blue Nile baseflow from Ethiopian tree rings. Holocene 20:837–848. https://doi.org/10.1177/0959683610365940

Woodhouse CA, Lukas JJ (2006) Multi-century tree-ring reconstructions of Colorado streamflow for water resource planning. Clim Change 78:293–315. https://doi.org/10.1007/s10584-006-9055-0

Worbes M (1995) How to measure growth dynamics in tropical trees. IAWA J 16(4):337–351

Worbes M (2002) One hundred years of tree-ring research in the tropics—a brief history and an outlook to future challenges. Dendrochronologia 2:217–231

Worbes M, Blanchart S, Fichtler E (2013) Relations between water balance, wood traits and phenological behavior of tree species from a tropical dry forest in Costa Rica—a multifactorial study. Tree Physiol 33(5):527–536. https://doi.org/10.1093/treephys/tpt028

World Bank (2017) Hydromet in Africa. Available at https://www.worldbank.org/en/region/afr/brief/hydromet-in-africa. Accessed 1 May 2020

Yuan X, Kaplan MR, Cane MA (2018) The interconnected global climate system-a review of tropical-polar teleconnections. J Clim 31(15):5765–5792. https://doi.org/10.1175/JCLI-D-16-0637.1

Chapter 16
Rehabilitation Sites Prioritization on Base of Multisource Remote Sensing Time Series, Erosion Risk, and Woody Biomass Modeling

Thomas Schneider, Mengistie Kindu, Martin Döllerer, Aleksandra Ivanova, and Thomas Knoke

Abstract Identifying the right sites for rehabilitation is an important step for successful restoration programs. This study identified rehabilitation priority sites using multisource datasets covering six decades from 1958 to 2016 in the Tis Abay area, as example for the Ethiopian Highlands. Two sets of B&W aerial photographs (1958 and 1984) and two sets of satellite images (MOMS-02/D2 for 1993 and RapidEye for 2016) were the main input data from which eight land use/land cover (LULC) types were classified. Combination of three cases, such as analysis of LULC changes, erosion risk as well as woody biomass demand and supplies, was considered and employed in a GIS environment for the rehabilitation site prioritizations. The change results show that area covered with woody biomass providing LULC classes decreased by 28% in 60 years. On the contrary, the demand increased from 2316 t in 1958 to 21,280 t in 2016 or more than 900%. The erosion risk analysis shows that over 2000 ha (33%) of the area were exposed to an erosion risk of >256 t ha^{-1} yr^{-1}. All in all, we found about 1000 ha (15%) of the study area as high to very high priority site for rehabilitation. We discussed options but limitations of concepts for landscape rehabilitations in response of global change, especially global warming, and a rapidly growing population.

Keywords B&W aerial photographs · MOMS-02/D2 · RapidEye · Changes · Fuelwood supply and demand · MUSLE · Restoration · Ethiopia

T. Schneider · M. Kindu (✉) · M. Döllerer · A. Ivanova · T. Knoke
Institute of Forest Management, TUM School of Life Sciences Weihenstephan, Technical University of Munich, Hans-Carl-von-Carlowitz-Platz 2, 85354 Freising, Germany
e-mail: mengistie@tum.de

T. Schneider
e-mail: tomi.schneider@tum.de

M. Döllerer
e-mail: doellerer@tum.de

T. Knoke
e-mail: knoke@tum.de

M. Kindu et al. (eds.), *State of the Art in Ethiopian Church Forests and Restoration Options*, https://doi.org/10.1007/978-3-030-86626-6_16

16.1 Introduction

The Ethiopian Highlands are one of the oldest known cultivated areas on Earth. In Ethiopia, over the last 3000 years, there has been progressive deforestation, which has accelerated tremendously during the last centuries. Within the last six decades, extensive forest clearing for cultivation, over-grazing, changes in political regime, and exploitation of forests for fuelwood and construction materials without considering sustainability resulted in considerable changes of land use/land cover (LULC) (Desalegn et al. 2014; Kindu et al. 2015, 2020; Demissie et al. 2017; Gashaw et al. 2018). Especially, the deforestations and conversions of large and often unsuitable sites into farmland, running hand in hand with the increase in livestock, magnified the pressure on the remaining woody biomass resources and promoted degradation (Paulos 2001; Kindu et al. 2013; Temesgen et al. 2013; Gashew et al. 2014; Gashaw 2015). The demand on fuelwood as the main energy source of the farmers in Ethiopia aggravates the problem (Kindu et al. 2015; Demissie et al. 2017).

The Ethiopian Highland Reclamation Study (EHRS) indicated that in the mid-1980s, 27 million ha or almost 50% of the highland area was significantly eroded, 14 million ha seriously eroded, and over 2 million ha beyond reclamation (FAO 1986). 'Generating three centimeters of top soil takes 1000 years, and if current rates of degradation continue all of the world's top soil could be gone within 60 years' a senior UN official is cited in 'Scientific American' in connection with the 'World Soil Day,' in December 2014.

The Ethiopian Governments were aware of the erosion problems since the early 1970s. The Soil and Water Conservation (SWC) program was launched with the aim to mitigate food insecurity and to sustain agricultural productivity (Kassie et al 2008; Getachew 2005). In the 1980s, the SWC program was implemented with the support of donors at large watershed level in the central, highly degraded areas of the Ethiopian Highlands. It was a top-down approach and could not meet the anticipated objectives (Tongul and Hobson 2013). As a result, it has undergone a series of changes in approach (bottom-up, community-based and in smaller units at sub-watershed level) and technical standards (Lakew et al. 2005). This was an essential step toward to integrated watershed management (Bewket 2007).

Nowadays, climate change is also superimposing the population growth, and erosion, food, and fuelwood provision challenges are becoming major issues in the sub-Saharan Africa. According to the United Nations Food and Agriculture Organization (FAO) (2014), Ethiopia is one of most affected countries in the region. Agriculture is the most vulnerable sector to climate change. That applies especially for the rain-fed crop and livestock-mixed systems of smallholder farmers in the Ethiopian Highlands (Deressa et al. 2011), because of their limited adaptive capacities (Keller 2009; Lewis et al. 2017). This emerging situation also pushed the efforts developing adapted holistic mitigation strategies in order to cope with the complex drivers and the different site conditions across the country. One of the fundamental issues seems to boost the efforts toward increasing carbon stock by afforestation and rehabilitation measures (Bewket 2007; Kindu et al. 2018; Knoke et al. 2020). Increase in forest-covered area will directly improve microclimate and hydrological regimes and will reduce erosion. On continental level, the African Great Green Wall (GGW) initiative is aiming to address the climate change issues as well. The GGW initiative was launched in 2007 by the African Union and ratified in 2014. GGW is part of the UN Sustainable Development Goals (SDGs) and supported by the United Nations Convention to Combat Desertification (UNCCD). The total area of the GGW initiative extends to 156 Mha, with the largest intervention zones located in Niger, Mali, Ethiopia, and Eritrea (https://www.unccd.int/actions/great-green-wall-initiative).

All initiatives need suitable areas for the intended afforestation and rehabilitation programs. Such areas need to be selected by considering multiple variables or drivers that triggered pressures to the existing LULC types (Kindu et al. 2018).

An effective geo-data-based land use management system with societal accepted concepts of how to stop negative trends is a necessity for handling such programs. Satellite remote sensing data is one of the appropriate sources to provide the basic information needed for the successful implementation of such instruments (Kindu et al., 2013; Demissie et al. 2017).

The main objective of the presented study is to demonstrate how to integrate multisource data collected over decades in a transdisciplinary framework with the aim to develop concepts for rehabilitation sites prioritization on base of remote sensing time series, erosion risk, woody biomass and fuelwood supply and demand modeling in the Tis Abay area, as example for the Ethiopian Highlands. The study is following a general approach developed in previous studies (Helldén 1987; Schneider et al. 1995, 1996; Reusing et al. 2000). Starting with trend analysis covering the last six decades, we identify restoration priority areas considering different situations, trying to compromise the diverse needs of a prospering society under the auspices of climate change situation. The general workflow starts with LULC classifications for the respective years providing the basic spatiotemporal basis for quantifying LULC changes (sub-task 1), fuelwood demand and supply (sub-task 2), and erosion risk (sub-task 3). In all cases, we rely on external, public accessible statistics for balancing and quantifying the target units. Finally, based on the outputs of sub-tasks 2 and 3, we develop our concepts for identifying potential restoration areas and provide a priority ranking scheme (sub-task 4).

16.2 Materials and Methods

16.2.1 Study Site

The study site is located in the Lake Tana area (Fig. 16.1) of the Western Ethiopian Highlands in the surroundings of the small village Tis Abay (Fig. 16.1). It covers an area of about 62 km^2 and is situated between the altitudes of 1530 and about 2200 m above sea level. The climate is tropical summer humid, with two rainfall seasons and precipitations of about 1400 mm/a. The land use in the area is dominated

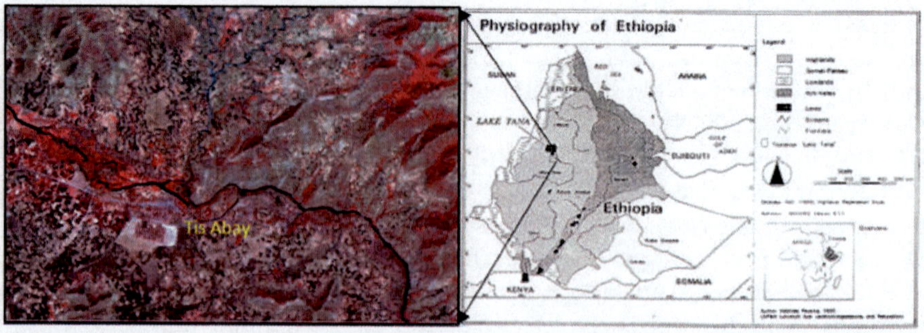

Fig. 16.1 Location of study site (Tis Abay), south part of Lake Tana. Left, with a color infrared composite from the 2016 RapidEye data. Vegetation covered areas appear in red tones

by agricultural crops and pasture farming. The remaining native forests composed mainly *Podocarpus falcatus* (Thunb.) R. B. ex. Mirb., *Juniperus procera, Ficus vasta, Croton macrostachyus* Del, and *Albizia gummifera*. The woodlands are dominated by Acacia species. Native forests and woodlands are degraded at a very high level.

16.2.2 Data Sources

Data from various sources was used in this study. Two sets of B&W aerial photographs from the years of 1958 and 1982 were utilized for classification and change analysis, both on a scale of about 1:45.000. The aerial photographs were provided by the Ethiopian Survey agency.

MOMS-02/D2: The German experimental system Modular Optoelectronic Multispectral Stereo Scanner (MOMS-02/D2) from the year 1994 was also used. MOMS-02 is a push broom CCD sensor equipped with four multispectral bands optimized for vegetation assessments combined with an on track stereo capability with two 22° fore and aft looking bands and a high-resolution nadir-looking band in the panchromatic region (Table 16.1). Due to data storage limitations, it must be decided in advance which of the seven modes should be operated. The presented results are based on MOMS-02/D2 mode 3 data of track 61, scene 8, recorded on April 30, 1994, 12.30 h local time at a sun elevation of about 86°. The imaging mode 3 of the MOMS-02/D2 sensor is designed to register simultaneously stereo (ST) and multispectral (MS) data. This mode combines the along track panchromatic bands 6 and 7 with the nadir-looking, multispectral bands 3 in the red and 4 in the near infrared wavelength region (Table 16.1) (Seige and Meißner 1993).

The RapidEye imagery data was used for the study year 2016. The RapidEye images were obtained from RapidEye Science Archive (RESA). The system was developed and initially operated by the RapidEye AG, Germany, since 2015 owned by Planet Labs Germany. RapidEye was a constellation of five identical satellites on slightly shifted sun-synchronous orbits (ESA 2020), launched in 2008 and out of

Table 16.1 Characteristics of the evaluated MOMS-02/D2 and RapidEye satellite data. MOMS-02/D2 mode 3 configuration used in this study is marked by italic letteres

MOMS-02/D2					RapidEye				
Band	Characteristic	Orientation	Band width (nm)	Pixel size (m)	Band	Characteristic	Orientation	Band width (nm)	Pixel size (m)
1	MS	Nadir	440–505	13.5	1	Blue	Nadir	440–510	5 (6.5)
2	MS	Nadir	530–575	13.5	2	green	Nadir	520–590	5 (6.5)
3	*MS*	*Nadir*	*645–680*	*13.5*	3	Red	Nadir	630–685	5 (6.5)
4	*MS*	*Nadir*	*770–810*	*13.5*	4	RE	Nadir	690–730	5 (6.5)
5	Pan	Nadir	510–760	4.50	5	NIR	Nadir	760–850	5 (6.5)
6	*ST fore*	*+21.40*	*520–760*	*13.5*					
7	*ST back*	*−21.40*	*520–760*	*13.5*					

operation since January 2020. The specifications of the five spectral bands are listed in Table 16.1.

In addition, other data was also utilized for geo-referencing and GIS analyses, including topographic maps scale 1:250 000 (based on the B&W aerial photographs from 1957/58) and 1:50 000 (based on the B&W aerial photographs from 1982/1984), soil maps, population census data (1984, 2007), and a digital elevation model (DEM) from the Japanese Advanced Spaceborne Thermal Emission and Reflection Radiometer (ASTER) (https://asterweb.jpl.nasa.gov/gdem.asp).

16.2.3 Methods

We integrated and analyzed primary and supplementary data sources in GIS environments. For 1958, 1982, and 1993, we used ARC-INFO GIS and ArcGIS for 2016. The data flow of our analysis is shown in Fig. 16.2.

16.2.3.1 Land Use/Land Change Analysis

Eight major land use/land cover (LULC) classes were considered to derive from aerial photographs (1958, 1984) and satellite data MOMS-02/D2 (Seige and Meißner 1993), RapidEye (ESA 2020). We selected the LULC types based on studies of the Food and Agricultural Organization of the UN (FAO) and to governmental requirements (MME 1986). A brief description of these LULC classes as defined for visual

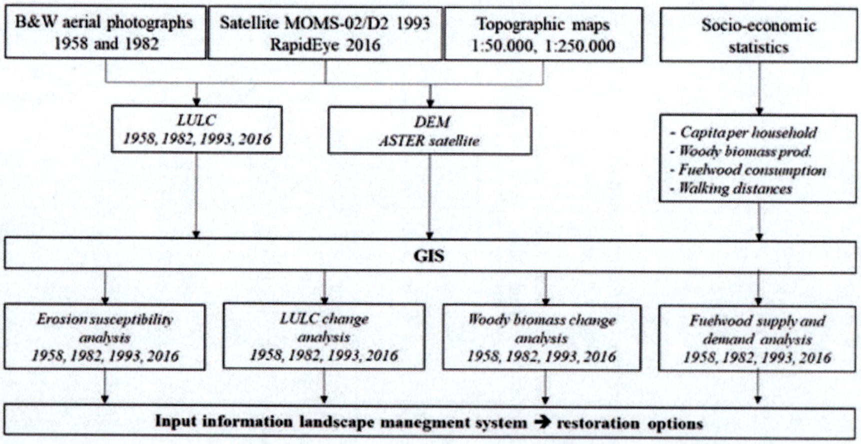

Fig. 16.2 Flowchart of the presented steps of a landscape management system for rehabilitation sites prioritization for restoration options in Tis Abay study site of Ethiopia

interpretation (aerial photographs) and automatic classification (MOMS-02/D2 and RapidEye) is given in Table 16.2.

Table 16.2 Land use/land cover classes distinguished by visual interpretation from B&W aerial photographs and image classification approaches from MOMS-02/D2 and RapidEye data for the demonstration studies in the Tis Abay study site

LULC classes	Class description	MOMS-02/D2; RapidEye	Aerial photographs
Forestland (FL)	Land covered by a close stand of trees of one or more stories with an interlaced upper canopy, rising 7–40 m or more in height	Forest	Forest
Woodland (WL)	Land supporting a stand of trees, up to 20 m in height, with an open or continuous, but not thickly interlaced canopy, extending over more than 20%	Woodland	Woodland
Bushland (BL)	Land supporting an assemblage of trees and shrubs, often dominated by plants of shrubby habit but with trees always conspicuous, with a single or layered canopy, usually not exceeding 10 m in height except for occasional emergents and total canopy cover greater than 20%	Bushland	Bushland

(continued)

Table 16.2 (continued)

LULC classes	Class description	MOMS-02/D2; RapidEye	Aerial photographs
Shrubland (SL)	Land supporting a stand of shrubs, usually not exceeding 6 m in height with a canopy cover greater than 20%. Trees, if present, contribute less than one-tenth of the canopy cover	Shrubland	Shrubland
Grassland wet (GLw)	More or like vital vegetation signal, mostly grazing land	Grassland	Grassland
Grassland dry (GLd)	Dry vegetation signal	Single trees	Grassland with single trees <20%
Cultivated land plowed (CLp)	Fresh plowed land without vegetation signal	Cultivated land	Cultivated land
Cultivated land harvested (CLh)	Dry remains of former vegetation	Single trees	Cultivated land with single trees <20%
Cultivated land fallow/degraded (CLf)	Sparse vegetation signal, irregular patches		
Cultivated land irrigated (CLi)	Vital vegetation signal, mostly vegetables or chad, coverage 65–90%	Cultivated land irrigated	
Water bodies (WB)	Blue Nil river, tributaries, hydropower channel	Water	Water
Wetland (WL)	Swamp areas, mostly papyrus and reed, vegetation coverage 80–100%	Wetland	Wetland

(continued)

Table 16.2 (continued)

LULC classes	Class description	MOMS-02/D2; RapidEye	Aerial photographs
Infrastructure/sealed/settlements/houses	Roads, houses, dam, bridges	Sealed	Sealed

The generated LULC classes were used as input for change detection, erosion risk, and woody biomass supply and demand as well as for the rehabilitation site prioritization.

For LULC identification and delineation in 1958 and 1982, we analyzed two B&W stereo pairs from each series, using a Wild Aviopret APT2 Stereoscope. For GIS integration and for comparison with MOMS-02/D2 and RapidEye outcomes, we digitized the delineated land use classes, transformed them into a raster format, and co-registered these data to the georeferenced MOMS-02 RGB image.

For LULC classification in 1993 on base of the MOMS-02/D2 data, we completed MS bands 3 and 4 with a relative vegetation index (RVI) (Reusing et al. 2000).

$$\mathrm{RVI} = ((\mathrm{band}\ 4/\mathrm{band}\ 3) - 255) * (-1) \tag{16.1}$$

We also used the pixel-based ties classifier (PCI 1993), a combination of parallelepiped and maximum likelihood method based on spectral differences (Schneider et al. 1996). Except for the homogeneous appearing Blue Nile class with six training areas, we established a minimum of 30 training sites per class on different altitudes, expositions, and soil type to train the algorithm.

We used a knowledge-based approach using the object-based image analysis (OBIA) software package eCognition Developer 9.5 from Trimble to classify the 2016 LULC on base of the RapidEye data set following the approach of Kindu et al. (2013). For the identification of LULC classes, we utilized the spectral ranges for the different bands, texture values, and indices, which we complemented by neighborhood relations for some classes. Again, we used the MOMS-02 data set for co-registering the data.

Accuracy assessments were done based on Lillesand and Kiefer (1987) for the classification from MOMS-02/D2 data and Kindu et al. (2013) for the classification derived from RapidEye image.

We used a map-based or post-classification approach for change detection quantification (Kindu et al. 2013). To facilitate the comparison and because our focus was on woody biomass classes, we aggregated the subclasses of cropland and grassland to one cropland and one grassland class each.

16.2.3.2 Fuelwood Supply and Demand

A special topic of the study was the identification of fuelwood supply and demand areas of the Tis Abay study site. The analysis is based on the quantification of the woody biomass of the classes forestland, woodland, bushland, and shrubland based on governmental studies (MME 1986; Addis Bar Forestry Project 1993) and the Swedish International Development Corporation (Sida) (Hellдèn 1987).

In the Tis Abay study site, forest areas are restricted either to the flood plains and islands inside the Blue Nile or to inaccessible steep valleys. We also considered detectable single trees on cropland and grassland, assuming as a woody biomass like the one for forest (Schneider et al. 1995).

The annual woody biomass increment can be calculated in two different ways. First, using the values as shown in Table 16.3 in the column {product.*1} (meaning productivity), as was done by (Schneider et al. 1995), or, by using the assumed mean annual increment (MAI) of 10% of the standing stock (MME 1986). For the ease of calculation in the present study, we calculated the fuelwood supply following the 10% increment assumption. The resulting differences by using these two approaches as well as the consequence of newly defining the woody biomass relevant LULC classes are briefly discussed later.

We calculated on the base of the visual interpretation of black and white (B&W) aerial photographs for the fuelwood supply in the years 1958 and 1982, while for 1993 and 2016 we calculated according to the woody biomass estimations from the MOMS-02/D2 and RapidEye satellite data analysis. Fuelwood supply is usually given as tons per hectare and year (t/ha/yr). The surplus and deficit areas visualized with the scaling factor based on the studies from Schneider et al. (1995, 1996).

The specific demand we calculated on base of the mean annual increment (MAI). For the conversion of the MAI into t/ha/yr, we followed the method used in the GIZ Technical Report (MME 1986) shown in Table 16.4. Multiplying the area for the respective class with the associated calorific value of the woody biomass in giga-calories per hectare [Gcal/ha] provides the woody biomass stock in [Gcal/ha].

Based on the calculation chain shown in Table 16.4, woody biomass in [Gcal] is transformed in metric tons [t]. Such unit is usually given for the annual consumption per capita, respectively, per household, which is in metric tons per year [t/capita/yr]. We calculated the fuelwood demand based on annual fuelwood need per capita of around 0.88 in 1993 and 0.76 t/year in 2016. By extrapolating these numbers for the consumption in 1958 and 1982, we calculated with a fuelwood demand for 1958 with 1.2 and for 1982 with 1 t/capita/year. The calculation is based on the numbers for the inhabitants of the study site, the fuelwood consumption per year [t/capita/year], and the annual fuelwood supply [t/capita/year] for the respective year.

We estimated the population density of the study site for 1993 and 2016 by digitizing the cottages from the topographic map 1:50.000 actualized on the base of the 1984 census for 1993 and BingMaps for 2016 combined with data for 2016 provided by the FactFish (2020) and Quandl platform (2015). According to interview data from the ADDIS BAR Forestry Project (1994), performed in an area nearby, the average number of inhabitants per hut was assumed to be 5.7 for 1993 and, according to FctFish (2020), of 4.8 for 2016.

Planning rehabilitation measures promoting forests or woodlands, the average daily walking distance as an important socioeconomic factor, must be considered. According to Helldén (1987), we calculated fuelwood deficiency and surplus areas

Table 16.3 Woody biomass estimation and productivity approach in the Tis Abay study site applied for 1993 MOMS-02/D2 and 2016 RapidEye data sets balanced with governmental statistics about standing stock and productivity (Mean Annual Increment, MAI) at the Ethiopian Highlands

	Ethiopian highlands, 1986[a]				Tis Abay study site, 1993[b]				
LULC	Area		*Growing stock*	Productivity[a]	Area		Growing stock	Provision 10%[b]	Provision[a]
	ha	%	*Gcal/ha*	Gcal/ha/yr	ha	%	Gcal (100%)	Gcal/yr	Gcal/yr
Forestland	2,879,000.0	8.80	*1123.10*	104.10	232.0	3.70	260,559.20	26,055.92	24,151.2
Woodland	80,000.0	0.20	*359.60*	20.00	194.0	3.09	69,762.40	6976.24	3880.0
Bushland	4,004,000.0	12.20	*87.10*	5.20	735.0	11.71	64,018.50	6401.85	3822.0
Shrubland	6,455,000.0	19.70	*2.20*	0.40	1223.0	19.49	2690.60	269.06	489.2
Total	13,418,000				2384.0	100.0	*397,030.70*	39,703.07	32,342.4
						[c]Inhabitants 1993			5557
						[c]Fuelwood need/person/year			0.9 t
						[c]Fuelwood need/Tis Abay/year			4763.0 t

Sources
[a] MME (1986)
[b] Schneider et al. (1996)
[c] Addis Bar Forestry Project (1993)

Table 16.4 Conversion of woody biomass from giga-calorie into tons (values per year calculated for Tis Abay 1993)

Calorific value, Tis Abay study site	Numbers [SI-units]	Sources
Woody biomass, total [Gcal]	*397,030.7 [GCal]*	MME (1986)
Assumed mean annual increment (MAI) (10%) [Gcal]	39,703.07 [GCal]	MME (1986)
Conversion factor Calorie/Joule	4.2	
Assumed mean annual increment (MAI)[%]	10	MME (1986)
Mean woody biomass broadleaved [MJ/kg]	12.54 [MJ/kg]	
Conversion calorific value [GCal] via [MJ] into tons:	*x**4.2 /12.54	
Assumed annual increment (10%) [t]	13,297.68 [t]	MME (1986), Schneider et al. (1996)

as well as walking distance zones as decision help for the definition of possible rehabilitation sites in the neighborhood of the settlements.

16.2.3.3 Soil Erosion Risk Analysis

Soil erosion is only one facet of land degradation which is controlled by natural and socioeconomic factors. For the identification of potential erosion risk areas within the study area, we applied a derivative of the universal soil loss equation (USLE) (Wischmeier and Smith 1978). The USLE is an erosion model designed to predict the longtime average soil losses in runoff from specific field areas in specified cropping and management systems (USA, Midwest states). Hurni (1986) modified and adapted the USLE to Ethiopian conditions based on recommendations of the Soil Conservation Research Project (SCRP). We followed the approach implemented by Stöhr et al. (1995) and Reusing et al. (2000) in the areas of our study site based on fieldwork between 1994 and 1995, known as modified universal soil loss equation (MUSLE). The MUSLE computes the soil loss for a given site as the product of six major factors whose most likely values at a particular location are calculated as follows:

$$\text{Mean soil loss rates: } A = R * K * L * S * C * P\left[t * ha^{-1} * yr^{-1}\right] \quad (16.2)$$

where A is the computed soil loss, expressed here as tons per hectare per year [$t*ha^{-1}*yr^{-1}$], *R* is rainfall and runoff factor, *K* is the soil erodibility factor, *L* is the slope-length factor, *S* is the slope-steepness factor, *C* is the cover and management factor, and P is the support practice factor.

We prepared the required factors for MUSLE in ArcGIS Pro. We derived *L* and *S* factors from the Aster DEM. For the *K* factor, we used the national soil map of Ethiopia. The precipitation distribution factor *R* we determined by interpolating records of nearby meteorological stations. We derived the LULC-related factors *C* and *P* factors from the classified map of 1993 and 2016. Detail description of MUSLE factors and methods of computation can be found in Reusing et al. (2000). We derived the assessment of mean soil loss rates by multiplication of all surveyed spatial erosion factors in ArcGIS.

16.2.3.4 Rehabilitation Area Prioritization

Finally, GIS-based evaluations of the extracted thematic parameters we used for identifying areas to be recommended for re-afforestation activities. The selection of periodization areas is based on the ecosystem service concept (ES) (Kindu et al. 2016; Knoke et al. 2016, 2020) and relies on balancing criteria for crop production (soil quality), erosion risk reduction, and distance to fuelwood demand areas. Thus, the main decision criteria we used are erosion risk, site conditions not appropriate for crop production (geology, soils, slope), and distance to settlements (reduced fuelwood provision pressure, distance to markets). For instance, highly productive soils in flat areas should be preserved for food production. Steep slopes associated with high soil erosion risk, in combination with long walking distance from demand hot spots without direct connection to the market, are prone to be considered as priority areas for rehabilitation. The analysis we made in ArcGIS Pro-based on a fuzzy logic approach.

16.3 Results

16.3.1 LULC Change Analysis

16.3.1.1 LULC Classification

The spatial distribution of eight LULC types for the years of 1958, 1982, 1993, and 2016 in each time step is presented in Figs. 16.3 and 16.4. In 1958, 43.3% of the land was covered with grassland followed by bushland (21.6%), woodland (11.8%), forestland (7.9%), and cropland (7.8%). Shrubland, water, and sealed accounted the smallest proportion of the study area. However, in 1986, croplands were dominant (34.2%), followed by grasslands (25.5%), bushland (14.7%), and woodland (9.6%). Cropland continued dominating in 1993 and 2016 with 45.2% and 46.7%, respectively. The remaining part is occupied by other LULC types in the two study periods.

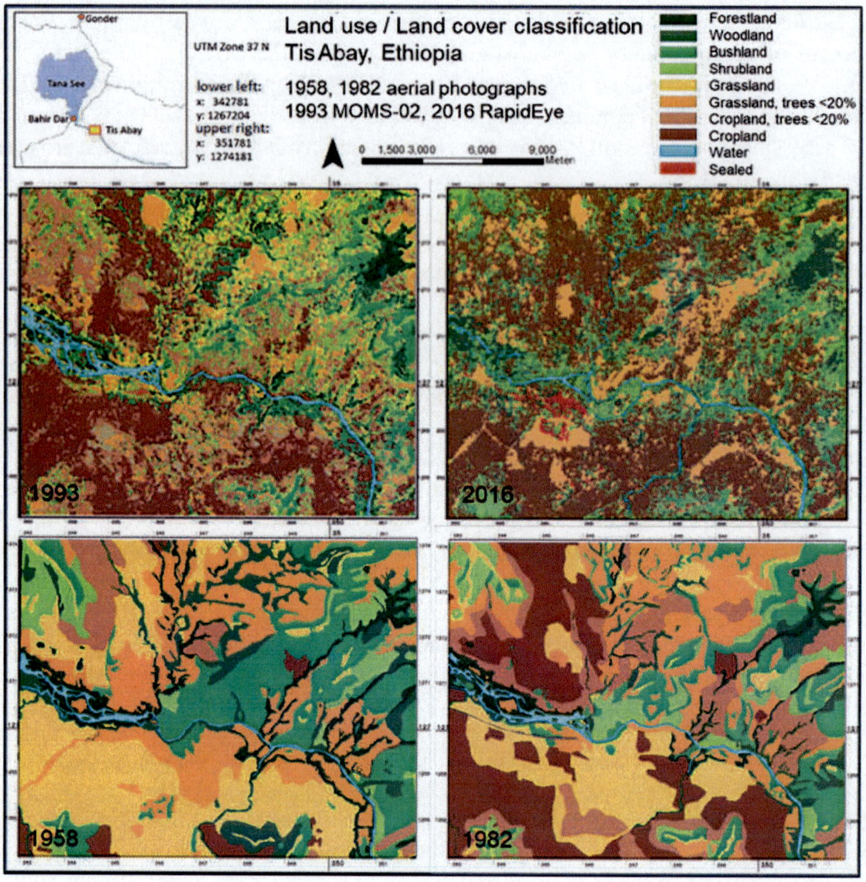

Fig. 16.3 LULC maps based on visual interpretation of B&W aerial photographs in 1958 and 1982 compared against the image processing classifications of the Moms-02/D2 (1993) by Ties/ML classification and RapidEye (2016) using OBIA classification

16.3.1.2 Accuracy Assessments

Tables 16.5 and 16.6 present the summary of the LULC classification accuracy for 1993 and 2016 reference years. The overall accuracies of the two reference years were between 88% and 89.2%, respectively. With 88.0% slightly lower overall accuracy, we achieved by the object-based classification of the RapidEye data set with 5 m pixel resolution.

16.3.1.3 Change Analysis

The changes in LULC are shown in Fig. 16.4. The change results showed continuous decline in forestland, woodland, bushland, and grassland throughout the study years. Analysis of the 60 years change showed that forestlands were decreased by 58%,

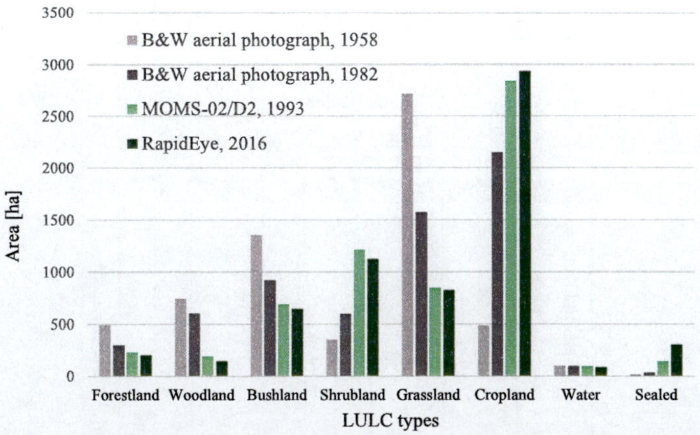

Fig. 16.4 Area covered by the classified LULC types of the Tis Abay study site as calculated from differing remote sensing systems in 1958, 1982, 1993, and 2016

woodlands by 80%, and grasslands by 70%. On the other hand, cropland, shrubland, and sealed showed an increasing trend. Of which, cropland gained an increase of 500% in 60 years.

16.3.2 Fuelwood Supply and Demand Analysis

The spatial distribution of fuelwood supply from the woody biomass estimation on base of the LULC classes forestland, woodland, bushland, and shrubland for the years 1958, 1982, 1993, and 2016 is shown in Fig. 16.5.

The area covered with woody biomass providing LULC classes decreased from 2946 ha in 1958 to 2122 ha in 2016, i.e., a conversion of 28% (Table 16.9).

The results also show that the growing stock decreased from 315,502 in 1958 to 114,405 t in 2016, which is a conversion by 64% of woody biomass in 1958 (Table 16.7). The distribution for the classes forestland (plus single trees 1958 and 1982), woodland, bushland, and shrubland is visualized in Fig. 16.6.

The amount of woody biomass which can be harvested without reducing the standing stock in the long run (sustainability principle), the mean annual increment (MAI), we assumed equivalent with 10% of the standing stock. The MAI decreased from 31,550.2 t/yr in 1958 to 11,440.5 t/yr or 36.3% in 2016 (Table 16.8). As shown in the table, column 'changes in %,' the highest loss we calculated for woodland which decreased to about 20% while shrubland expanded to more than 320% in 2016 compared to the year 1958.

In Table 16.9, we show the calculation of the annuall per capita fuelwood supply for each of the years 1958, 1982, 1993, and 2016. While for 1958 a provision per capita of 13.62 [t/capita/year] was available, the supply dropped down to 0.54

Table 16.5 Confusion matrix of the MOMS-02/D2 1993 data classification accuracy from the Tis Abay study site, Ethiopia

	Visual interpretation Tis Abay										Classified sum = 685	Users acc. %
LULC classes	FL	WL	BL	SL	GL	GLt	CLp	CLh	sl	W		
Forestland (FL)	47	2	3								52	90.38
Woodland (WL)	3	41	1	4							49	83.67
Bushland (BL)		5	48	3							56	90.57
Shrubland (SL)		1	1	119		1		1	1		124	95.97
Grassland (GL)	1				33						34	97.06
Grassland <20% tree (GLt)				1	1	32					34	94.12
Cropland plowed (CLp)			3	7			137	10			157	87.26
Cropland harvested (CLh)				11			5	113			129	87.6
Sealed (road, villages) (sl)								5	24	2	31	77.42
Water (W)	2									17	19	89.47
Visual interpretation: sum = 685	53	49	56	145	34	33	142	129	25	19	611	
Prod. acc. %	96	81	91	89	97	97	96	90			Overall acc. %	**89.2**

Table 16.6 Confusion matrix of the RapidEye data 2016 classification accuracy from the Tis Abay study site, Ethiopia

	Visual interpretation Tis Abay										Classified sum = 900	Users acc. %
LULC classes	FL	WL	BL	SL	GL	GLt	CLp	CLh	sl	W		
Forestland (FL)	78	5	2				7		1		93	83.87
Woodland (WL)	6	77	3	1		1	1				89	86.52
Bushland (BL)	1	3	76	9	0	2	1				92	82.61
Shrubland (SL)	2	4	6	71	2	3		2			90	78.89
Grassland (GL)	1				81	4		3			89	91.01
Grassland <20% tree (GLt)		1	1	4		78		4	2		90	88.64
Cropland plowed (CLp)	1						81				82	98.78
Cropland harvested (CLh)	1		2	5	7			81	4	1	101	81.82
Sealed (road, villages) (sl)						2			81	1	84	96.67
Water (W)									2	88	90	97.78
Visual interpretation: sum = 900	90	90	90	90	90	90	90	90	90	90	792	
Prod. acc. %	87	86	84	79	90	87	90	90	96.7	97.8	Overall acc. %	**88**

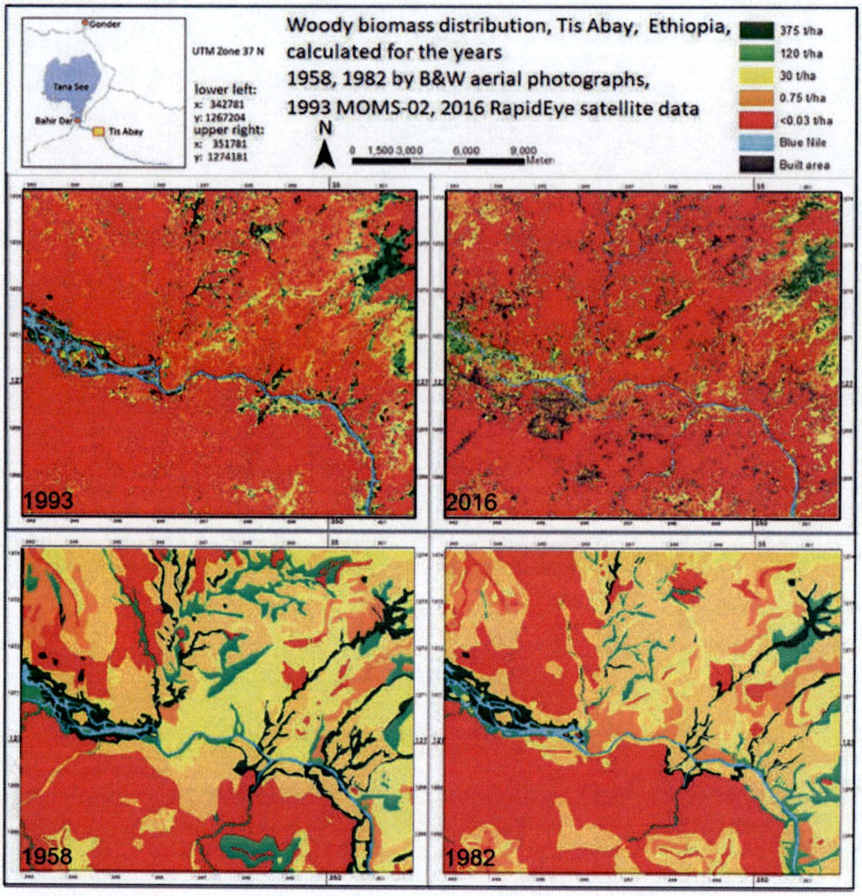

Fig. 16.5 Distribution of woody biomass as estimated on base of the LULC classes forestland, woodland, bushland, and shrubland

[t/capita/year] in 2016, or, from an oversupply of 1135% in 1958 to an undersupply of only 71% of the needed fuelwood in 2016.

The spatial distribution of fuelwood supply areas (provision) calculated for the study area Tis Abay for 1993 and 2016 is shown in Fig. 16.7a (upper row) in metric tons per hectare (t/ha). The surplus and deficit areas are visualized with a provision 'surplus' starting with 2.7 t/ha. The demand areas are calculated on base of the population density per hectare and the annual fuelwood consumption per capita in metric tons per capita and year [t/capita/year) as given in Table 16.9. As indicated in the table, the demand increased from 2316 t in 1958 to 21,280 t in 2016. This is over 800% as compared to the demand in 1958. The population distribution visualized as inhabitants per hectare is shown in Fig. 16.7b (mid row). We visualized the walking distances in 100 m steps up to more than 900 m distance from the demand areas

Table 16.7 Areas and corresponding woody biomass in tons [t] and giga-calories [GCal] of LULC classes in Tis Abay, calculated for 1958, 1982, 1993, and 2016. Gray shaded are the numbers used for calculating the woody biomass and derived the supply as shown in Table 16.8

LULC	*Grow.stock*	Area with woody biomass [ha]				Woody biomass stock [t]				Woody biomass stock [Gcal]			
	[GCal/ha]	1958	1982	1993	2016	1958	1982	1993	2016	1958	1982	1993	2016
Forest + s.t	*1123.1*	495.0	298.0	232.0	205.0	186,198.2	112,095.1	87,268.6	77,112.4	555,934.5	334,683.8	260,559.2	230,235.5
Woodland	*359.6*	743.0	602.0	194.0	147.0	89,487.1	72,505.0	23,365.4	17,704.7	267,182.8	216,479.2	69,762.4	52,861.2
Bushland	*87.1*	1356.0	923.0	735.0	643.0	39,557.6	26,926.0	21,441.6	18,757.8	118,107.6	80,393.3	64,018.5	56,005.3
Shrubland	2.2	352.0	598.0	1223.0	1127.0	259.4	440.6	901.2	830.4	774.4	1315.6	2690.6	2479.4
Total		2946.0	2421.0	2384.0	2122.0	315,502.2	211,966.7	132,976.8	114,405.3	941,999.3	632,871.9	397,030.7	341,581.4

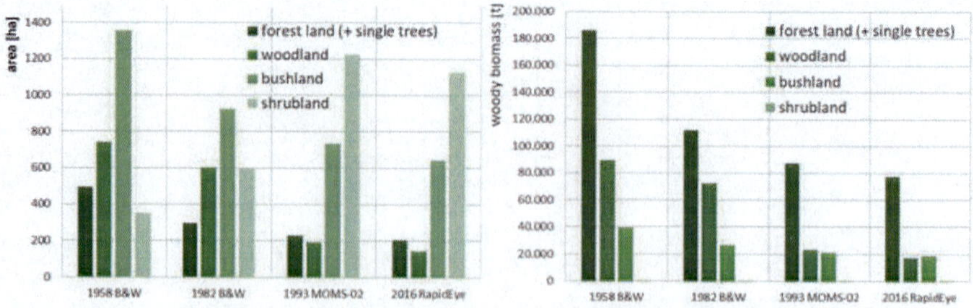

Fig. 16.6 Area covered [ha] (left) and woody biomass [t] (right), calculated for the LULC classes forestland, woodland, bushland, and shrubland for 1958, 1982, 1993, and 2016, Tis Abay study site

(Fig. 16.7c). The walking distance zones are a metric, was later used as decision criteria for the identification of potential rehabilitation areas.

16.3.3 Erosion Risk

The spatial distribution and areas of annual soil erosion rates and severity classes are shown in Fig. 16.8 and Table 16.9. In 2016, for over 2000 ha (33%) of the area very severe soil loss risk of more than >256 t ha^{-1} yr^{-1}, was calculated, of more than >256 t ha^{-1} yr^{-1}, which is an increase of over 40% of the area exposed in 1993 (Table 16.10).

16.3.4 Rehabilitation Areas

The spatial distribution and areas of priority sites for rehabilitation are shown in Fig. 16.9 and Table 16.11. The priority map clearly shows that the northern part of the study site requires implementation of rehabilitation or protection for reducing the risk of erosion while keeping the demand of fuelwood for the community. Based on the spatial distribution of priority areas, three priority classes,—high, very high, and medium high—account for about 34% of the total area of the study site. Only 18% of the area shows a very low priority class.

16.4 Discussion

Identifying the right sites for rehabilitation is an important step for successful restoration programs. In this study, using multisource data sets covering six decades, we demonstrated on behalf of retraceable criteria how to identify rehabilitation priority

Table 16.8 Annual woody biomass supply in tons/hectare/year [t/ha/yr] and giga-calories/hectare/year [GCal/ha/yr] calculated for 1958, 1982, 1993, and 2016, Tis Abay assuming a mean annual increment [MAI] of 10% as well as changes in percent [%] of the initial year of comparison

LULC	Supply (at 10% MAI) [GCal/yr]				Supply (at 10% MAI) [t/ha/yr]				Changes in % initial year		
	1958	1982	1993	2016	1958	1982	1993	2016	1958/2016	1982/2016	1993/2016
Forest+s.t	55,593.45	33,468.38	26,055.92	23,023.55	18,619.82	11,209.51	8726.86	7711.24	41.41	68.79	88.36
Woodland	26,718.28	21,647.92	6976.24	5286.12	8948.71	7250.50	2336.54	1770.47	19.78	24.42	75.77
Bushland	11,810.76	8039.33	6401.85	5600.53	3955.76	2692.60	2144.16	1875.78	47.42	69.66	87.48
Shrubland	77.44	131.56	269.06	247.94	25.94	44.06	90.12	83.04	320.017	188.46	92.15
Total	94,199.93	63,287.19	39,703.07	34,158.14	*31,550.22*	*21,196.67*	*13,297.68*	*11,440.53*	36.26	53.97	86.03

Table 16.9 Woody biomass supply and demand of Tis Abay site for the year 1958, 1982, 1993, and 2016 as well as respective changes

Calorific value, Tis Abay	1958	1982	1993	2016
Area with woody biomass [ha]	2946	2421	2384	2122
Inhabitants (rounded)	1930	4005	5600	28,000
Fuelwood demand [t/person/year]	1.2	1	0.88	0.76
Annual fuelwood demand, total [t]	2316	4005	4928.00	21,280.00
Annual fuelwood supply [t]	*31,550.22*	*21,196.67*	*13,297.68*	*11,440.53*
Annual fuelwood supply [t/capita/yr]	13.62	5.29	2.38	0.54
Annual per capita fuelwood supply [%]	1135.23	529.26	270.45	71.05

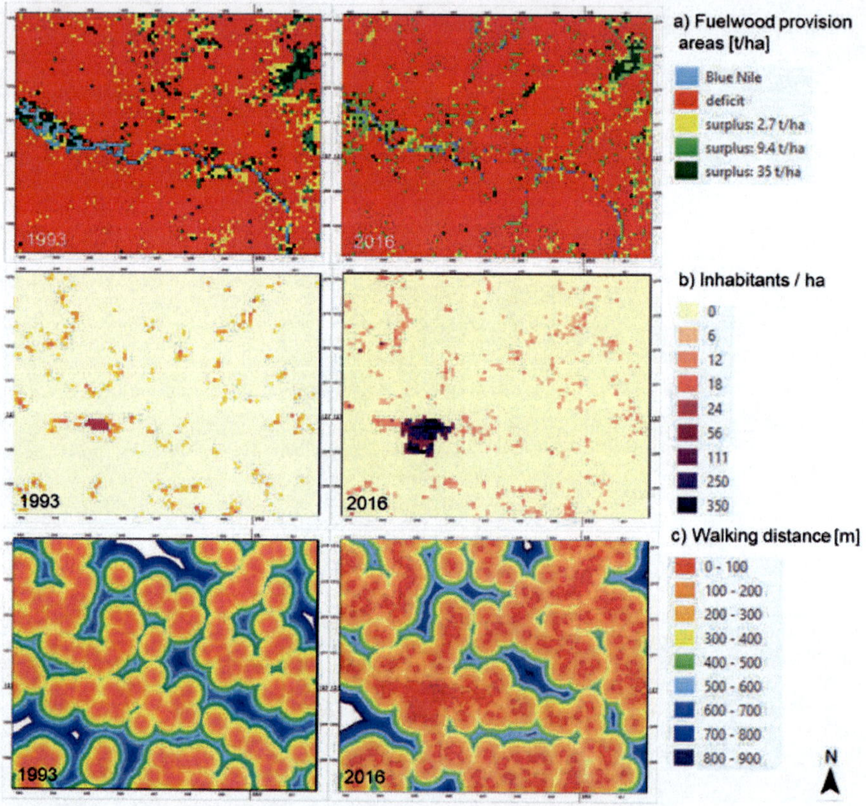

Fig. 16.7 Fuelwood supply, population distribution, and per capita provision in percent of the demand [%] calculated for the Tis Abay study site for the year 1993 (left column) and 2016 (right column)

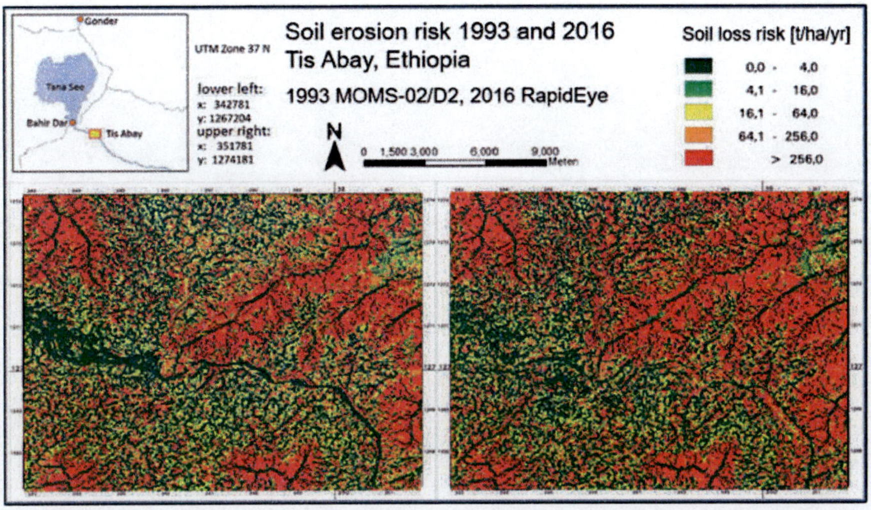

Fig. 16.8 Map of annual soil erosion rates and severity classes for 1993 (left) and 2016 (right)

Table 16.10 Annual soil erosion rates and severity classes for the years 1993 and 2016

Soil loss (t/ha/yr)	Erosion risk class	1993		2016	
		Area (ha)	%	Area (ha)	%
0–4	Low	2535.8	40.4	2484.3	39.6
4.1–16	Moderate	278.5	4.4	229.4	3.7
16.1–64	High	632.1	10.1	513.3	8.2
64.1–256	Severe	1131.0	18.0	953.1	15.2
>256	Very severe	1697.6	27.1	2095.0	33.4

sites, mainly, by employing analysis of LULC changes, erosion risk, and woody biomass demand and supply.

Comparing multi-temporal and multi-sensoral data sets, the ranges of possible misinterpretation sources have to be estimated. This is especially demanded for investigations focused on changes of dynamic systems like vegetation coverages and land use. The most disturbing error sources are related to registration issues, especially different phenological stages and atmospheric transmission conditions during data takes but on sensor characteristics and analyzation methods (extraction methodology).

A fundamental request is the comparability of the phenological stages of multi-temporal data takes. In our case, we compare aerial photographs from 18.01.1958 and 01.12.84, with digital scanner data from 30.04.1993 and 30.01.2016. The aerial photographs are from the main dry season, and there is no doubt about comparability. The MOMS-02 data is from the March/April rain period. Due to very little

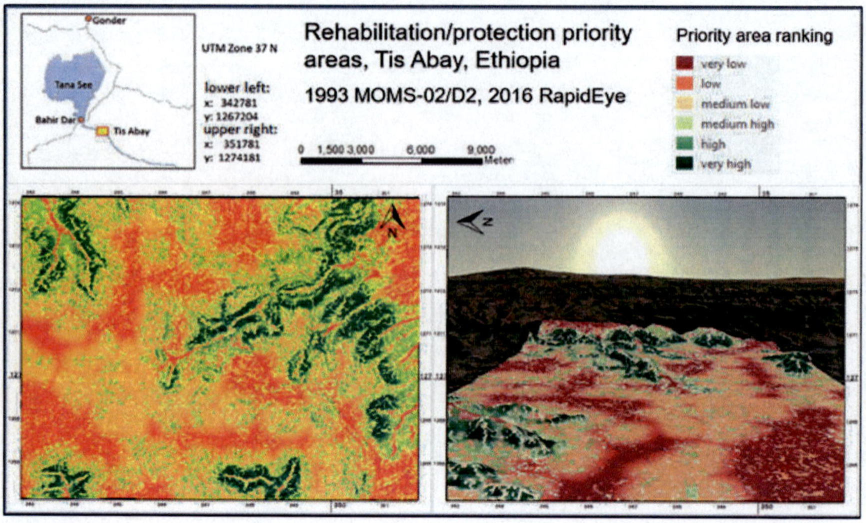

Fig. 16.9 Visualization of rehabilitation/protection priority areas as calculated for 2016

Table 16.11 Rehabilitation site priority classes as calculated for 2016

Priority class	Area (ha)	%
Very low	1131.8	18.0
Low	2042.8	32.6
Medium low	963.4	15.4
Medium high	1187.4	18.9
High	440.8	7.0
Very high	508.9	8.1

precipitations 1993, the phenological changes are negligible. The RapidEye data was recorded end of January 2016 during dry season. Considering these facts, a comparison seems to be possible. Information about the atmospheric conditions during the data takes was not available. On the other hand, the general atmospheric conditions in that region are very favorable for remote sensing data acquisition. Therefore, this error source seems to be low.

The next group of possible information bias is the data extraction method. We compared results produced by visual interpretation (B&W aerial photographs), pixel- and object-based digital image analysis (satellite data). The analyzation of B&W aerial photography is based on operator's interpretation of gray value, structure, texture, topology, size, and shape of objects and their shade. Depending on image scale, used film material and developing process, but also on conservation status of the historic film material, interpretation errors may happen. For LULC mapping purposes adapted interpretation keys are to be developed, which define the association criteria for each class. The delineation of classes depends on the subjective interpretation of the operator. The high geometric resolution of the aerial photographs and the stereoscopic interpretation opportunity enables the operator to detect single objects, like trees, bushes, huts, etc. Due to practical, economic reasons, it was not possible to map each of the single objects. Because of the importance for the ecological evaluation of the respective class for our investigations, we additionally defined the classes cultivated land with trees <20% and grassland with trees <20% by using an area grid and 'expert opinions decision' of the interpreter.

In case of the digital source provided by the satellite data, the different spatial, spectral, and radiometric resolutions of the MOMS-02 (1993) and Rapid Eye (2016) systems but the available radiometric correction algorithms at the time of acquisition affect the accuracy of information extraction and of course comparison. The suboptimal radiometric correction of each CCD sensor element in case of MOMS-02/D2 is obvious by striping effects in band 3 (red). This affected especially the differentiation of grassland wet and grassland dry, normally distinguished due to the darker gray tone of grassland wet. Therefore, we decided to distinguish only one grassland class in both data sets. The vegetation classes forestland, woodland, bushland, and shrubland differ in gray value and texture. The higher the woody biomass fraction, the darker the gray tone, the lower the signal. Due to the reasons mentioned above, also the differentiation of these classes was often difficult. Especially in hilly terrain, topography caused shading effects altered the gray value, and biased the interpretation.

Our classification showed an overall accuracy of over 88%, which is within the range of accuracies found in previous similar studies (Kindu et al. 2013; Desalegn et al. 2014; Demissie et al. 2017). In tropical regions, classification results depend very much on the acquisition date (season) of the satellite data. In the presented case, it worked out quite well, despite the fact, that data was recorded at the end of the dry season when usually the rain period starts and greening of the herbaceous vegetation makes a differentiation from woody vegetation more difficult. One explanation for the better accuracy levels achieved might be that the rainy season did not start that year before the data take (Schneider et al. 1996).

The decreasing amount of fuelwood available, paired with an increasing population that already demands more fuelwood than could be sustainably harvested from the current supply, suggests that the Tis Abay area is in desperate need of some solutions to increase the availability of fuel in the area. This is due to their exclusively reliant on fuelwood for their energy. Purely from the perspective of fuelwood, however, planting trees or restoring and expanding smaller patches of existing forest close to communities to create communal wood lots is an efficient answer for the population of the Tis Abay as it brings the resource closer to where it is consumed. This touches sociocultural (women, who are responsible for collecting the household need) as well as socioeconomic aspects (time may be used for other activities). As proxy for the time lost for fuelwood provision, we calculated the average daily walking distance.

The rate of erosion risk obtained in this study is in line with previous studies in other parts of the country (Hurni 1988; Temesgen et al. 2013; Gashaw et al. 2014; Gashaw 2015). It has to be mentioned that the MUSLE can simulate the highly complex processes involved in soil erosion only in a very simplified procedure (Reusing et al. 2000). The accuracy of the predicted soil loss rates depends on how exactly the erosion parameters are described but will never be absolute. Under specific situations like years with heavy rain events and droughts, soil losses may differ substantially from the long-term average estimated by the MUSLE for specified physical conditions. Even if the quantitative estimations are contestable, the qualitative results concerning the spatial distribution of erosion susceptibility are correct.

The priority classes considered in this study are, mainly, from the very steep slopes and high erosion risk classes. Walking distances from the demand spots did not affect the calculation that much. The identified priority areas may be covered by remnant woody biomass or completely deforested one. Accordingly, we suggest the right approach of protection or rehabilitation as double strategy, as recommended in (Schneider et al. 1996; Kindu et al. 2018).

Bishaw (2001) summarized the recommendations of diverse national and international organizations as the 'need of conservation-based integrated development strategies.' These strategies should secure the provision of the people with food, fuelwood, and fodder on sustainable bases while considering the ongoing climate changes. The author suggested the implementation of agroforestry and social forestry in the rural areas with subsistence farming, the expansion of plantation forestry on currently uncultivated and sloping lands, and the strict protection of the remaining natural forests to conserve species and biodiversity.

As to species selection, the effect of already expanding Eucalypt species needs to be taken into account. We suggest planting them as green fences or plantations close to houses, while using endemic species for rehabilitation and reforestation measures on steep and inaccessible areas (Schneider et al. 1996).

Statistical values provided by different sources made it difficult for the approach of using multiple data sources in this study. For example, there are discrepancies in estimates of MAI for woody biomass, soil erosion rates, and the changes in definitions of forest, woodland, bushland, and shrubland and over time. Depending on such definitions, the area covered by the respective LULC class may change as well. In this case, depending on the data input source, the magnitude may differ but not the relative differences between neighboring areas over time.

16.5 Conclusions

The study presented an important contribution to the use of multisource data sets and disclosed the ability of using a combination of techniques for prioritization of rehabilitation sites for successful restoration in the Tis Abay area, as example for the Ethiopian Highlands. LULC classifications based on aerial photographs from 1958 and 1984 were compared in a stack with MOMS-02/D2 mode 3 LULC classifications from 1993 and RapidEye data from 2016, and their associated dynamics for the last six decades were analyzed. Erosion risk as well as fuelwood demand and supply analysis were conducted using the data sets as input. Problems associated with the comparability of information from B&W aerial photographs and digital data sets from satellite are discussed. The fuelwood demand and supply study is based on the woody biomass estimations of forestland, bushland, shrubland, and woodland, as well as from single trees sprinkled over grassland and cultivated land, classified from the MOMS-02/D2 data from 1993 and the RapidEye data from 2016. Linking governmental statistics, results from FAO studies and RapidEye, MOMS-02/D2 mode 3, historical aerial photograph derived thematic and relief parameter in a GIS, a database for rehabilitation/afforestation planning was developed, which is not perfect, but at present offers the best information options the best for the region under investigation. We demonstrated the value of historic data sets and the ability to integrate new data, derived from different sensor types and differing methodology. The overall approach allowed us to make analyses of site priority based on our input criteria.

Acknowledgements This work is supported by the Bavarian State Chancellery, within the project 'Exploring Church Forests as ecosystem baseline for restoration concepts in the Ethiopian Highlands.' RapidEye data was provided by the RapidEye Science Archive (RESA) of the German Space Agency (DLR). Our special thanks are for Susanne Lichtenwald and Mathias Reusing who established the data basis for this demonstration study by their work in the mid 1990th.

References

Addis Bar Forestry Project (1993) Technical report, Addis Ababa, Ethiopia

Bewket W (2007) Soil and water conservation intervention with conventional technologies in Northwestern highlands of Ethiopia: acceptance and adoption by farmers. Land Use Policy 24(2):404–416

Bishaw B (2001) Deforestation and land degredation in the Ethiopian highlands: a strategy for physical recovery. Northeast Afr Stud 8(1):7–25

Demissie F, Yeshitila K, Kindu M, Schneider T (2017) Land use/land cover changes and their causes in Libokemkem district of South Gonder, Ethiopia. Remote Sens Appl: Soc Environ 8:224–230

Desalegn T, Cruz F, Kindu M, Turrión MB, Gonzalo J (2014) Land-use/land-cover (LULC) change and socioeconomic conditions of local community in the central highlands of Ethiopia. Int J Sust Dev World 21(5):406–413

Deressa T, Hassan R, Ringler C (2011) Perception and adaptation to climate change by farmers in the Nile basin of Ethiopia. J Agric Sci 149(01):23–31

European Space Agency (ESA) (2020) Online document. https://spacedata.copernicus.eu/data-offer/missions/rapideye. Accessed Aug 2020

Factfish (2020) Online database. Ethiopia: fuelwood, consumption by households. http://www.factfish.com/statistic-country/ethiopia/fuelwood,+consumption+by+households. Accessed Aug 2020

FAO (1986) Highlands reclamation study Ethiopia final report (Vol I & II). Rome, Italy

FAO (2014) Forestry statistics. http://www.fao.org/forestry/statistics/84922/en/. Accessed 2017

Gashaw T, Bantider A, Silassie HG (2014) Land degradation in Ethiopia: causes, impacts and rehabilitation techniques. J Environ Earth Sci 4(9):98–104

Gashaw T, Tulu T, Argaw M, Worqlul AW, Tolessa T, Kindu M (2018) Estimating the impacts of land use/land cover changes on ecosystem service values: the case of the Andassa watershed in the Upper Blue Nile basin of Ethiopia. Ecosyst Serv 219–228

Getachew A (2005) Determinants of land degradation in the Lake Tana basin and its implications for sustainable land management: the case of Angereb and Gish Abbay Watersheds. MSc. Thesis, Alemaya University, Alemaya

Gashaw T (2015) The implications of watershed management for reversing land degradation in ethiopia. Res J Agric Environ Manag 4:5–12

Helldén U (1987) An assessment of woody biomass, community forests, land use and soil Erosion in Ethiopia—A feasibility study on the use of remote sensing and GIS-analysis for planning purposes in developing countries; SIDA contract 30/86, 1 ETI 52, final report, Sweden. Lund University Press, Lund

Hurni H (1986) Guidelines for development agents on soil conservation in Ethiopia, ministry of agriculture, community forests and soil conservation development department. Addis Abeba

Hurni H (1988) Degradation and conservation of the resources in the Ethiopian highlands. Mt Res Dev 8:123–130

Kassie M, Pender J, Yesuf M, Köhlin G, Mulugeta E (2008) Conservation on mean crop yield and variance of yield. Evidence from the Ethiopian highlands. Environment for development. EFD DP discussion paper 08–08

Keller M (2009) Climate risks and development projects assessment report for a community-level project in Guduru, Oromiya, Ethiopia. 'Bread for All' and 'HEKS' report

Kindu M, Schneider T, Teketay D, Knoke T (2013) Land use/land cover change analysis using object-based classification approach in Munessa-Shashemene landscape of the Ethiopian highlands. Remote Sensing 5(5):2411–2435

Kindu M, Schneider T, Teketay D, Knoke T (2015) Drivers of land use/land cover changes in Munessa-Shashemene landscape of the south-central highlands of Ethiopia. Environ Monit Assess 187(7):60

Kindu M, Schneider T, Teketay D, Knoke T (2016) Changes of ecosystem service values in response to land use/land cover dynamics in Munessa-Shashemene landscape of the Ethiopian highlands. Sci Total Environ 247:137–147

Kindu M, Schneider T, Döllerer M, Teketay D, Knoke T (2018) Scenario modelling of land use/land cover changes in Munessa-Shashemene landscape of the Ethiopian highlands. Sci Total Environ 622–623:534–546

Kindu M, Angelova D, Schneider T, Döllerer M, Teketay D, Knoke T (2020) Monitoring of Urban growth patterns in rapidly growing Bahir Dar City of Northwest Ethiopia with 30 year Landsat imagery record. ISPRS Int J Geo Inf 9(9):1–19

Knoke T, Paul C, Hildebrandt P, Calvas B, Castro LM, Härtl F, Döllerer M, Hamer U, Windhorst D, Wiersma YF, Curatola Fernández GF, Obermeier WA, Adams J, Breuer L, Mosandl R, Beck E, Weber M, Stimm B, Haber W, Fürst C, Bendix J (2016) Compositional diversity of rehabilitated tropical lands supports multiple ecosystem services and buffers uncertainties. Nat Commun 7:11877

Knoke T, Paul C, Rammig A, Gosling E, Hildebrandt P, Härtl F, Peters T, Richter M, Diertl K-H, Castro LM, Calvas B, Ochoa S, Valle-Carrión LA, Hamer U, Tischer A, Potthast K, Windhorst D, Homeier J, Wilcke W, Velescu A, Gerique A, Pohle P, Adams J, Breuer L, Mosandl R, Beck E, Weber M, Stimm B, Silva B, Verburg PH, Bendix J (2020) Accounting for multiple ecosystem services in a simulation of land-use decisions: does it reduce tropical deforestation? Global Change Biol 26:2403–2420

Lakew D, Carucci V, Asrat W, Yitayew A (2005). Community based participatory watershed development: a guideline. Ministry of agriculture and rural development, Addis Ababa, Ethiopia

Lewis K (2017) Understanding climate as a driver of foodinsecurity in Ethiopia. Clim Change 144:317–328

Lillesand TM, Kiefer RW (1987) Remote Sensing and Image Interpretation. John Wiley & Sons, New York-Chichester-Brisbane-Toronto-Singapore

Ministery of Mines and Energy of the Provisional Military Government of Socialist Ethiopia (MME) (1986) Biomass energy resources, GTZ. Technical Report 1, Addis Ababa, Ethiopia

Paulos D (2001) Soil and water resources and degradation factors affecting their productivity in the Ethiopian highland agroecosystems. Michigan State University Press 8(1):1–18

PCI (1993) Pace Multispectral Analysis Package; chapter 1.17.2, Maximum Likelihood Classification, PCI users handbook

Quandl (2015) Online financial, economic and alternative data. https://www.quandl.com/data/CITYPOP/CITY_TISABAYAMETHIOPIA-Population-of-Tis-Abay-AM-Ethiopia. Accessed Aug 2020.

Reusing M, Schneider T, Ammer U (2000) Modelling soil loss rates in the Ethiopian highlands by integration of high resolution MOMS-02/D2-stereo-data in a GIS. Int J Remote Sens 21:1885–1896

Schneider TH, Stöhr H, Lichtenwald S, Ammer U (1995) Land cover mapping approaches with MOMS-02/D2 mode 3 data and the application in a fuelwood demand and supply analysis in the Lake Tana region of Ethiopia. Proceedings of MOMS-02/D2 Symposium 5–7 July 1995. Köln (Cologne), Germany, pp 185–192

Schneider, TH, Lichtenwald S, Ammer U (1996) Woody biomass change detection in the Tana Lake district of Gojam, Ethiopia—study based on MOMS-02/D2 mode 3 data from 1993 completed with B&W aerial photographs from 1958 and 1982. Proceedings of 5th scientific workshop synergy of remotely sensed data, Dundee, Scotland, 6–7 May

Seige P, Meißner D (1993) MOMS-02: an advanced high resolution multispectral stereo scanner for earth observation. GIS 1(1993):4–11

Stöhr H, Reusing M, Schneider TH, Kornus W, Lehner M (1995) Mapping of erosion risk zones in the Lake Tana region, Proceedings of MOMS-02/D2 Symposium 5–7 July 1995. Köln (Cologne), Germany, pp 193–200

Temesgen H, Nyssen J, Zenebe A, Haregeweyn N, Kindu M, Lemenih M, Haile M (2013) Ecological succession and land use changes in a lake retreat area (main Ethiopian Rift valley). J Arid Environ 91:53–60

Tongul H, Hobson M (2013) Scaling up an integrated watershed management approach through social protection programmes in Ethiopia: the MERET and PSNP schemes

Wischmeier WH, Smith DD (1978) Predicting rainfall erosion losses: a guide to conservation planning. U.S. Department of Agriculture. Agriculture Handbook No. 537. Washington: U.S. Government Printing Office

Chapter 17
Towards Ethiopian Church Forests and Restoration Options—Synthesis and Conclusions

Mengistie Kindu, Thomas Schneider, Alemayehu Wassie, Mulugeta Lemenih, Demel Teketay, and Thomas Knoke

Abstract The chapters in this book present a wide variety of evidence-based studies about the Ethiopian church forests and potential restoration options. These are pooled by the editors in this synthesis and concluding chapter of the book "State of the Art in Ethiopian Church Forests and Restoration Options".

Keywords Sustainability · Church forests · Ecosystem services · Restoration · Prioritizations · Ethiopia

17.1 Overview of Synthesis

17.1.1 Contributions

Ethiopia's native forests are rich in biodiversity. For they occupy a very mountainous and rugged topography they offer ecosystem services of paramount significance such as watershed protection. Centuries of mismanagement are resulting in their

M. Kindu (✉) · T. Schneider · T. Knoke
Institute of Forest Management, TUM School of Life Sciences Weihenstephan, Technical University of Munich, Hans-Carl-von-Carlowitz-Platz 2, 85354 Freising, Germany
e-mail: mengistie@tum.de

T. Schneider
e-mail: tomi.schneider@tum.de

T. Knoke
e-mail: knoke@tum.de

A. Wassie
Organization for Rehabilitation and Development in Amhara, P.O. Box 204, Bahir Dar, Ethiopia

M. Lemenih
Unique Forestry and Land Use GmbH, Addis Ababa, Ethiopia

D. Teketay
Department of Range and Forest Resources, Botswana University of Agriculture and Natural Resources, Private Bag 0027, Gaborone, Botswana
e-mail: dteketay@buan.ac.bw

M. Kindu et al. (eds.), *State of the Art in Ethiopian Church Forests and Restoration Options*, https://doi.org/10.1007/978-3-030-86626-6_17

continued decline leaving behind pockets of remnants around churches. Their decline and degradation are intensifying proportional to growth in human population in rural parts of the country where land is the main production factor and agriculture is the mainstay of the economy. There is a growing recognition at all levels to conserve the remnant forests and complete them with restoration in the landscapes containing the remnants. This, however, requires full understanding of the problems that drove past and present declines and degradation from a multidisciplinary perspective and identification of effective policies, strategies, practices, and incentives that support their restoration.

Chapters included in this book have synthesized and shared varieties of state-of-the-art knowledge and understanding of Ethiopian church forests (ECFs). The main messages show that as much church forests have religious values they also provide ecological and economic values. They provide habitat as critical sanctuaries for many endangered and endemic plants and animals especially birds as well as invertebrate taxa of Ethiopia (e.g., Goodin; Kifle et al.; Kindu et al.; Mequanint et al.; Zegeye). They are important reserves of propagules for restoring landscape with native species and biodiversity conservation (Alem et al.; Mequanint et al.; Stimm et al.) as well as major stocks of above- and below-ground carbon (Alebachew et al.; Assefa et al.; Tolla et al.). The church forests also serve as a source of climate information by using tree-rings of old trees as proxy data sources of past hydroclimate information (Mokira et al.).

The studies revealed high number of floristic compositions with varied structures in the various church forests studied. For example, Mequanint et al. documented as high as 115 woody species representing 53 families and 97 genera at 24 church forests in southeast of Lake Tana. Similarly, from four selected church forests, Alem et al. showed that church forests harbored a variety of woody species, ranging from 16–40 species, and their diversity ranged from 0.52–2.15, with an average density of woody species 1446 stems ha^{-1}. Kifle et al. also investigated woody species diversity and species richness, of three church forests. They revealed the presence of 34, 17, and 27 woody species in Assela Teklehymanot, Etisa Teklehymanot, and Saramba Kidanemhret, respectively, of which, nine represent IUCN Red List of Threatened Species. From his study sites, Zegeye also documented a total of 263 vascular plant species belonging to 198 genera and 79 families. These results validate the potential and actual roles of ECFs for biodiversity conservation. In most cases, they represent the last option to preserve the endangered plant and animal species of the country.

These forests are also important storage of soil and biomass carbon; hence play role in climate mitigation. In their study that examined soil carbon storage of church forests, Assefa et al. reported a higher soil organic carbon storage of church forests than the adjacent land use systems (eucalypt plantations, grazing land, and cropland). Alebachew et al. and Tolla et al. also reported similar results from Asebot Monastery and Abune Teklehyimanot Church forests, respectively. The contribution, Mokira et al. revealed the opportunity church forests offer for dendroecological study and understand of past and present climate and hydroclimatic reconstructions. They confirmed that church forests are valuable archives of regional hydroclimate information and important to understand the growth dynamic of old-growth Afromontane forests. They further stressed that church forests are storehouses of long-term and past climate information, particularly in areas where data about long-term climatic and environmental changes are scarce and urgently needed.

17.1.2 Overview of the Challenges

The studies also revealed a wide variety of challenges facing the church forests and their surrounding landscapes. The church forests are continue declining from both natural and anthropogenic drivers/factors, all of which necessitate immediate invention to persevere these biodiversity hotspots. Kindu et al. indicated that conversion to croplands, population pressure, livestock grazing, inappropriate government policies, isolation, and their small sizes were the most common drivers/causes for declining trends of forest coverage. Shifts in the values of church forests, emphasis on built structures around churches instead of the forests, replacement by exotic species (eucalypts) for economic optimization, and forest fire were also reported as drivers for church forest conversions. Most of those identified drivers are also listed in other chapters in this book that are, mainly, based on case studies (Alebachew et al.; Alem et al.; Zegeye).

The negative effects of such drivers/factors in church forests are also reported in different chapters. Conversion of church forest to cropland or grazing land can reduce soil organic carbon stocks such as by 58–69% in <50 years (Assefa et al.). Mequanint et al. found the demographic structure for majority of the species showed I-shape, irregular shape, and J-shape patterns, suggesting poor regeneration status in the church forests of southeast of Lake Tana. They reported that human influence, such as plantation of exotic species, affects the structural composition and regeneration status of church forests. Alem et al. also reported expansion of eucalypt plantations in most of their studied church forests. Similarly, Kifle et al. also observed the tendency to replace the indigenous species with exotic plantation species, such as *Eucalyptus globulus* Labil. and *Cupressus lusitanica* Mill.

The replacement of church forests by exotic species has further consequences. For example, the expansion of eucalypt species in the church forests could hinder the regeneration of woody species due to the high nutrient and moisture demanding nature of the species (Alem et al.). In addition, the regeneration status of each species in the studied church forests showed a worrying trend (Alem et al.; Mequanint et al.; Kifle et al.). For example, Mequanint et al. identified several species at different church forests that had lower numbers of seedlings and saplings than mature trees, which confirm the low regeneration and recruitment status of the forests.

17.1.3 Possible Solutions

For the multiple challenges identified multiple actions or measures have been identified. Policies and effective actions identified in the chapters of the book need to be implemented at scales appropriate to restore and conserve church forests. Starting point should be the protection of existing church forests from mounting pressures using stone walls or fencing (Alem et al.; Goodin; Kindu et al.; Mequanint et al.). Second step is the application of appropriate silvicultural treatments and management practices, such as weeding, hoeing and loosening the ground under the forest canopy, removing the thick litters that cover the ground under the forests, thinning, pruning, and enrichment planting with indigenous woody species by ensuring species-site matching (Alem et al.; Kifle et al.). Other conservation actions such as in situ and ex situ conservation measures should also be urgently considered for the ecologically and culturally most important rare indigenous species and families that exhibited low importance value indices in all of the church forests (Mequanint et al.; Kifle et al.; Zegeye). Fire prevention and control systems should be established with the necessary facilities for those church forests prone to forest fires (Zegeye). If there exist degraded lands in the surroundings of church forests, these can be incorporated into church forests through planting fast growing species as buffer zone plantations (Alem et al.). However, the land should be secured through participatory and inclusive negotiation and full consent of the local community. The consent should be ratified by signature collected from all community members.

Evidence also shows that successes of conservation and protection measures depend on the presence or absence of the wandering monastics who make the church forests their home (Goodin). For this reason, it is essential to restore those biotic communities, which are a vital part of the local communities, both through serving as intercessors before God through prayer and as a welcoming vocation for the landless to have respect and community support (Goodin).

In some areas where church communities live within the church forests and in more recently established church forests, eucalypt species are planted to gain income (Alem et al.; Mequanint et al.). This may have a long term effect in terms of altering species composition of church forest unless policies and measures are put in place that controls such planting of exotic tree species in the vicinity of church forests (Mequanint et al.). Alternative income generating activities such as apiary, fattening, and NTFPs could be considered to augment incomes of households that rely on plantings of exotic species (Alem et al.). Church forest managers therefore should strategically think of how to integrate local communities in the management of the forests and how they can ensure incentives for communities in their surroundings. Payment for ecosystem services including carbon payment could also be a possible alternative in this regard (Alebachew et al.; Assefa et al.). Choice of options that bets integrate local community in management, conservation and restoration of church forests need to be made in consultation and full participation of the community and considering local economic, social, and ecological contexts (Goodin).

In the context of degraded lands interspaced between church forests exclosures, intense rehabilitation measures and protection activities are key options proposed to ensure fast restoration (Kindu et al.). To effect these measures understanding of the soil seed banks, ecology of the indigenous forests and tree species as well as species site matching is needed. In measures that involve planting, Stimm et al. suggests the use of the church forests in the vicinity as sources of propagules, hence these remnant church forests should always be treated as repositories for sourcing reproductive materials such as seeds and locally adapted vegetative materials.

Ecological corridors should be negotiated at landscape level to connect the church forests, any remnant patches of natural woodlands interspaced between adjacent church forests to provide landscape connectivity and enhance resilience of the genetic material of species in the church forest (Demissie et al.). In this regard, Gudeta et al. stressed the need for identifying and prioritizing areas of intervention for reducing further degradation and reclaiming the degraded areas. In line with this, they recommended various erosion models and/or multi-criteria evaluation approaches using remote sensing and GIS for identifying and prioritizing areas of intervention (Gudeta et al.; Schneider et al.). Specifically, remote sensing is considered the most efficient technology to handle these problems since it can explicitly reveal spatial patterns of the landscape over a large geographic area in a regular and consistent way. Although remote sensing-based inventories within and surrounding areas of church forests have improved considerably in recent years, there are remote sensing based methodological challenges in highly structured church forests for generating and delivering high quality information, which should be further improved in this area (Kindu et al.).

In addition, through using dendrochronological data for reconstructing multi-century climate information, church forests are also important as evidence-based climate-smart restoration efforts. Such an approach is crucial as predictions of climate change in Africa are characterized by a high degree of uncertainty due to the limited availability of long-term and high-quality climate data. They help to better understand forest growth dynamics, tree to forest level responses to changing climate, and to reconstruct multi-century climate information and characterize extreme climate events beyond instrumental periods (Mokira et al.). The authors also stress that the integration of tree-ring with other ecological information may provide an extraordinary opportunity for a better understanding of forest productivity and visualizing the consequences of forest degradation through the calculation of the required time to recover degraded landscapes. Church forest islands are, thus, towers for monitoring and reconstructing past climate change information. Moreover, church forests provide unique archives and important references on the environment, which enhance informed decision-making to support designs and scaling-up of massive climate-smart restoration initiatives across and beyond the region.

17.2 Conclusions

The approaches used to investigate ECFs and the suggested restoration options synthesized in this book provide key building blocks for a new form of sustainable resource management that:

- Considers the underlying causes of changes within and surrounding areas of ECFs,
- Recognizes contribution of ECFs to the conservation of biodiversity and ecosystem services; specifically, their contribution as critical sanctuaries for many endangered and endemic plants as well as invertebrate taxa,
- Understands biotic diversity gives resilience to a forest, but the ECFs get their resilience, in large part, from the diverse religious history that comes together in the Tewahedo eco-theology,
- Preserves, or in some cases recreates, the protective sense of sacredness of ECFs, which is vital for restoration projects to allow natural regrowth of the forests through enclosure walls rather than mass plantings of seedlings that are not seen as sacred in the eyes of the people,
- Features the potential of ECFs, such as apiculture/beekeeping, tourism/ecotourism, scientific research, and educational field trips,
- Addresses the protection of church forests from external pressures through stone walls or fencing,
- Considers the use of ECFs as a blueprint for restoration of degraded landscapes in the county; especially, strongly requires the use of church forests as repositories for forest reproductive material of native species as important for restoration of degraded areas as well as carbon stock conservation,
- Urgently considers conservation activities, such as in situ and ex situ measures, for the ecologically and culturally most important rare indigenous species and families that have low importance value indices and family importance values in all church forests,
- Uses remote Sensing, GIS, and ground inventory-based approaches as essential tools for change detection studies within church forests and surroundings, and for identifying and prioritizing areas of intervention,
- Uses possible options to create forest corridors intended to expand coverage of forests and increase habitats of the associated wild animals as well as reduce the magnitude of forest fragmentation and
- Features ECFs as sources of climate information by using tree-rings from church forests for generating multi-century climate information that is critical to improving the understanding of long-term patterns of climate changes.

However, there cannot be a single stakeholder, who is solely responsible for the implementation of such sustainable resource management approach. Forest conservation and landscape restoration are long-term processes, which have a focus on conservation of important species and restoration of ecological functionality of degraded lands for enhancing human wellbeing using a variety of land uses and diverse plant species. As a result, all those who have a direct or indirect influence on church forests and degraded landscapes shall take greater responsibility for the development and implementation of the suggested approaches. More rigorous discussion is needed on the specificity of their linkages. Models that capture key land use/land cover dynamics in and the surroundings of ECFs with their associated services and socioeconomic drivers need to be developed under scenarios of local, regional, and global changes. There is a need to strengthen the development of local technical capacities for the establishment of conservation strategies as well as collection, production, and distribution of important tree species that can facilitate the goal of landscape restoration. The Editors simultaneously view these as a form of practical implementation that needs a multidisciplinary perspective.

Index

M. Kindu et al. (eds.), *State of the Art in Ethiopian Church Forests and Restoration Options*, https://doi.org/10.1007/978-3-030-86626-6